Visual Basic (Bootcamp) Certification Exam Guide

New Technology Solutions
Dan Mezick
Scot Hillier

McGraw-Hill
New York • San Francisco • Washington, D.C. • Auckland
Bogotá • Caracas • Lisbon • London • Madrid • Mexico City
Milan • Montreal • New Delhi • San Juan • Singapore
Sydney • Tokyo • Toronto

Library of Congress Cataloging-in-Publication Data

Mezick, Dan.
    Visual Basic bootcamp certification exam guide / Dan Mezick, Scot Hillier
        p.   cm. — (Certification series)
    Includes index.
    ISBN 0-07-913671-0
    1. Microsoft Visual BASIC. 2. BASIC (Computer program language)
    3. Electronic data processing personnel — Certification.
    I. Hillier, Scot. II. Title. III. Series.
    QA76.73.B3M492 1998
    005.26'8—dc21   97-41798
                                                     CIP

McGraw-Hill

A Division of The McGraw·Hill Companies

1 2 3 4 5 6 7 8 9 0 DOC/DOC 9 0 2 1 0 9 8 7

PN 046609-2
PART OF ISBN 0-07-913671-0

The sponsoring editor for this book was *Judy Brief* and the production supervisor was *Claire Stanley.* It was designed, edited, and set in Century Schoolbook by *TopDesk Publishers' Group.*

Printed and bound by *R. R. Donnelley & Sons Company.*

McGraw-Hill books are available at special quantity discounts to use as premiums and sales promotions, or for use in corporate training programs. For more information, please write to the Director of Special Sales, McGraw-Hill, 11 West 19th Street, New York, NY 10011. Or contact your local bookstore.

 This book is printed on recycled, acid-free paper containing a minimum of 50% recycled, de-inked fiber.

CONTENTS

iii

Contents

Contents

Contents

Contents

PREFACE

It is safe to say this book was one of the easier books we have written. That is mostly because the information on the pages, the questions, and the overall approach to the topic is taken directly from the way we teach VB at the VB BOOTCAMP seminar and hands-on training that we do across the USA. We've trained tens of thousands of developers in this topic of VB development. It's a blast to teach the power of this product. In your hands is the text version of our approach to VB training. We believe—and hope—you will find it very useful as a tool for learning.

It's doubtful anyone can learn VB in 21 days, and we don't pretend that you will. There are many ways to learn VB. At least initially, reading the product documentation is not one of them. This book makes our best possible attempt to teach you the VB product in the shortest possible amount of time. If you will take the time to go through the entire book, at the end you will be ready to dig in and become a reasonably competent programmer of business applications using VB. You will also be ready to take a stab at the certification exam.

You'll find questions and answers at the end of major sections of this book. The questions assume that you have examined the help file and worked a little extra with the product, beyond the material presented in the chapter. On the CD, you will find a timed exam that explores your ability to remember trivia, much like the certification exam. No, no—just kidding. There is no replacement for actually working with the product, which is why you will find numerous hands-on exercises in this book. The test we provide is intended to give you a flavor of the actual, timed VB certification exam.

Regardless of how you learn Visual Basic, you need to understand: VB is a rather large product. No one person can actually know the whole thing. The approach with this book is to show you a key topic with examples, and then have you do a hands on exercise that focuses on that topic. There is no attempt to build a huge monolithic application that contains a little of everything. In our opinion that would just make learning VB harder. This book is about learning VB, but it is up to you to go on and actually apply what you learn. From this book you will learn the things you need to know to go on and decompose and digest the entire product.

If you already know a little VB, you can just index in to a chapter with material new to you, and start from there. For example if you know VB4 you can dive in at around Chapter 8 or Chapter 9. You'll find the rest of the book a good read! You may want to check out the beginning chapters anyway, to make sure you have all the topics mastered.

So dig in. And, be sure to visit our web site http://www.vb-bootcamp.com to stay current, because this topic moves really, really fast. And remember we are available to your organization if your team wants a customized, hands-on, instructor-led training program.

We both hope you enjoy learning VB with our book.

Scot Hillier and Dan Mezick
Fourth Quarter, 1997

ACKNOWLEDGMENTS

Each time we write a book, we continue to be amazed at how many people are actually required to make the effort a success. This is no less true with this work, and we want to take the time to offer our sincere thanks to those that deserve it.

As always, we thank our families first. Without the support and love of these most-important people, we would certainly be less successful.

From McGraw-Hill we would especially like to thank Acquisitions Editor Judy Brief for initiating the project and Alice Tung for helping make it a reality. McGraw-Hill also has a strong partner in TopDesk Publishers' Group where Chuck Wahrhaftig and Judy Allan put this book together.

At New Technology Solutions, we wish to thank Gary Shomsky who helped create testing materials and reviewed our early revisions. We also want to thank all the other staffers at NTSI that support the business behind the scenes. Thanks to everyone.

Once again, to Nancy.   —Scot Hillier

To Roberta, my best friend and wife,
the highest form of life on the planet.   —Dan Mezick

CHAPTER 1

Design Time Fundamentals

What You Need to Know

You need to be familiar with using Windows applications like Excel or Word before you start this topic. You also need to be a programmer.

What You Will Learn

- Project Window
- Properties Window
- Toolbox
- Code Window
- VB's Most Important Menu Commands
- Adding Forms, Modules, and Classes
- Starting a New Project from Scratch

Understanding VB means completely understanding the primary interfaces that are used to manipulate objects. In this book, we define an object as anything that has a property or a method or an event. Understanding these interfaces is what makes you a VB programmer. VB's Toolbox, for example, is populated with many controls, a special kind of object you can drop on a VB form (see Figure 1-1). Before you can begin with VB you need to know how to move around and what the items are that make up a VB project. After that, you need to know how to manipulate the properties, events, and methods of objects in the VB environment. That comes in the next part of the book.

A Quick Tour of VB Design Time

Properties, Project, and Code Windows

Manipulating VB's various design time windows is how you get things done when using VB. The more important windows include the Project window, also known as the Project Explorer, and the Properties window (see Figure 1-1).

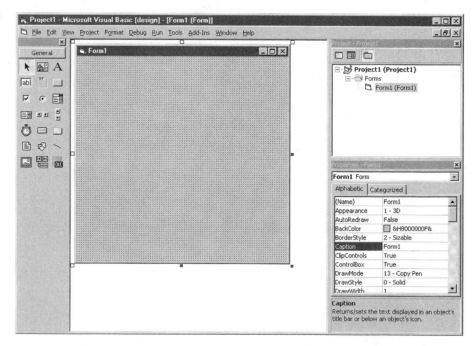

Figure 1-1
Important elements of the Visual Basic (VB) design environment include the Toolbox, the Properties window, and the Project Explorer.

The Properties Window

The Properties window is where you alter the settings for the appearance or behavior of an object.

Object: Any item in a VB program that possesses a property or a method. A property is a setting that describes appearance or behavior. Properties and methods are part of an object's interface and are covered in depth shortly.

Check It Out: Changing the Properties of a Form

You can change a form's appearance at design time. Try this:

1. Start up VB. If you are prompted to select a project template, select a Standard EXE.

2. Select the View/Properties menu command. The Properties window will appear if it is not visible now.

3. Select the BackColor property of the form by clicking on the name of the property. Click on the drop-down button that appears to be pointing down. You will see a tabbed dialog with two tabs: System and Palette. Click on Palette. Note the background color of the form. Also note the hexadecimal value for the BackColor. (The hexadecimal value represents the Red-Green-Blue mix of colors that make up the shade you have selected.)

4. Now change the BackColor by clicking a color on the palette. Note how both the background color of the form and the hexadecimal value of the property change.

The Project Window

The Project window is where all of the elements of your project are contained and displayed. For example, in the Project window you can view all of the items you have added to the project. Double clicking on any program object will bring up a window for that object. Double clicking on a module will bring up its code window. Double clicking on a form will bring up its Form Designer window. To insert new items in your project, select the Project menu command, where you will find further choices for inserting various program objects such as forms, modules, MDI forms, and class modules. (There are more items you can add, but we'll skip them for now.)

A brief explanation is in order. Forms are windows that will appear in your application's user interface, or UI. Modules are special files that hold data and procedures that are typically accessible from your entire application. These data items and procedures are said to be globally accessible from the entire application. A class module is a special kind of module that serves as a kind of plan or blueprint for a homegrown, custom object. This will be discussed in great detail later, and is the basis for object-based programming in VB. Finally, an MDI (Multiple Document Interface) form is a special kind of form that can contain subordinate "child" forms. In this book we will use all of these project components. Please note that, when saved, all of these items exist on your hard drive as text files. Later in the book we'll return to the topic of project management and explore how VB saves your work.

Check It Out: Inserting Several Key Program Items into Your Project

You can insert many project components in the Project window. Here we will look at three key ones: a form, a module, and a class module.

1. Select the Project/Add Form menu command. The form will appear inside VB's Form Designer, in the main area of the VB display. This main area of the VB display is a child window in the VB MDI design environment.

2. Close the form now. If it is maximized, you close it by clicking on the icon that appears next to the File menu, and select Close from the system menu that drops down. If the form is normalized, you close it by clicking the icon that appears in the upper left corner of the Form Designer, and select Close from the system menu that drops down. See Figures 1-2 and 1-3.

3. Select the View/Project Explorer menu command. You will see Form2, the form you just created, inside the Project window. It is under the folder called Forms. You can open and close this folder. Try it with the mouse now.

4. Select the Project/Add Module menu command. A code window will appear. (We describe code windows later.) Close it just as you closed the form you created previously, by clicking on the icon that appears next to the File menu and selecting Close from the system menu that drops down. Look in the Project window, where your module named Module1 appears under a folder named Modules. You can add more modules. You can also open and close the Modules folder by double clicking on it.

5. Select the Project/Add Class Module menu command. A code window will appear. Close it just as you closed the form and module you created

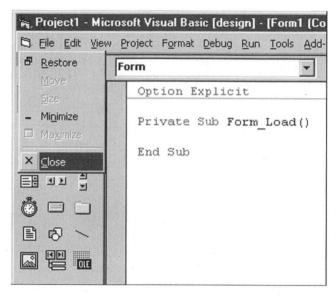

Figure 1-2   Closing the code window of a form if the window is maximized.

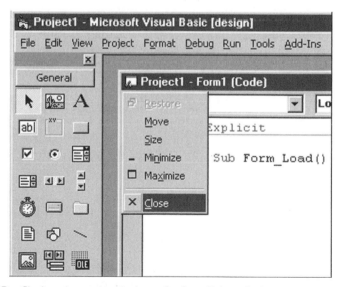

Figure 1-3   Closing the code window of a form if the window is normal.

previously. Look in the Project window, where your class module named Class1 appears under a folder named Class Modules. Class modules are covered in depth later in this book. More class modules can be added in the same way as you added Class1. VB will name the next class modules Class2, Class3, and so on. Later you'll learn how to override the default name. Just accept it for now. You can also open and close the Class Modules folder inside the Project window by double clicking on it just like the other folders in this example.

The Code Window

The code window is where you program Visual Basic. For the form object, you can display its code window by double clicking on the form in the Form Designer. You can also click on the item in the Project window (also called the Project Explorer) and press the View Code button that appears at the upper left of the Project window. The code window will be explored in detail throughout this book.

▨ Check It Out: Viewing the Code Window for a Form and Module

Forms and modules can contain Visual Basic code. You edit this code from the code window. In this window, you can edit VB code for the object that contains it. Try these steps:

1. Double click on the name Form1 in the Project window. It will appear in the Form Designer. Double clicking on the form will bring up a code window for the form. There are two drop-down combo boxes that appear at the top of the window. The left one contains the objects that are in this form, including the form itself (more on this later). The Form object appears by default. The right combo box contains built in event procedures for the Form object. The Form_Load event procedure appears by default. This is known as the default event. Click on the drop-down combo boxes now.

2. Close the code window. Close the Form window.

3. You can also display a code window for a Form by clicking the leftmost button at the upper left of the Project window. Point the mouse there now to display a ToolTip labeled View Code. Click on the name Form1 in the Project window, and then click on the View Code button (see Figure 1-4). This series of actions will also display a code window for a selected Module.

4. Select the Project/Add Module menu command. A code window will appear for the new module. The [general][declarations] section is displayed. A module is a library that contains VB code and data variables. You will learn more about modules in the topics that follow.

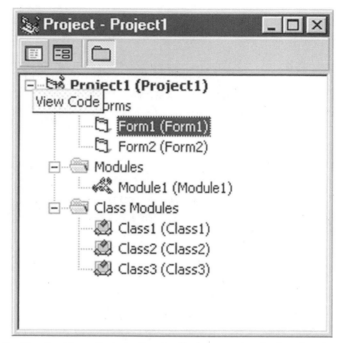

Figure 1-4 Use the View Code button to display a code window for the object that is selected in the Project window.

The Toolbox, Toolbar, and Menu

The Toolbox is where the reusable, visual components called ActiveX controls are displayed for use. You can display the Toolbox by selecting the View/Toolbox menu command of Visual Basic. This window is used to select (ActiveX) controls for use on a VB form object. When the Toolbox is visible and a form is displayed, you can drop CommandButtons, ListBoxes, and other visual objects onto the form directly from the Toolbox.

Check It Out: Dropping Controls

You can add controls to your forms like this:

1. Display Form1. (Click on Form1, then click on the View Object button at the top of the Project window, or just double click on Form1 from the Project window.)

2. Select the View/Toolbox menu command, making the Toolbox appear.

3. Double click on the Label control (the one with the capital letter A on it). This will place a Label control of a default size, centered on Form1.

4. Click on the Label Control icon in the Toolbox, then left-click on the form, dragging your mouse from upper left to lower right, holding the mouse button down. This will "draw" a Label control on Form1, of whatever size and location you choose.

The Toolbar

The Toolbar is a panel of buttons that is displayed beneath VB's menu. These buttons form a subset of VB's menu functionality and can be used in place of the corresponding menu command. For example, on the toolbar is a button for saving the project. Clicking on this button is the equivalent of selecting the File/Save Project menu command. The Toolbar can be toggled off and on by selecting the View/Toolbars/Standard menu command. Turning it off will get you more screen real estate. This is recommended, since the menu provides all of the Toolbar's functionality.

Adding and Removing Controls from the Toolbox

You can add new custom controls to your Toolbox by selecting the Project/Components menu command. This will display a list of installed controls from which you can select new ones for use in your project. (See Figure 1-5.)

To add a new control, simply click the checkbox next to the control name and click on the OK button. The new control will be added to your project. You can remove controls from your Toolbox in the same way. If a control is currently being used, just click on it to clear the checkbox and press OK. The control will disappear from your Toolbox.

Check It Out: Adding and Removing Controls from Your Toolbox

You can customize the Toolbox to include your favorite controls. Try this:

1. Select File/New from the VB menu. Do not save the previous project. Click on No when prompted about saving your work.

2. You'll be presented with a set of standard project templates. Select Standard EXE and press OK. Now you are working with a new project, consisting of a single form named Form1. If the Toolbox is not visible, select the View/Toolbox menu command to display it.

3. Select the Project/Components menu command. When the dialog appears, scroll to the Microsoft Data Bound Grid Control, and click on the checkbox that appears next to it.

Figure 1-5
The Toolbox may be customized by selecting from the list of installed controls, making them available for use in your project.

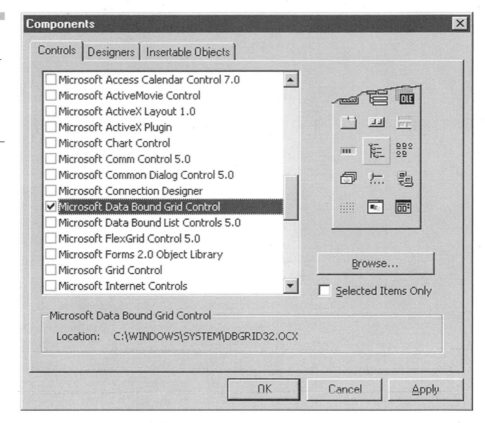

4. Move the dialog so you can see both the Toolbox and the Components dialog. When you click on the OK button on the Components dialog, you will see the Data Bound Grid appended to your Toolbox. You can remove it the same way you added it.

You can add and remove controls from your Toolbox using the Project/Components dialog. Later in the book, you will learn how to create your own controls using VB. These controls, when properly constructed, will appear in your Components dialog, right alongside all the Microsoft controls that ship with VB.

You might want to try deselecting all of the controls in the Components dialog and see what happens. You will find that the Toolbox does not become empty—some controls remain in the Toolbox, even though you attempted to remove all of the controls. These remaining controls (such as the CommandButton, ListBox, and Label) are known as intrinsic controls. They are always there because they reside in the VB runtime Dynamic Link Library named MSVBVM50.DLL. These controls have been optimized for speed and are always part of any VB application you create. For these reasons, it's smart to use them whenever possible.

The Menu and Tools/Options

VB provides a Preferences dialog for handling developer settings. This dialog is located under VB's menu as the Tools/Options... dialog. This tabbed dialog provides you with many settings you can customize to set up VB to look and work the way you want. For example, you can set the font name and size for the code window, and you can specify the colors for code and comments.

Check It Out: Setting VB Design Environment Options

VB provides you with a way to set the environment to your preferences. Try this:

1. Select the Tools/Options... menu command.

2. The Options dialog will appear. It has several tabs. Select the Editor Format tab.

3. From the Editor Format tab, you can change the font name and font size for the code that will appear in your code windows when you are editing. Alter the font name and size to suit yourself.

4. Click on the OK button. Now display Form1, and double click on it to display its code window. When you do this, your editor settings should be in effect.

5. Again select the Tools/Options... menu command.

6. Explore the other tabs on the Options dialog, but do not change anything. When you are done, press Cancel, in case you changed some setting while exploring. Later sections of the book will show what most of the other settings are and how to use them.

Closing All Windows

It's a useful exercise to close all of the windows in VB. Double clicking on the Project1 title in the Project Explorer will close all of the Form Designer and code windows in this project.

Starting a New Project

When you start a new VB project, you are presented with a list of templates. These provide a convenient way to get started with the many types of specialized projects you can build in VB. For now, you'll be selecting the Standard EXE template for all examples and exercises. From VB you can create standalone EXEs, components called ActiveX servers, ActiveX controls, and other types of output, such as ActiveX documents. All of these will be covered in due time.

Using Search Strings

No discussion of using VB would be complete without some mention of the Help file. Clicking on the Help/Microsoft Visual Basic Help Topics menu command will reveal a tabbed dialog. Select the Index tab and type in the topic you are seeking. The Help file will perform an incremental search, showing you all of the topics available based on the text, as you enter it. (See Figure 1-6.) Later in the book you'll learn more about the VB Help file.

Figure 1-6
The Index tab of VB's Help dialog does an incremental search for your topic as you type.

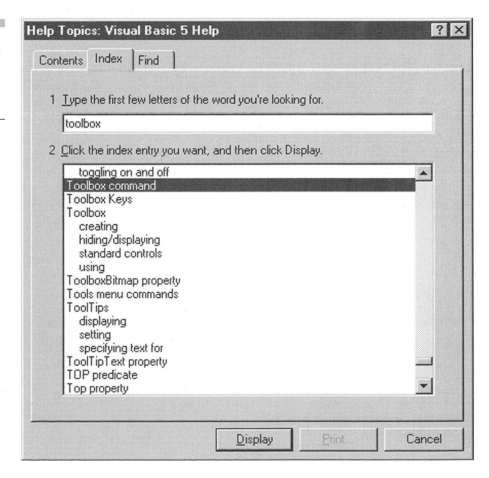

If you have performed a full installation of VB, you will have the books online available. You can access it by selecting the Help/Books Online menu command. This resource makes all of the VB documentation available online. When doing a Find for a topic that is a multiword phrase, remember to use quotation marks. If you forget, Books Online will return to you a list of all topics that contain all of the words that you included, in any order.

SUMMARY

If you understand the Project, Properties, and Toolbox windows, and have a clue about code windows and the VB editor, then you are ready to begin learning the whole product. Understanding how to add forms and modules to the Project window is important. With this limited but important information as background, you are now ready to begin learning VB more completely.

We have covered some important design time basics. We have seen the Project, Properties, and Toolbox windows. We have worked with code windows and the VB editor. We have learned how to add forms and modules to the Project window. We have also found some important menu commands under the menu bar headings of View, Project, Tools, and of course, Help. If you understand this limited but important background information, you are ready to begin learning, ready to learn the entire VB product.

Increasing Your Knowledge

Pertinent Help File Search Strings

- Toolbox
- Properties Window
- Project Explorer
- Form Window
- Editor Tab
- Editor Format Tab
- Code Window

Pertinent Books Online Search Strings
Remember: use quotation marks when searching for multiword topics.

- "Elements of the Integrated Development Environment"
- "Adding Controls to a Project"

Mastery Questions

1. The Properties window is used to:
   a. alter the appearance of an object such as a form.
   b. change settings related to an object's behavior.
   c. view the current settings of the properties of an object.
   d. All of the above

2. Which main menu command leads to submenu commands that add new program objects such as forms, modules, and class modules?
   a. File
   b. Edit
   c. View
   d. Project

Answers to Mastery Questions

1. d

2. d

Properties, Events, and Methods

What You Need to Know

You need to be a programmer who is familiar with the Visual Basic design environment.

What You Will Learn

■ Programming the ListBox properties, events, and methods

■ Attaching code to event procedures

■ Starting, running, and saving a complete VB project

Properties. Events. Methods. These are your interfaces to objects in Visual Basic. Understand these interfaces, and there is nothing you cannot do. If, however, you choose to just dive in and not learn these fundamental topics, then you may never really understand VB completely.

In VB, everything revolves around properties, events, and methods. Everything! If you are brand new to VB development, you will want to carefully study this section before proceeding. Lets get started!

Properties

A property is a characteristic or attribute of an object that typically describes the object's appearance or behavior the object can display. Think of a property as a piece of data or a variable associated with an object.

Almost everything you manipulate in VB possesses properties. The ListBox control, for example, has a property named BackColor, which describes the appearance of the ListBox. It also has a property named MultiSelect, which describes certain behavior. Besides controls, there are other "system objects," such as the App object, which also have many properties. (In this book, until further notice, an object is any program item that has a property or method.) Understanding the various properties of each object you manipulate in VB code is absolutely essential to competency in Visual Basic.

Properties at Work

VB has three modes of operation: Design mode, Run mode, and Break mode. Design time is when you establish forms, add controls to forms, and get the visual appearance right for your application. This is also when you define code for execution at Run time. Run time, or Run mode, is the mode in which you test your application. In this mode, you can test your app before compilation. A third mode, called Break mode, is used for debugging the VB application. This mode and the debugging process will be addressed later in the book.

Why discuss this now? Because some properties of some objects, such as the ListBox object, are "Run time only." This means you will not see the property at design time at all! Other properties are "read-only at run time." As you will see in the upcoming sections, all properties are not created equal. Some are not available at design time, and others cannot be set at run time. Without a mental model of how properties work, you can get frustrated with error messages VB issues when your program runs. Understanding how properties work is central to understanding almost everything you'll do with Visual Basic.

Changing Properties at Design Time

Most properties can be changed via the Properties window. (See Figure 1-7.) The Properties window lists the properties of an object in alphabetical order, or arranges them by category, depending on the tab you select. At the top of the Properties window is the Object box. This drop-down combo box displays the name and class name of the currently selected object.

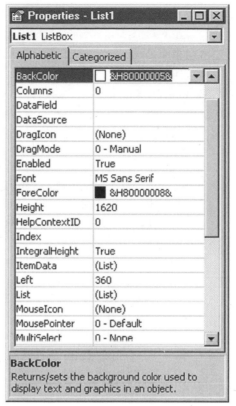

Figure 1-7 The Properties window lists some of the properties of an object: all those that are available at design time.

The class name is the control or object type. You can select the Properties of an object from the object box, or you can click on an object on the form to view its properties.

Check It Out: Working with Design Time Properties

The Properties window displays the design time properties of an object. Start getting familiar with the properties of some commonly used objects.

1. Start a new project and drop a Command Button, a ListBox, and a Label control on Form1.

2. Bring the Properties window into view by selecting the View/Properties Window menu command.

3. Click the CommandButton. Note the display in the Properties window. It shows the properties of the CommandButton named Command1.

4. Click on the ListBox. Note the display in the Properties window. It shows the properties of the ListBox named List1.

5. Click on the Label. Note the display in the Properties window. It shows the properties of the label named Label1.

6. Click the form. Yes, once again, note the display in the Properties window. It shows the properties of the form named Form1.

7. Now check out the object box of the Properties window. Drop the list down by clicking on the down-arrow button. You'll see four objects: Command1, Form1, Label1, and List1. Select List1. Note how the focus on the form shifts to the ListBox and its properties are displayed when you select List1. Selecting List1 from the object box of the Properties window produces the same results as clicking on the ListBox.

8. Select Command1. Click on the Categorized tab of the Properties window. The same properties are listed under both the Alphabetic and the Categorized tabs, but in a different order.

9. Notice the bold headings of the categories (**Appearance, Behavior, DDE**, etc.). These headings are not properties. Click on the minus sign in the box to the left of each heading, collapsing the list down to view only the headings.

10. Select Label1. Which category is still expanded? Click on the minus sign next to its heading to collapse the Properties window down to only headings.

11. Select List1. Which category is still expanded? Collapse the Properties window down to headings only.

12. Select Form1. Which category is still expanded?

13. Change the BackColor property of Form1 to white.

Changing Properties at Run Time

You can change most properties of controls and forms at design time. Most properties of most controls can also be changed at run time. Changing a property at run time requires code, and code typically is placed under event procedures. For example, the form object has an event procedure named Form_Load. This event procedure executes when the form is loaded into memory. You can place code under this event procedure. You will now experiment with changing a property of the Label object at run time. When you make a change to a property of an object at run time, you do it with code you place under event procedures. For example, the Label object has a property named Caption. It contains the text that appears on the Label. You can alter the displayed text in the Label with the following line of VB code:

```
Form1.Label1.Caption = "VB BOOTCAMP"
```

This code specifies that the Caption property of the control named Label1 located on Form1 will be changed from its current value to "VB BOOTCAMP". Specifying a property in this way (i.e., FormName.ObjectName.PropertyName) is known as "full qualification." When this code executes, the Caption property of Label1 changes.

Check It Out: Changing Properties at Run Time

Properties that may be changed at run time can be manipulated in VB code, typically inside of an event procedure. Try this:

1. Bring Form1 into view (if it is not already) by selecting Form1 from the Project window and clicking on the View Object button of the Project window. (You can also double click on Form1 to make it visible.)

2. Now double click on Form1. A code window will appear displaying the Form_Load event procedure of Form1.

3. Now enter the following line of VB code inside the Form_Load event procedure:

```
Form1.Label1.Caption = "VB BOOTCAMP"
```

4. Now select the Run/Start menu command. Tell VB No you don't want to save the changes. Now you move from design time into run time, and VB displays Form1.

5. Label1 should be displaying the text VB BOOTCAMP because you changed the value of Label1's Caption property at run time. Click the Run/End menu command to return to design time. You will note that the caption of Label1 will revert back to the original design time value.

The event procedure named Form_Load executed when you started the app. This is because Form1 is the default Startup Object. You will learn how to change it later. Event procedures execute when certain events take place, such as a form loading into memory. Other examples of events include mouse clicks and mouse moves by the user. Event procedures are covered in much more detail in the upcoming sections. For now, understand that event procedures are part of most controls and objects such as forms, and you can place code under the procedure to respond to events and user actions. The code we inserted under the Form_Load event of Form1 changed a property at run time. The Caption property of the Label object is an example of a property that can be changed both at design time and, with code, at run time.

The Help File

You can learn about all of the properties related to the Label object by referencing the VB Help file. To do this, click on the Label object on Form1 and press the F1 key. You will see the Help topic for the Label object, and at the top of the listing you will see a hotspot in green labeled Properties. If you click on that hotspot, you'll get more information on all of the properties related to the Label object. You need to examine the Help file *often* when you start learning VB, or you will miss important information.

Read-Only at Run Time Properties

Some properties are read-only at run time. This means that you can sample the value, but you cannot alter it in any way. Attempting to change a property that is read-only at run time will get you an error message. For example, the ControlBox property of the Form object can be set to either True or False at design time, specifying whether or not a system menu appears in the form to the upper left corner of the form. The ControlBox will always appear at design time. Making the ControlBox property False will take effect at run time.

When you click on the system menu of a Form, you can close the form by selecting the menu's Close command. However, if the ControlBox property is set to False, the system menu will not appear. You can change the ControlBox property of the Form object at design time. However, if you attempt to change the ControlBox property of a form at run time, you get a compile error as depicted in Figure 1-8. This is because the ControlBox property of the form object is read-only at run time. If you search the Help file by selecting the Help/Microsoft Visual Basic Help Topic menu command and search on "ControlBox", you will access the Help file for this property. Note the help file indicates the property is read-only at run time.

Unlike the Caption property of the Label object, the ControlBox property of the Form object cannot be changed both at design time and at run time. The ControlBox property can only be sampled at run time, it cannot be altered. It is read-only at run time. There are many properties of numerous controls that fit into this category. When you explore the properties of a control you are learning more about, start thinking about how

Figure 1-8
This is the VB error you get when you attempt to change a read-only at run time property. This message will pop up at compile time, preventing this type of code from running.

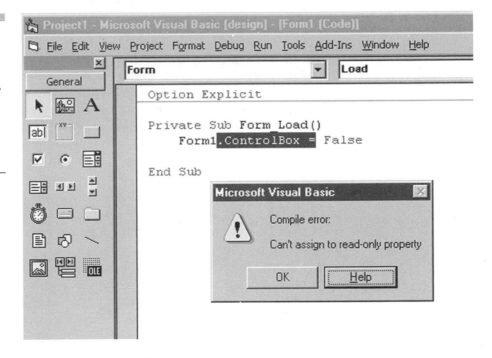

the properties behave at design time and run time. Unfortunately, the important fact that a property is read-only at run time is usually just a small detail in the Help topic. You need to understand the significance of any limitations that are imposed on the run time behavior of a property you are exploring when using Visual Basic.

The Default Property of a Control

As you know, the following line of code will change the caption of a label at run time:

```
Form1.Label1.Caption = "VB BOOTCAMP"
```

What you might not know is that this line of code will do the same thing, and without an error, even if the Caption property is not specified. The following two lines of code are equivalent:

```
Form1.Label1.Caption = "VB BOOTCAMP"
Form1.Label1 = "VB BOOTCAMP"
```

Notice that, in the case of the second line of code, no property is specified. When you code in this way, the default property of the Label object will get the assignment. This means that the Caption property of Label1 will get assigned "VB BOOTCAMP", even though no property has been specified.

Coding in this manner, using implied functionality such as the default property, should be avoided. Why? Because anyone who maintains this code will have to (a) know the name of the default property of the Label object or (b) have to stop and slog through the documentation or Help file looking for this detail. Avoid implicit coding and keep what you are doing obvious to all, unless you seek to write what we call "career secure" code. Career secure code is typically written by consultants; the excuse commonly used is "performance." Career secure code is code that no one can comprehend-except you. Career secure code is evil. We shall have much more to say about maintainability and good coding practice as this book unfolds. For now, avoid the default property.

As a matter of historical note, the default property of most controls has been either poorly documented or undocumented altogether. This is because custom control writers could not define a default property when custom controls first came into vogue. Microsoft subsequently introduced the ability to define a default property as an optional feature that custom control writers could choose to implement at their discretion. Some did, some didn't. Most did not update their Help files to call out the property with this special status. One notable custom control vendor, Videosoft, always made sure the default property was called out in the Help file for their products. As it stands now, some controls have a default property, and some don't. Those that do rarely document the default property as they should. This has only made matters worse for new VB programmers. Don't use the default property of any control, even if you actually know what it is.

Dependent Properties

Some controls have dependencies established across two or more properties. The Timer object, for example, has a property named Interval and a property named Enabled. The Interval property is used to specify how many milliseconds should elapse before the Timer's only event procedure, called the Timer event procedure, executes. The timer control is great for things you need to do periodically. One good use for it is to check the file system for the existence of a particular filename. Say, for example, that your VB app needs to know when some other app concludes processing. One solution is to have the other app write a certain file to a certain directory to indicate it is finished. Your VB app can then check every Interval milliseconds to see if the file is there. If it is there, you can delete it and perform other processing. All of this can happen under the Timer event procedure using standard VB code. You can shut the Timer off by setting the Enabled property to False at run time. If you do this, the Timer will not tick. Even if the Interval property is set to a valid value, it does not matter, because the Interval property of the Timer object is utterly dependent upon the value of the Enabled property.

Many third party custom controls possess property dependencies. Again, like the default property, this stuff is typically poorly documented. Be on the alert for dependencies across the properties of third party custom controls.

Design Time Only Properties

Design time only properties are another property type. For some controls, you will see properties such as "(About)" or "(Custom)" displayed in the properties window. These properties typically lead you to a dialog box when you click on them, and as such, they are design time only properties. Attempting to access these properties at run time will get you an error. Try this:

Check It Out: Design Time Only Properties

A design time only property like "(Custom)" has obvious value at design time. Try this:

1. Add the "Microsoft Data Bound Grid" to your Toolbox using VB's Project/Components menu command. Click on this menu command and examine the dialog. (See Figure 1-9.)

2. The list of components that appears is actually the list of all custom controls installed on your machine. Note how many are yours for free with Visual Basic. (It was not always this way.) Scroll down the list until you hit the listing titled Microsoft Tabbed Dialog Control 5.0. Click on the checkbox to its left, and click on the OK button on the dialog. You will see the SSTab control add itself to VB's Toolbox. (It's called SSTab because the vendor that provides the control is named Sheridan Software.)

3. Select the SSTab from VB's Toolbox and drop one of these controls on your form. Bring up the Properties window. Click on the Alphabetic tab, and find the "(About)" property. Double click on it, and watch what happens.

4. Dismiss the About dialog and double click on the "(Custom)" property in the Properties window. The window that comes up is called a property sheet. This dialog provides a point-and-click interface for the setting of properties on the control.

Figure 1-9
Use this dialog box to
select additional con-
trols that are installed
on your system to be
added to your Tool-
box.

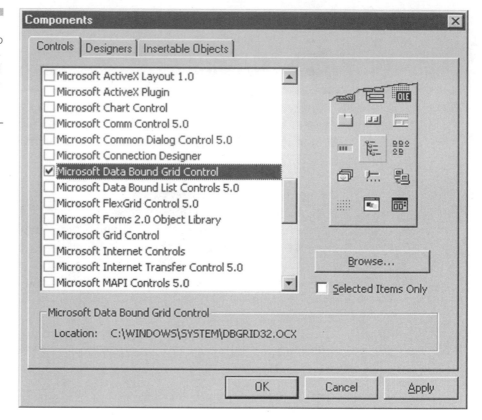

5. Set the Tab Count to 4 and click on OK. What happens to the look of the SSTab1 on your form?

6. Experiment with other settings on the property page that appears when you select the "(Custom)" property of SSTab1.

Summary of Concepts: Properties

Understanding how to use custom controls means understanding the interfaces that the control exposes. Properties are a big part of the total interface exposed by a control. Right now, you are learning how to use objects. Later, you'll learn to build objects for use by others. Examples include ActiveX EXEs, ActiveX DLLs, and ActiveX controls, all of which can be constructed in VB. When you build reusable components and controls, you will define properties as part of the interface. Understanding design time and run time properties, the default property, dependent properties, and property pages becomes critical, because you will later be incorporating these features into the properties of your own homegrown components.

Leveraging the VB Help File

You can use many approaches to leveraging the Help file. Besides clicking on a form and pressing F1, there are other ways to gain access to the Help file.

Check It Out: Leveraging the Help File During Development

There are several ways to get to the information in VB help. Try these:

1. Start with a new Standard EXE project by selecting File/New Project from VB's menu. Drop a ListBox on Form1.

2. Click on the ListBox and expose the Properties window. Click on the List property and then press F1. What happens? You can click on any property in the Properties window and then press F1 to get Help on that property. Close VB Help.

3. Next, simply click on the CommandButton on the Toolbox. Now press F1. What happens? You can click on any control in the Toolbox and press F1 to get help on that control. Close VB Help, again.

4. Now, go back to the ListBox. Double click on it and a code window will appear, displaying the Click event procedure. Inside this Click event, type in the word "MsgBox". Next, click on that keyword, and then press F1. What happens? You can click on any code in VB with a valid keyword in the code, and VB will provide the Help topic for that keyword.

If you are a developer, you don't read documentation. It's a characteristic of the developer personality type to avoid documentation whenever possible. But the Help file is your buddy. The Help file is your pal. Use the Help file. If you do not pick up the VB manuals, and if you do not skim the Help file, the entire right side of your VB brain will be missing. You will be a lopsided VB programmer, thinking you know the product when in fact you don't. VB's design time environment makes it easy to skip reading the documentation and Help file. If you succumb to this obvious temptation, you'll have no one to blame but yourself for your complete ignorance of things that are useful but invisible at design time. For example, in upcoming sections we'll explain system objects like the App, Screen, and Printer objects. You will just never hear about these important objects if all you do is play around at design time without at least reading the Help file.

We strongly urge you to get in the habit of examining the Help file and picking up the documentation. VB ships with Books Online, and we strongly recommend you install them in total. VB's Help menu command provides access to Books Online as a menu option, and a search engine is provided. This is a great resource, a required resource in your bag of tools for learning VB. If you are like one of us (guess which one), you may just go off and start looking for jobs in VB, and maybe even immediately get one after you digest this book. If you do, there will be a strong temptation to "go build something" and skip the ongoing study we are recommending here. If you skip the study of the Help file and other documentation now, you will eventually have to go back and do it anyway, because your colleagues will be showing you VB stuff you've never seen before when you review their code. So get it over with now and develop a balanced VB brain. Install Help and Books Online. Then use them daily. OK, enough. Let's do event procedures.

Events and Event Procedures

It's important to note that the whole purpose of VB is to create Windows programs fast and without a fuss. You are writing a Windows program when you program in VB. Remember that. Also remember that Windows is an event-driven, message-based programming environment. We won't get into this in detail now, but realize that writing Windows programs in VB is simple largely because VB reduces the complexity of Windows programming significantly by providing event procedures where you can place custom code to make your program come to life.

Event Procedure: *A predefined, empty procedure where you put custom code.*

During the lifetime of your program, there are many events. The first form loads. Then the first control on that form gets the focus. Then the user moves the mouse and clicks it on a button. These are all examples of events that occur in your program. In VB, you are provided with empty event procedures that will execute when specific events take place like the ones we just described. If you care about the occurrence of the event, you write code in the corresponding event procedure, which is initially empty. Event procedures are provided to you by VB so that you can react to specific events during the execution of your application.

For each item you program, there may be a predefined set of event procedures, or there may be none at all. The form object, for example, has many event procedures, while the timer control (discussed earlier when we talked about property dependency) has only one: the Timer event procedure. As we will see later, some controls have very few event procedures, while others have lots of them. Event procedures are optional-do not expect every object you program to have them. Also, in other books on VB, you will see the terms "event" and "event procedure" used interchangeably. We will not do that to you. Instead, we define an event as something that occurs, and we define an event procedure as a predefined procedure that executes *in response* to a specific event, such as the user clicking the mouse. Events happen; event procedures execute in response. Let's look at some examples.

The Default Event Procedure

Start a new project and drop a ListBox, a Timer, and a Label on Form1. Next, double click on the Form. This will bring up a code window. The code window will display the Form_Load event procedure. This is because the Form_Load event procedure is defined as the default event procedure. If you then double click on the ListBox control, you'll see the List1_Click event procedure displayed. Guess what? For any control, there is a default event procedure defined. This is the procedure that you see first when you display a code window for the control.

If you double click on Form1, you will see Form_Load displayed—the default event procedure. But wait, there are many more. Click on the drop-down combo to the upper right of the code window. All of the event procedures that are defined for the Form object will display in the drop-down. All of these procedures are empty, waiting for your custom code. Think of event procedures as notifications. When an event happens, the corresponding event procedure executes, providing you with the opportunity to react to the event.

■ **Check It Out: Event Procedures of the Form Object**

Let's look at some of the most fundamental event procedures found in Visual Basic: the event procedures of the Form object. The Form object has many event procedures of interest (see Figure 1-10). Let's look at two: the Load event procedure and the Unload event procedure.

1. Start a new project and double click on Form1 to display a code window. The Form_Load event procedure will appear. It is the default procedure for the Form object.

2. Add the following code to the Form_Load event procedure:

```
MsgBox "FORM LOAD"
```

Figure 1-10

The right side drop-down box of the form's code window displays the event procedures of the Form object.

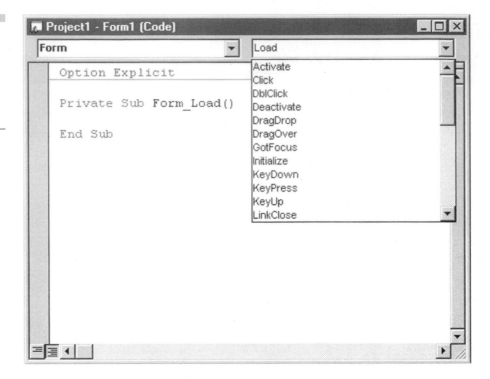

3. Now select the Unload event procedure by clicking on the Procedure dropdown that appears to the upper right of the code window, and click on Unload from the list. Add the following code in the Unload event procedure:

```
MsgBox "FORM UNLOAD"
```

4. Now click on the Run/Start menu command. You will see the output from the Form_Load event procedure's MsgBox statement: a modal dialog box displaying the text "FORM LOAD". What you are observing is the Load event procedure executing. This happens when the form comes into memory, but before it goes on screen.

5. Click the OK button of the MsgBox dialog. Now the form is visible. Click the ControlBox on Form1, and from the system menu, select the Close command. When you do this, the Form_Unload event procedure executes and displays a dialog with the text "FORM UN-LOAD". The Unload event procedure of the Form object executes before the form leaves memory, and before it goes off-screen.

6. The Form object has many other interesting and important event procedures. These are well documented in the Help file. (How would you find this important online documentation for the Form object? *Hint:* see "Leveraging the VB Help File.")

Event Procedure Firing Order

The event procedures of the Form object are well documented. What is not well documented is the relationship between all the individual event procedures that are part of the Form object.

Check It Out: Analyzing Event Firing Order

To understand the order in which various event procedures execute, you need to do some experimentation.

1. Double click on Form1 to bring up its code window.

2. Add the following MsgBox statement under the Activate event procedure:

```
MsgBox "FORM ACTIVATE"
```

3. Add similar MsgBox statements with appropriate descriptive text under the Initialize, Paint, and Resize event procedures.

4. Now run the project. What happens? What is the event firing order?

5. Now experiment. Resize the form smaller. Resize the form larger. Minimize, and then restore the form. What happens?

The Initialize, Load, Resize, Paint, and Activate event procedures are interrelated: When the form loads, they all execute in a certain order. Knowing the order is important if you are going to be a competent VB developer. This important knowledge concerning the event firing order of event procedures is very poorly documented in the VB manuals and Help file. We expect that will change now that this book is published.

Using Debug.Print

For many event procedures, you can experiment with placing MsgBox statements inside and then fiddle with the control or form and see what happens. You probably won't read the documentation before you do this. That could cause problems. For example, the Form_MouseMove event procedure will get you in trouble if you put MsgBox statements under it to see what it does. That's because the MouseMove event procedure executes every time you move the mouse one pixel. If you put a MsgBox statement under the MouseMove, it will execute and appear on screen. Next, you'll click on the OK button to dismiss it-but that will move the mouse a pixel, which will (you guessed it) execute the MouseMove event procedure again. And again. And again.

Clearly, the MsgBox statement is not your all-purpose solution when experimenting with event procedures. You want to use Debug.Print instead. A Debug.Print statement will write text to the Immediate window (also known as the Debug window by crusty, veteran VB1.0 developers).

Check It Out: Printing to the Immediate (Debug) Window

Using the same project with the MsgBox statements, see what happens when sending output to the Immediate window.

1. Double click on Form1 to bring up its code window.

2. Click on the procedure dropdown to the upper right. The event procedures will appear in the dropdown. Note that, for each event procedure under which you have placed code, that event procedure name is represented as bolded text. This indicates that it contains something, probably some code, but at least a comment.

3. For each event procedure with a MsgBox statement under it, replace it with a Debug.Print statement instead, such as:

```
Debug.Print "FORM LOAD"
```

4. Run the app, and then select the View/Immediate Window menu command. This will bring up the Immediate (or Debug) window. What do you see?

5. Form1 is still running, as a separate app. Switch the window to Form1 and close it. Which event procedure executes when you do this? You may need to use the View/Immediate Window menu command again, now that you are back to design time. (*Note:* If you clicked inside the Immediate window, you may have repositioned the cursor within that window. This would affect the position of the output of any Debug.Print statements that followed.)

6. Now put similar Debug.Print statements under these events: Deactivate, QueryUnload, and Terminate.

7. Observe what happens, and in what order, when you start and stop the application. To clear the contents of the Immediate window and start over, select its contents (in design time) and delete it.

Read the Help file for each of the event procedures of the form object you have tested. This technique of exploring the event firing order of event procedures is valuable to you when you begin to write production VB applications. For example, if you assumed that the Resize event procedure executed only when the form was resized, you'd be wrong, even though that assumption would be a reasonable one to make. By doing this little experiment, you know now that the Resize event executes right after the form loads. That is valuable information! Later, you will learn how to short-circuit the default behavior and make VB do what you say. For now, understand the relationships between event procedures of the form object. They execute in a predefined order, which you need to know *before* you start doing serious work in Visual Basic.

Things You Can't Do with Event Procedures

Event procedures impose a rigid set of rules on you as a developer. You cannot delete an event procedure. You cannot add a new event procedure. You cannot change the name of an existing event procedure. You cannot change the return value (if any) of an event procedure. Most event procedures do not include passed parameters, but some do. (Check out the Unload and QueryUnload event procedures for examples of this.) All event procedures consist of a name and a set of parentheses, which are typically empty, having no parameters passed in by VB or Windows. You cannot insert new passed parameters between the parentheses, nor can you remove any passed-in parameters that may appear between the parentheses. You also are prohibited from changing the name or data type of any passed-in parameters.

In short, you have two choices: Place code under event procedures of interest, or ignore the event procedure and do nothing with it whatsoever. Do event procedures execute whether you place code under them of not? *Yes,* they absolutely do! Event procedures make it easy to write event-driven Windows applications, but you do have to obey the rigid rules. That's the price you pay for the convenience of programming in Visual Basic.

Summary of Concepts: Events

An event procedure executes when a specific event occurs. Visual Basic objects provide you with empty event procedures. If you choose to respond to that particular event, you may put your code inside that event procedure. Events have a specific firing order. Understanding the firing order is key to successful VB development. VB has reduced the complexity of Windows programming, and with that come certain restrictions concerning the use of event procedures.

Methods

A method is a routine that has a name and is part of the object you are programming. It's a procedure that performs an action upon the object to which it belongs. For example, the AddItem method of the ListBox control appends a new item to the visible display of the ListBox object. By way of comparison, the RemoveItem method of the ListBox object deletes text items from the ListBox.

A method is a callable procedure that is part of an object such as a form, List-Box, or scroll bar. The method can be thought of as a kind of verb, or action, that can be performed upon the object.

Method Examples

Methods often come in pairs. For example, the ListBox object contains AddItem and RemoveItem methods. These are opposites that complement one another. The Form object supports a Show method, which displays the form, and a Hide method, which hides it from view. For a complete list of the methods of the form or ListBox, you can examine the Help file for each object.

The syntax of a method call is very similar to the syntax of a property of an object. Compare the following two lines:

```
ObjectName.PropertyName
ObjectName.MethodName
```

How do you tell the difference between a property and a method? The first thing to notice is that the property name will be a description of an attribute of the object. The method name will identify an action to be performed on the object. It will usually make obvious sense for the property to have a value that represents a quantity (a number) or a state (True or False, for example). In grammatical terms, a property is like an adjective that describes; a method is like a verb that does an action.

For now, you can experiment with the methods of the ListBox.

▆   Check It Out: Methods of the ListBox control

It's time to experiment with the ListBox control. Here you will work with some typical examples of methods.

  1. Start with a new project. Drop a ListBox on Form1.

2. Double click on Form1. Add the following code to the Form_Load event:

```
Form1.List1.AddItem "Thing1"
Form1.List1.AddItem "Thing2"
Form1.List1.AddItem "Thing3"
```

3. Run the app. List1 will display the text added to List1 with the AddItem method calls, which are coded under the Form_Load event procedure.

4. End the app by clicking on the Run/End menu command.

5. Drop a CommandButton on Form1. Attach the following code to the Click event of this button:

```
Form1.List1.RemoveItem 0
```

6. This button when clicked will remove the first item from List1. The RemoveItem method requires an integer value specifying the ordinal number of the item to remove. Since the first item in a ListBox starts with the number zero, this is what we have specified to delete the first item. Run the app. Click on Command1. What happens?

7. Click on Command1 to delete all of the items, then click once more. You get an error because you are trying to delete the first item, and there is no first item to delete. Congratulations! You are now in VB's Break mode. We will get to that later. For now, click on End form in the modal error message dialog box to return to design time.

8. Drop another button on Form1, and attach this code under the Click event procedure:

```
Form1.List1.Clear
```

9. Run the app. Click on Command2. What happens?

10. End the app. Click on List1, and press F1 to bring up the Help. For further study, check out all of the methods of the ListBox object.

The AddItem and RemoveItem methods of the ListBox object are typical of the way methods operate on an object. Once you learn the methods of one control, you will find that the knowledge may be applicable to many controls. The ComboBox control, for example, also has AddItem, RemoveItem, and Clear methods, just like the ListBox control. As you learn more about each control, you will see the common methods, and quickly move past the ordinary and mundane, easily identifying the very unique. With each new control that you learn about, you will recognize a method or two that are unique to that particular control.

Summary of Concepts: Methods

A method is a callable procedure. Methods do the work on an object. They often come in pairs, like the AddItem and RemoveItem methods of the ListBox control. You have seen a typical use of method calls, used from event procedures to make the object do its actions. In the exercise to follow, you will seem more of how properties, events, and methods interact.

Exercise: Properties, Events, and Methods

In the following exercise you will practice using properties, events, and methods for ListBox manipulation. Notice how a method is called from within an event procedure, but is conditional upon the value of a property.

1. Start a new Standard EXE project. Immediately save the project to a directory named VB BOOTCAMP/PEMS. The VB BOOTCAMP directory will hold all of the projects you will create when working with the exercises in this book. PEMS (Properties, Events, and Methods) is your first project. All of the elements of this first project will be saved to the PEMS directory.

2. Make Form1 visible if it is not already. Double click on it to reveal a code window and add the following code under the Form_Load event:

```
Form1.List1.Additem "Thing3"
Form1.List1.Additem "Thing2"
Form1.List1.Additem "Thing1"
```

3. Run the app by selecting the Run/Start menu command. List1 should display the three items you have added in the previous step. End the app by clicking on the Run/End menu command, returning you to design time.

4. Drop two ListBox controls on the form. Drop a CommandButton between them. Double click on the command button to reveal the Click event of Command1. You'll be placing code under the Click event of Command1.

5. The purpose of Command1 is to move the selected item from List1 to List2. Add the following code to the Click event procedure of Command1:

```
'CHECK TO SEE IF AN ITEM IS SELECTED:
If Form1.List1.ListIndex = -1 Then Exit Sub
'COPY THE SELECTED ITEM FROM LIST1 TO LIST2:
Form1.List2.AddItem Form1.List1.List(Form1.List1.ListIndex)
```

6. Run the app. Click on the first item in List1, and then click on Command1. What happens? The selected List1 text should appear in List2. End the app and return to design time.

7. Search the Help file for the following properties of the ListBox control:
   a. ListCount
   b. List
   c. ListIndex

8. Also, search the help for the following methods of the ListBox:
   a. AddItem
   b. RemoveItem
   c. Clear

9. Now search the help for this VB command:
   a. Exit sub

10. The code you attached to the Command1_Click event procedure performed the task of copying the selected text from one list box to another. One additional step remains. You must remove the text from List1, the source list box. Double click on List1 and append the following code lines to the very end of the event procedure:

```
'DELETE THE SELECTED ITEM FROM LIST1
Form1.List1.RemoveItem Form1.List1.ListIndex
```

11. Run the application and click the first item on List1. What happens?

12. With nothing selected in List1, click on Command1. What happens? Why?

13. The app you are coding is common to setup and initialization programs, where you can select from a list of possible choices to arrive at the final list of selections. You may have seen this type of user interface in the past. If you have, you know that usually, there is not just one button between the list boxes. There are usually four. One

button moves selected items from left to right, and another moves selected items from right to left. Two more buttons are used to move everything in one direction or the other. In this exercise, we'll create buttons only to move a single selected item at a time. We just created the first button that moves a selected item from left to right. Now we'll move it back.

14. Drop another button on List1. It will get the name Command2 by default. We will copy code from Command1 and change it slightly.

15. Command2 will do everything Command1 does, but it will move selected text form List2 into List1. The code in Command1 will be useful to do this. All we have to do is make a few changes, right? Double click on Command1, then select all of the code you entered in the previous steps. Click on the Edit/Copy menu command. This will copy the selected code to the Windows clipboard.

16. Now close the code window and double click on Command2. The Command2_Click event will display in the code window that appears. Click anywhere *inside* this event procedure and select the Edit/Paste menu command. This (in effect) copies the code from Command1 to Command2, by way of the clipboard. Save your work by clicking on the File/Save Project menu command. Remember to save often!

17. Now, edit the code you pasted. Carefully change it so that this code moves the selected text in List1 to List1. *Be careful* and deliberate. Change the comments to reflect what you intend the code to do. Change every reference to List1 to List2, and vice versa. The effect of these edits will be to enable Command2 with the functionality to move the selected item in List2 directly into List1. Save your work by clicking on the File/Save Project menu command.

18. Run the app. Click on the first item in List1 and click on Command1 to move it to List2. Now select the first item in List2, and click on Command2. Is everything working? If Command2 doesn't work, return to design time. Go back to the code under Command2_Click and check it over. Is it a good idea to copy code this way?

19. You may have noticed that items in both ListBoxes to do not appear in sorted order, which is something that most users have come to expect. Click on List1 and change the Sorted property to True. Do the same for List2.

18. Save your work and run the app. All items should appear sorted in both ListBoxes, regardless of the order in which they were added to the ListBox.

SUMMARY

A large part of the interface of a control that you use is its properties. These are attributes of the object. Objects usually have some methods, which are callable procedures that do its actions. Events fire event procedures in a specific order. You may call methods from within event procedures that you care about to get the work done.

Properties, events, and methods are the interfaces that you have to each control. The interface that is exposed to you is the control. Mastering the fundamentals of the properties, events, and methods of each control is essential to use a control properly. If you just skim over these fundamentals and move on, then you may never really understand VB completely.

Increasing Your Knowledge

Help File Search Strings

In the last exercise, you looked up just some of the properties and methods of the ListBox control. Starting with the Help topic on the control itself, look up all of the properties, events, and methods of each of the following controls:

- ListBox Control
- Label Control
- CommandButton Control
- TextBox Control
- Form Object

Books Online References

- "Understanding Properties, Methods and Events"
- "The Basics of Working with Objects"

Mastery Questions

1. Which property sets a value that determines whether a form can be moved?
   a. MDIChild property
   b. FixedMode property
   c. Moveable property
   d. ScaleMode property

2. The StartUpPosition Property identifies where
   a. a form will be displayed when first shown.
   b. a control will be displayed when first shown.
   c. the data control will set its first record at start up.
   d. the current cell of the MSFlexGrid will start off.

3. The DisabledPicture Property sets a reference to a picture to display in which of the following controls when the control is disabled?
   a. CheckBox control
   b. OptionButton control
   c. CommandButton control
   d. PictureBox control

4. The Style property of a ListBox control determines
   a. whether it will be a pull-down list box or a standard list box.
   b. whether graphical images can be displayed instead of a list of text.
   c. whether graphical images (icons) can be displayed next to each text item.
   d. whether a checkbox will appear to the left of each item to show if it is selected.

5. Which properties may be changed for a form from the Form layout Window?
   a. Top and left
   b. Width and height
   c. StartUpPosition
   d. WindowState

6. Which of the following statements are true about the ToolTips property?
   a. It may be set to True in order to enable, or False in order to disable the ToolTip feature.
   b. You must set the ShowTips property to True to display ToolTips.
   c. Any value of text in the ToolTips property will enable the ToolTip feature in most controls.
   d. The maximum length of a ToolTip text is 255 characters.

7. The ItemCheck Event of the ListBox control
   a. occurs when a list item is highlighted.
   b. occurs when a list item is selected.
   c. occurs when the checkbox of the list item is selected.
   d. occurs when the checkbox of the list item is cleared.

8. The MultiLine property applies to which of the following controls?
   a. Line control
   b. ListBox control
   c. Label control
   d. TextBox control

9. The MultiLine property is
   a. not available at run time.
   b. read-only at run time.
   c. read/write at run time.
   d. None of the above

10. The Change event applies to
    a. the ComboBox control.
    b. the DirlistBox and DriveListBox controls.
    c. the Hscrollbar and Vscrollbar controls.
    d. the Label control.

11. The Activate and Deactivate events occur
    a. only when moving the focus from another application.
    b. only when moving the mouse from another application.
    c. only when moving the focus within an application.
    d. every time the mouse moves to another form in the application.

12. The left, right, and middle mouse buttons may be distinguished in which events?
    a. Click event
    b. MouseDown event
    c. MouseUp event
    d. MouseMove event

13. KeyDown and KeyUp events never occur for which of the following keys?
    a. Enter
    b. Delete
    c. Tab
    d. Escape

14. Which of the following event procedures have parameters that indicate the physical state of the keyboard?
    a. KeyDown
    b. KeyUp
    c. KeyPress
    d. KeyPreview

15. Which of the following methods may invoke a Paint event?
    a. Paint method
    b. Repaint method
    c. Show method
    d. Refresh method

16. Which event(s) occurs as a user repositions the scroll box on a scroll bar?
    a. Scroll event
    b. Change event
    c. Reposition event
    d. All of the above

17. The Clear method
    a. clears all property settings of the Err object.
    b. clears the contents of a ListBox, ComboBox, or the system Clipboard.
    c. clears all objects in a collection.
    d. clears graphics and text generated at run time from a Form or PictureBox.

18. The Circle Method applies to which of the following objects?
    a. Form object
    b. PictureBox control
    c. Image control
    d. Shape control

Answers to Mastery Questions

1. c

2. a

3. a, b, c

4. d

5. a, c

6. c

7. c, d

8. d

9. b

10. a, b, c, d

11. c

12. b, c

13. c
14. a, b
15. c, d
16. a
17. a, b, c
18. a, b

CHAPTER 2

Multiple Forms

What You Need to Know

You need an understanding of VB design time use and the basics of properties, events, and methods to get the most from this section of the book.

What You Will Learn and Why You Care

This part of the book shows you how to connect forms together to craft a complete application. You need to know about this because applications that consist of a single form are quite useless.

Overview

The Form object has some special methods you'll be using a lot: the Show method and the Hide method. Guess what they do? There are also some associated commands that are intrinsic to VB: Load and Unload. Loading a form fires the Load event of the Form object. Unloading a form fires the Unload event of the Form object. Let's go to work by linking two forms together.

▨ **Check It Out**

1. Start a new VB project. Add a new form, resulting in two forms for the project: Form1 and Form2.

2. Both forms will display in the same location at run time, so you need to move Form1. Click on the View/Form Layout Window menu command to reveal the Form Layout window. Click on the form that appears (this will be Form2) and drag it to the lower right. You should see Form1 remaining to the upper left, which is what you want. (See Figure 2-1.) For more information on Form Layout, search Help for "Form Layout".

3. Drop a button on Form1. Add this line under the Click of Command1:

```
Form2.Show
```

This code will bring up Form2 when you click on Form1.Command1.

Figure 2-1
The Form Layout window.

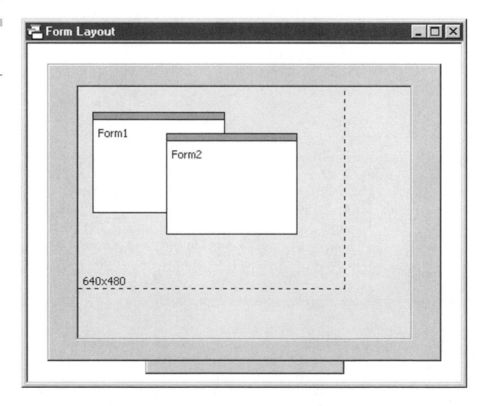

4. Run the app and click on Command1. See Form2? Size Form2 smaller and position it on Form1. Now click anywhere on Form1. Form2 disappears. To where? Move Form1, and you will see Form2 behind Form1, which obscured Form2. Form2 allowed this because it was not shown "modally."

5. End the app and search Help on Show.

A modal form is a form that halts execution of the application when it is displayed. In truth, the Windows operating system defines windows and something called a modal dialog box. For example, the MsgBox statement displays a modal dialog box with your message constrained within it. When the Msgbox displays, execution of your application halts. This is known as application-modal behavior because you can still start and interact with some other application while your application is in a modal state. (The Msgbox statement can be utilized to display a "system-modal" dialog, which locks the entire system.) In the same manner as the Msgbox statement displays modal behavior, so can your forms. We will now go back and make this happen in the small two-form application you have going now.

Check It Out

1. Double click on Form1.Command1 to reveal a code window with the Command1_Click event displayed for editing. Change this line:

```
Form2.show
```

to this:

```
Form2.Show 1
```

2. Run the app. Size and position Form2 to sit on top of Form1. Now click on Form1, attempting to obscure Form2. What happens? Form2 is now being displayed modally. You must close the window for execution to resume. Click on the ControlBox (located in the upper left corner of Form2) and click on the Close system menu command to dismiss Form2.

You may connect as many forms as you want by using the Show method of the Form object. You may not know this, but the Show method will load the form if it is not already loaded—this is the default behavior of the Show method. This can be easily proved and will be in the next step.

When you click on the system menu (ControlBox) of a form and click on the Close menu command, you fire the Unload event procedure. This also can be proven by experimentation, as follows:

▮ Check It Out

1. Add this code to the Load event procedure of Form2:

```
Msgbox "FORM2 LOADING"
```

2. Add this code to the Unload event of Form2:

```
Msgbox "FORM2 UNLOADING"
```

3. Run the app. Click on Form1.Command1. What happens? The Show method is automatically loading Form2, firing the Load event procedure, and executing your code.

4. Click on the system menu of Form2 when it appears. Click on the Close menu command. What happens? Clicking on the Close menu command automatically fires the Unload event procedure and executes your code. Now the form is no longer in memory.

5. End the app and return to design time.

This approach is fine for simple apps, but realize that code to add to event procedures of the form increases the size of the form. Also note that any code you attach to event procedures of controls located on the form also bloats the size of the form. A form can grow quickly in size, which will slow the load time. You therefore want to preload your key forms into memory when the entire application loads, and keep them in memory for the lifetime of the application. This way, the form will appear faster when you execute the Show method.

The way to preload forms is to perform the action at startup. Visual Basic provides a special way to initialize your entire application by using a special procedure. This procedure is called Sub Main(). Sub Main must be located in a module and must be declared Public. When your app has a Sub Main, and you indicate to VB that Sub Main should be used as the Startup object, VB will run the procedure Sub Main() and execute all of your VB program statements found there. You can, therefore, preload your most important forms when the entire app loads for the first time, by performing the loading action in the Sub Main() procedure. This is exactly what we'll do next.

Check It Out

1. Click on the Project/Add Module menu command. Visual Basic will add a module named Module1 to your project.

2. Now click on the Insert/Procedure menu command. The Add Procedure dialog will appear. Specify Main as the name of the procedure, and make sure it is specified as Public (not Private) and as a Sub (not a Function). Click on OK when you are done.

3. Add these lines to Sub Main:

```
Load Form1   'Preload Form1
Load Form2   'Preload Form2
Form1.Show   'Show startup form
```

4. You have defined Sub Main and included your statements for application startup, but there is one more thing to do. You must specify to VB that Sub Main should be used as the Startup object.

5. Click on the Project/Project1 Properties menu command; a modal dialog will appear. (See Figure 2-2.)

6. Click on the Startup Project drop-down combo. Specify Sub Main as the Startup object.

7. One last thing: Add this line of code under the Load event procedure of Form1:

```
Msgbox "FORM1 LOADING"
```

8. Run the app. What's going on? Form1 is loading from Sub Main. Also, Form2 is loading from Sub Main. Finally, Form1 is showing from Sub Main. The Msgbox statements under the Load event procedures of Form1 and Form2 execute as a result of Sub Main() executing as the Startup object.

9. Now, if you click Command1 on Form1, Form2 will not load, because it is already in memory.

10. Dismiss Form2 by closing it. What happens? Form2 unloads and leaves memory.

Figure 2-2
The Project Properties
dialog.

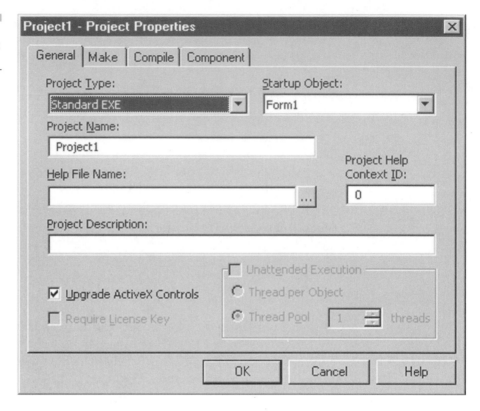

11. Click Command1 on Form1 again. Why does Form2 load? Because it
is no longer loaded, and the Show method therefore loads it automatically.

12. End the app. Return to design time.

The only thing missing here is a way to make Form2 leave the screen
while keeping it in memory. This is accomplished with the Hide method.
If you drop a button on Form2 and add this code under the Click event
procedure, you will have a way to accomplish the task:

```
Form2.Hide
```

The right way to manage forms is to understand that a few of your
forms need to be preloaded in memory all the time. Your startup form, the
forms that implement your major function points, and so on need to be preloaded from Sub Main() to ensure fast and responsive performance. Forms
such as Preference dialogs and About boxes need not be in memory all of

the time. In other words, 20% of your forms are typically used a whole lot, while the remaining 80% are barely used at all. It's the 20% that need to be preloaded from Sub Main() and then hidden and shown using the Hide and Show methods.

SUMMARY

You have now used your knowledge of Show and Hide methods of the form object to accomplish something useful. You understand how to connect forms together; this understanding becomes the basis for creating prototypes and setting up the overall structure of your applications. You also now have worked with Sub Main() and understand how and why to preload your most frequently used forms.

At this point, you've done quite a few Check It Outs, and you must be thinking: What about the VB language? What about data variables and data types? What about functions and subroutines? What about any built-in functions VB might provide? What about managing my projects? What about error handling and debugging? What about building and distributing an executable? What about menus and online Help? And most important, what about database connectivity? All of these things and more will be covered as discrete topics as your move through the book. As of now, you are in a good position to learn all of these things.

Increasing Your Knowledge

Help Search Strings

- Hide method

- oad event

- Unload event

Books Online Search Strings

- "Show"

- "Hide"

- "Load "

- "Unload"

- "QueryUnload"

- "Visible Property"
- "Initialize"
- "Terminate"

Mastery Questions

1. Another way to show a form without using the Show method is to
   a. Call the Activate event in code
   b. Set the Visible property to True
   c. Call the Paint event in code
   d. None of the above

2. Since loading forms can be time-intensive, it is best to:
   a. Load forms on demand to reduce perceived loading time
   b. Load all forms in advance while displaying a 'splash' screen
   c. Use the Load keyword and then the Show method to display each form as demanded by the user
   d. Use the Load command to load the most-often-used forms at startup so they are preloaded when demanded by the user

3. The Unload event
   a. Follows the QueryUnload event
   b. Follows the Terminate event
   c. Occurs before the Terminate and after the QueryUnload event
   b. None of the above

Answers

1. b

2. d

3. c

System Objects

What You Need to Know

You need an understanding of VB design time use and the basics of properties, events, and methods to get the most from this section of the book.

What You Will Learn

This part of the book reveals an aspect of VB that you will not be aware of unless you read the Help file in sorted alphabetical order. You could also get this information from skimming the VB documentation. You care plenty about system objects because they come with every VB project and the set of system objects you receive is a useful resource is building robust VB applications.

Overview

Every time you start a new VB project, that project automatically receives a set of system objects. These objects, such as the Screen and Clipboard objects, can help your application in many ways. There is a complete set of system objects that typically have useful properties and methods. For example, to get the horizontal width of the screen at run time, you could simply use this line of code:

```
Msgbox "The screen is " & Screen.Width & " twips in width."
```

We shall describe what a "twip" is later. For now, notice and understand that every application written in VB has access to the Screen object. You can learn more by searching on "Screen object" in the Help file. In this section of the book we cover all the system resource objects you get for free with every VB app:

- **App Object:** Provides access to system information
- **Screen Object:** Provides information on the Screen state
- **Clipboard Object:** Reads from and writes to the Windows clipboard
- **Error Object:** Provides info on run time errors
- **Debug Object:** Allows you to write program output to the Debug window
- **Printer Object:** Writes output to the printer
- **Forms() Collection:** Provides access to every loaded form in the project
- **Controls() Collection:** Provides access to every control on the active form
- **Printers() Collection:** Provides access to installed printers on your system

All of these objects are available to all VB applications all the time. We will now cover each in turn. You will note the last three items are not objects at all, but collections of objects. These will be covered last as an introduction to the concept of a Visual Basic collection, a kind of list of objects you will see in VB programming.

Please keep in mind that this discussion of the system objects is an introduction designed to acquaint you with these VB application resources. You will need to examine the Help file concerning each object, study the example code in the Help file, search the VB Books Online, and do the exercise that follows to get a good grip on how to use these objects. The Printer object, for example, is very large, and a thorough treatment would take a great many pages. An exhaustive treatment of each object is beyond the scope of this book. You are expected to explore, experiment with, and understand these objects through hands-on exercise and a complete reading of the applicable VB documentation.

The App Object

The App object has many useful properties; it has just two methods. One of the more useful properties of this object is the Path property. It reveals where the application is located in the file system on the computer. For example, if your application (or its executable) is located in the directory named C:\VBBOOTCAMP\MYAPP\, then the following line of code will produce the output shown in Figure 2-3.

```
Msgbox "This app was launched from: " & App.Path
```

Figure 2-3
Msgbox showing the message "This app launched from C:\VBBOOTCAMP\MYAPP".

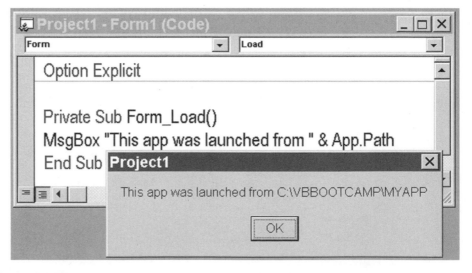

The general syntax of system object usage follows this pattern:

```
object.property
```

or

```
object.method
```

All of the system objects including the App object follow this general syntax. There is nothing special you have to do in code to obtain access to the App object or any of the other system objects. They are available automatically in every VB application. The system objects are extremely useful to your applications! Table 2-1 shows a partial list of what the App object offers in terms of useful functionality.

The Screen Object

The Screen object provides useful information about the screen's resolution and state. It provides access to the form and control with the focus, and other information such as the fonts that are supported. The following example reports the resolution of the screen in pixels:

```
MsgBox _
"Screen is " & Screen.Width / Screen.TwipsPerPixelX & " wide in
pixels."
```

This example introduces some new information. First, the continuation character in VB is a single blank followed by an underscore. The "space

	Property Name	Comment	Usage
TABLE 2-1:			
Useful members (properties and methods) of the App object	Path	Provides the working directory for the application.	Stringvariable = App.Path
	TaskVisible	Specifies whether the app is visible in the Windows Task Manager	App.Taskvisible = [True I False]
	StartMode	Indicates how the application was started, either as a standalone app or as a component used by another application.	Integervariable = App.Startmode

underscore" tells VB that this line of code continues on the following line in the code module. This example also introduces a new measure of screen resolution called the twip. The terminology is from the world of typography, where a point is defined as 1/72nd of an inch. The twip is 1/72nd of a point, or 1/1440th of an inch. The default unit of measure for VB is the twip. For example, the screen and the form object use the twip as the default unit of measure.

The screen provides the TwipsPerPixelX and TwipsPerPixelY properties. These allow you to derive the number of pixels on the screen by dividing the twip-measured width by TwipsPerPixelX. A similar calculation can be done to determine the vertical pixels of the screen.

ActiveForm and ActiveControl

The active form and the active control are supported as properties of the Screen object. This code will report the window title of the form in your project that currently has the focus:

```
Msgbox _
"The Caption of the form with the focus now is " & _
Screen.ActiveForm.Caption
```

Note how Screen.ActiveForm is equivalent to specifying the name of the form. When you specify Screen.ActiveForm, all of the properties and all of the methods of that form are available to your program. You can even determine the name of the active form by sampling the name property, like this:

```
Msgbox _
"The Name of the form with the focus now is " & _
Screen.ActiveForm.Name
```

You can also find which control has the focus by simply sampling the value of the ActiveControl property. This property provides access to the properties and methods of the control that has the focus on the active form. For example, this code will report on the name of the control with the focus on the active form:

```
Msgbox "The active control right now is: " &
Screen.ActiveControl.Name
```

You'll want to examine the Help file to learn more about the properties of the Screen object.

The Clipboard Object

The Windows clipboard is a shared area of memory that all applications on the Windows desktop can read from and write to. You have probably used the clipboard many times from Word, Excel, or other applications. The basic idea with the Clipboard object is that you can copy text and even images to this area of memory, and other applications can paste it into their own documents or forms. Here is a simple example for you to try:

Check It Out: Writing to the Clipboard

1. Start a new Standard EXE project. Drop two textboxes and two buttons on Form1. The control names should be Text1, Text2, Command1, and Command2 by default. Do not rename the controls, just stick with the VB-assigned names.

2. Under the Click of Command1, add this code, which will copy the textbox text from Text1 to the Clipboard:

```
Private Command1_Click()
  Clipboard.Clear 'Clear the Clipboard contents
  If Text1.Text <> "" Then 'if Text1 contains something…
    Clipboard.SetText Text1.Text  'Copy Text1.Text to the Clipboard
    Msgbox "Text1.Text contents is now in the Clipboard."
    Msgbox "You can start NotePad or Word and select Edit/Paste!!"
  Else
    Msgbox "NOTHING in Text1 to copy to clipboard."
  End If
End Sub
```

3. Run the application. Enter some text into Text1 and click on Command1. The contents of Text1 will be copied to the Windows clipboard. Next, run a word processor such as Write, NotePad, WordPad, or Word and select the Edit/Paste menu command. The contents of the clipboard will be copied into your document. You should see the data from your VB application in the word processing document. You have copied data from your VB application into another Windows application with just a few lines of code.

4. Pasting from the clipboard does not clear it out. You can paste from the clipboard into any application that supports such data transfers. You can paste the same text 8 times to the same document if you wish. You can even paste the clipboard data into your own VB application. This is the next step. Add this code under the Click of the second button named Command2:

```
Private Sub Command2_Click()
  Text2.Text = Clipboard.GetText(1)  'get text from clipboard
  'GetText(1) indicates the format of the text to copy
End Sub
```

5. Run the application. Add some text to Text1, then click on Command1 to copy it to the Clipboard. Next, click on Command2, which will copy whatever is in the Clipboard to Text2.Text.

As you can see, it is simple to write to, and read from, the Windows clipboard. Keep in mind that the Clipboard object is simply the VB programmer's interface to what is in essence a Windows desktop facility, shared by all applications.

The Error Object

The Error object is very interesting and is covered completely in the section on error handling and debugging. For now, understand that the Error object is used to manage errors raised by VB. You can also "raise" an error in VB code, but when you do this you must be sure to use an error code that is not reserved by VB itself. There is a way to print out all of the error codes reserved in VB; you will learn to do this shortly in the sample that follows.

The most important properties of the Error object are the Number and Description properties. Err.Number gives you the code for the error, and Err.Description provides the associated descriptive text that matches the code. Table 2-2 lists some of the properties of the Error object.

Error information is typically sampled when a trappable error occurs. You do this by reading the Err.Number and Err.Description properties of the Error object. You can create and then raise your own custom error code by using the Err.Raise method. You do not need to be too concerned about this right now, because this book contains a section that deals with these issues completely. Later, you will learn the ins and outs of trapping errors. For now, be aware that the Error object exists and that you can exploit it to good effect in your VB applications later in the learning program, when you learn how to trap errors.

The Raise method sets the Err.Number to the number you specify. This has the effect of allowing you to add new errors to the syntax and processing errors that VB already supports. For example, you could use one of the unreserved error codes (listed in the next Check It Out) to force an error that will be trapped by your program. The idea here is that your application-defined error can be responded to by your error trapping procedure,

TABLE 2-2:

Selected properties
of the Error object

Property Name	Comment	Usage
Number	Indicates the error number of the most recent error	Integervariable = Err.Number
Description	Provides the text associated with the last error number	Stringvariable = Err.Description
LastDLLError	Indicates the last error code pertaining to the last call to a procedure in a Dynamic Link Library. For 32-bit Windows only. (VB cannot trap DLL errors since they occur in the address space of another process.)	Longvariable = Err.LastDLLError
Source	Indicates the project where the error originated, useful when using components.	Stringvariable = Err.Source

just like the standard trappable VB runtime errors. Don't be too concerned about trapping errors at this point, as the subject is covered in a complete manner in a subsequent chapter. For now, understand that the Error object exists, it comes for free with each VB application, and it is very useful when the time comes to handle trappable runtime errors in your VB programs.

The Debug Object

The Debug object has two methods: Assert and Print. We will cover Assert later, since it is new to VB with VB5. The Print method writes text output to the Immediate window (which was formerly called the Debug window). An example of the syntax is:

```
Debug.Print "This is a sample of writing to the Immediate window."
```

It uses the standard Object.Method syntax and is very useful in the process of debugging your programs. What you print may be a status message or the value of a variable at a key point in your code where you insert the Debug.Print statement.

Another valuable feature of this handy statement is that VB will execute it in the design environment, but will completely ignore it when your application is compiled into an executable. Also, the Immediate window may be used as a quick and easy way to view program output that might otherwise go to the printer. Use the following exercise to list the error codes in Visual

Basic, with just a few simple lines of code that uses the Error() function that is built into the VB language. The following little exercise lists out error numbers and descriptions to the Immediate window.

Check It Out: Debugging Basics

1. Start a new Standard EXE project. Drop a button on Form1; it will be named Command1 by default.

2. Place the following code under the click of Command1. This code will list out the first 100 reserved VB error codes to the Debug window:

```
Private Sub Command1_Click()
  Dim i As Integer
  For i = 1 To 100
    Debug.Print i, Error(i)
  Next i
End Sub
```

3. Now run the application. Select the View/Immediate Window menu command. This will display the Debug (or Immediate) window. Size and position the Debug window and Form1 to be side by side.

4. Click on Command1 and 100 lines of output will come spewing out of the Debug window. When the 100th item prints, check out the contents of the Debug window. These are error codes reserved by VB. Note that some of these may say "Application-defined or object-defined error." These are codes that are unreserved and not currently used by VB.

The Error object also has a set of associated constants. These will be discussed when the entire topic of error handling and debugging is covered in a subsequent chapter.

The Printer Object

The Printer object is the largest system object available to you. It has more properties and more methods than any of the other system objects.

The basic idea with the Printer object is that it is an output-only object; all you can do is write to it. Many of the Printer object's properties and methods are similar to the properties and methods of the Form object. For example, the Printer object has Circle and Line methods just as the Form object does.

To send output to the printer using the Printer object, all you do is this:

```
Printer.Print "YOUR MESSAGE HERE"
```

Windows will make an attempt to send output to the printer designated as the default printer. If that printer is properly connected, the printer will receive the output.

Along with the printer object, you get the Printers() collection. This will be discussed under the section that follows on the Forms and Controls collections. Table 2-3 lists some of the more interesting properties and methods of the Printer object.

Printing Your Source Code

At this point, you might be asking: "Speaking of the printer, just exactly how do I get a printed program listing?" Although not directly related to programming the Printer object, there is a facility for printing out your source code. The File/Print menu command provides a small dialog that allows you to specify the current module or form. There is also an option for printing the entire project. Figure 2-4 depicts the Print dialog.

TABLE 2-3:
Selected, Useful Properties and methods of the Printer object.

Properties	Description	Usage
Copies	Specifies how many copies	Read-write runtime only
Devicename	Returns the name of the printer	Read only
Orientation	Specifies which way print appears, landscape or portrait	Read-Write
Papersize	Specifies the paper size to be used.	Read-write

Methods	Description
Print	Prints to the Printer object
KillDoc	Terminates the current print job
NewPage	Ejects the current page and also a blank page

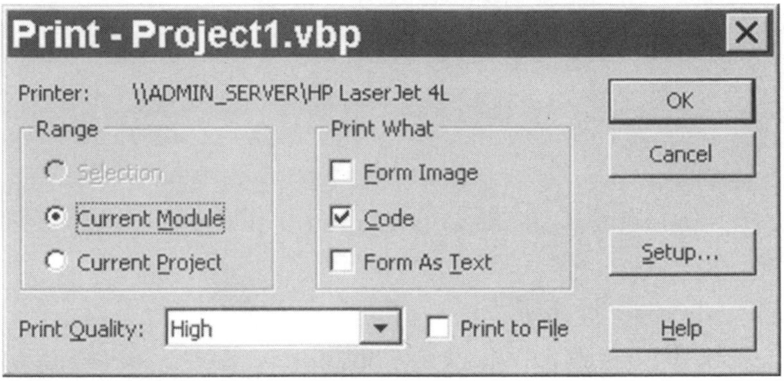

Figure 2-4 The Print dialog.

The Forms() Collection

In addition to the system objects that come with every VB project, there is also a group of resources known as collections. These are lists of useful items you may wish to exploit in your VB programs. For example, there is a list of every loaded form in your project, maintained automatically by the system in the Forms() collection.

To examine the contents of the Forms collection, use the For Each…Next looping construct that is part of the VB language. It works like this: For any array or collection, the For Each…Next construct allows you to iterate over the entire list without knowing the number of items it contains. Here is an example:

```
Dim FormItem as Form
For Each FormItem in Forms()
  MsgBox "Form " & FormItem.Caption & " is in the Forms() collec-
tion."
Next
```

Since the entire VB language is covered in a subsequent chapter, a brief discussion of this little bit of code is in order. This line:

```
Dim FormItem as Form
```

reserves storage for a variable named FormItem. This variable can contain a reference to a form object because it is declared as type Form. (You could name the variable whatever you want. You do not have to use this name.) The variable is used for referring to a particular Form object. As you know, the Forms() collection contains all of the forms in the project that are currently loaded. To access the entire list, the For Each...Next statement is used to iterate over the entire list without the need to know how many items are in the list. This line:

```
For Each FormItem in Forms()
```

sets up the loop so that each item in the list will be processed. With each iteration through the collection, FormItem is assigned a reference to a form in the collection. The first, then the second, then the next form is referred to by the variable FormItem. This allows you to address each item in turn until all the items are processed.

For example, this line:

```
MsgBox "Form " & FormItem.Caption & " is in the Forms() collection."
```

uses FormItem.Caption to refer to the Caption property of each item in the collection, one per iteration, and report the value of the Caption property via a simple MsgBox statement.

Check It Out: The Forms Collection

1. Start a new VB project. Add three more forms. They will be named Form2, Form3, and Form4.

2. Add this code to the Form_Load event procedure for Form1:

```
Load Form2
Load Form3
Load Form4
```

3. Close the code window and display Form1. Drop a button named Command1 on Form1. Add this code to the Click event procedure of Command1:

```
Dim FormItem as Form
For Each FormItem in Forms
  MsgBox "Form " & FormItem.Caption & " is in the Forms()
collection."
Next
```

4. Start the application, and click on the Command1 button. What happens? You should see a MsgBox with the form caption for each form

loaded in your project. (*Note:* The order that VB reports these forms
is not predictable. Which item is first, for example, can vary. Usually
VB reports the forms in reverse order, that is, last item first.)

5. End the application with the Run/End menu command.

6. Now go back to the Form_Load of Form1 and erase or remark out
this line:

```
Load Form4
```

7. Run the project again. What happens? Form4 is in the project, but
not loaded, so it is not part of the Forms() collection.

8. End the application and return to VB design time.

Collections like the Forms() collection are yours free with each VB project. For example, you also get the Fonts collection, the Printers collection, and the Controls collection of the Form object. Each collection works
in the same way as the Forms() collection we just discussed. You just iterate over all the items with the For Each...Next loop construct.

The Controls() Collection

The Controls collection is a part of each form. You can iterate over the
Controls() collection in the same way as you can over the Forms() collection, but it is important to note that the controls collection is part of a
form, not a part of the entire project. Each form has a Controls() collection. For example, to report all the controls that are on a form, place this
code under a button on the form, and click on the button:

```
Dim ControlItem As Control
For Each ControlItem In Form1.Controls()
  MsgBox "Form1 contains this control: " & ControlItem.Name
Next
```

In this example, we report the Name property rather than the Caption
property. The Name property is used in code to refer to the object, while
the Caption is simply descriptive text that is displayed on the object as a
kind of label.

There is a way to dynamically add new controls to a VB form at run
time. When you do this, you'll want to examine the Controls() collection
to refer to each control you have dynamically added. (There is also a way

to add new forms to the application dynamically at run time. More on this later.)

The Printers() Collection

The Printers collection provides a simple way for you to get access to all of the printers installed in Windows, not just the one that is currently set as the default. You iterate over the Printers collection the same way you would iterate over the Forms() and Controls() collections, by using the For Each...Next looping construct.

Searching the Help file on Printers will reveal more information. Here is a simple code sample that reports the names of all the installed printers and specifies which one is the default:

```
Dim PrinterItem As Printer   'This is a VB reserved word and object
type
For Each PrinterItem In Printers
  MsgBox PrinterItem.DeviceName & " is installed."
  'The Printer object points to the default printer…
  'PrinterItem is the current printer item in the Printers
collection.
  'Question: Are they one and the same?
  If PrinterItem.DeviceName = Printer.DeviceName Then
    MsgBox PrinterItem.DeviceName & " is also the DEFAULT."
  End If
Next
```

You can attach this code to the click of a button on a form and see the results for yourself in just a couple of minutes. Check it out. It will list every installed printer and tell you which one is the default right now.

The Help file has more to say about the Printer object and the Printers collection, but we will not address these other details now. If you are comfortable with what has been presented so far concerning the Printer object and the Printers collection, you are in great shape. Ditto for all the other system objects presented in this chapter.

SUMMARY

All of these system-supplied programmable objects are extremely useful. The new VB user typically discovers these assets after 6 months or more of VB use. The main reason for this is that no self-respecting developer examines the printed documentation. This is unfortunate, because the documentation has a ton of useful information that you need to really exploit the full power of the product.

EXERCISE: A SIMPLE WORD PROCESSOR

This exercise will explore multiple forms and the Printer object to create a small but functional word processor. You will learn how to read and write text files from the disk as an additional lesson in this exercise.

Hardware and Software Requirements
Visual Basic Version 5.0 Pro or Enterprise

Required Files
None

Assumed Knowledge
Knowledge of menu creation

Knowledge of properties, events, and methods concepts

Knowledge of App and Printer objects

What You Will Learn
In this exercise, you will learn how to do file input and output in VB and how to use Multiple Forms, Menus, and System Objects.

Summary Overview of Exercise

Creating the User Interface
In this section of the exercise, you will set up a TextBox on a form to be a word processing document and set up a menu to access the functionality that will be created.

Writing Text Output to a File
In this section you will learn how to open a file and write to it, and then test the results of your output.

Reading Text Input from a File
In this section you will learn how to open a file and read from it. You will also put a Do...Loop construct into action.

Printing Text to the Printer

In this section you will use a very powerful and yet simple code statement to send your text to the printer.

Copying Application Data to the Clipboard

Creating the User Interface

Step 1: Setting Up

Create a directory named TINYWP (Tiny Word Processor). Start a new Standard EXE project and immediately save everything to this directory. Next drop a textbox on Form1. A textbox can be used as a kind of word processor, if you change certain properties. Size Text1 to occupy nearly the entire form. Then click on the TextBox and go to the Properties window. Set the MultiLine property of the TextBox to True, and set the ScrollBars property of Text1 to 3-Both. These two settings will allow multiline input and provide vertical and horizontal scrollbars, much like Word or NotePad.

Step 2: Resizing at Run Time

When the form is resized, you will want Text1 to occupy the entire "client area" of the form. (In Windows parlance, the client area of a window excludes the border, menu bar, and title area, but includes the interior of the window.) To make Text1 occupy the entire client area of the form at run time, you will need to attach code to the Resize event procedure of Form1 that will automatically resize Text1 to occupy the entire Form1 each time Form1 is resized. Here is the code you need to add to the Resize event procedure of Form1:

```
Text1.Top = 0    'Upper left vertical
Text1.Left = 0    'Upper left horizontal
Text1.Width = Form1.ScaleWidth   'Resize Text1 horizontal
Text1.Height = Form1.ScaleHeight   'Resize Text1 vertical
'Check Help file for info on ScaleHeight, ScaleWidth properties of
Form
```

Run the application and resize, minimize, and maximize Form1. Text1 should occupy the entire form at all times.

Step 3: Creating a Menu

The next step is to be able to enter some text and then to write that text to a file. For simplicity, we will always write to and read from the same text file. The name of the file will be SAMPLE.TXT, and it will be written to your application's project directory. To write out the text to a file, we'll use a menu structure and create a menu item to handle the writing of the text to a file. Add the following menu structure to Form1 using the Menu Editor:

APPEARANCE OF MENU ITEMS:	NAME OF MENU ITEMS:
File	mnuFILE
New Text File	mnuNEW
Open Text File	mnuOPEN
Save Text File	mnuSAVE
Print Text File	mnuPRINT
Copy	mnuCOPY
————	mnuSEP1
Exit	mnuEXIT

This menu structure will allow us to manipulate text in our word processor. Make sure you properly name these menu items in the Menu Editor and also properly indent them. If you are unsure, you need to go back and read up on how to define a menu using the Menu Editor.

Step 4: Ending the Application
We will need a way to end the application. Add this code to the mnuEXIT_Click event procedure:

```
MsgBox "Tiny Word Processor Is Terminating…."
End  'This reserved VB word kills the application.
```

Run the application and then click the File/End menu command. The application should terminate.

Writing Text Output to a File

Step 5: Writing the File
To write out the text you enter as a file, you must perform text file operations. Under the mnuSAVE_Click event procedure, add this code:

```
'WRITE OUT THE TEXT TO A FILE:
Close #1
'OPEN THE FILE FOR OUTPUT, OVERWRITE EXISTING DATA IF ANY:
'Note the use of App.Path
Open App.Path & "\SAMPLE.TXT" For Output As #1
'WRITE THE TEXT TO THE FILE
Print #1, Text1.Text
```

```
'CLOSE THE OPEN FILE HANDLE
Close #1
'ISSUE A MESSAGE
MsgBox "Text saved."
```

Simple file I/O is based on a file "channel" or "handle". This is nothing more than a number that specifies to VB an open connection to a specific file for input or output. You are using "#1". The Print statement uses this file handle to do the job of writing out the text to the file. Once this is done, the file handle is closed using the Close keyword. You should now look over the Help for these search items: "Open Statement", "Close Statement", "Write Statement", and "Print Statement".

Step 6: Checking the Results

Run the application. Type some text. Click on File/Save. If everything went OK, you should be able to open this file using Windows NotePad. Be sure to search for the file in your project's directory, since you used App.Path to specify the location. (*Note:* if you did not save the project, then App.Path will refer to the directory that the VB EXE is installed in.) Start NotePad (Windows 95 and NT client both contain this application in the Programs/Accessories menu command) and be sure the file SAMPLE.TXT exists and can be opened. If you can see the file contents using NotePad, you did it the right way.

Reading Text Input from a File

Step 7: Reading the File

Reading the file from the disk is very similar to writing. In fact, it is the exact opposite, since we want to get the text and insert it into Text1 for editing. Place this code under the mnuOPEN_Click event procedure of Form1:

```
Text1.Text = ""  'Clear Text Box text
Close #1 'no error if already closed, so always do this.
Open App.Path & "\SAMPLE.TXT" For Input As #1
Dim FileText As String  'Temp variable for each line
Do While Not EOF(1)  'Until no more lines in file
  Input #1, FileText   'Get the next line
  'Rebuild the text, tack on a carriage return/linefeed
  Text1.Text = Text1.Text & FileText & vbCrLf
Loop  'While there is still stuff in the file to read
Close #1  'Close the file
```

Run the application. Click on File/Open to copy the contents of SAMPLE.TXT into the text box.

Step 8: How It Works

An explanation is in order! We dimension a variable with this statement:

```
Dim FileText as String
```

This creates a variable that we need to hold each line in the file as we process it. For each line in the file, we move it to the variable, and then assign it to Text1.Text. This is because the Input statement will not simply accept Text1.Text as the place to store the file text. We need to use a variable temporarily, and then copy the variable contents to Text1.Text.

This code also demonstrates the use of the Do...Loop construct. We are using Do While, which specifies a condition. While EOF(1) is not True, we continue to process the file. This line accomplishes that:

```
Do While Not EOF(1)
```

The EOF(1) is a function that is built into VB. The "(1)" indicates you are testing for End of File on the file indicated as open on file handle #1. So long as this file still has contents to process, the iterative loop continues, processing each item in the file until EOF(1) evaluates to True.

This statement:

```
Input #1
```

reads the text until a carriage return/line feed is encountered. This end-of-line marker is dropped by the Input statement during processing. This means our program needs to append it to each line as we process it and assign it into the variable FileText. The VB constant vbCrLf contains the ASCII character sequence for a carriage return/line feed combination. We tack this value onto the end of each line processed so that the restoration of the text into Text1.Text is complete.

Printing Text to the Printer

Step 9: Checking for a Printer

If you have a printer attached, the default printer should be active and working. You can test this by firing up NotePad and then trying to print a sample page. If that test works, you can proceed to trying to print some text under VB program control.

Step 10: Printing

The Printer object will point to the default printer. Under the mnuPRINT_Click event procedure insert the following code:

```
Printer.Print Text1.Text
```

This one simple statement will spew the contents of Text1.Text to the default printer. This statement may appear to be quite simple, and it is. Powerful printing techniques may be built around this simple statement.

The default behavior is for VB to send it to the printer when your application ends. Since this is rarely the desirable behavior, add the following line of code after the Print statement to make it print immediately:

```
Printer.EndDoc
```

Copying Application Data to the Clipboard

Step 11

In many applications, you will want to implement some simple cross-task data transfer abilities. For example, your users may request that portions of your form's displayed data be accessible to any word processor they are using. This is best accomplished with the Clipboard object. It lets you read from and write to the Windows clipboard. This last part of the exercise shows you how to write to the Windows clipboard from your application, thus making your application's data available to all applications.

Switch to Form1 and add the following code to the mnuCOPY_Click event procedure:

```
Clipboard.Clear
Clipboard.SetText Text1.Text
```

Close the edit window and run the application. Enter some text into Text1 and then click on the File/Copy menu command. Then minimize the application and fire up NotePad or Word. From the word processor, use the Edit/Paste menu command to get a copy of the current Clipboard contents into the word processor you are using.

Exercise Extra Effort

1. Write a corresponding Paste menu command that copies the contents of the clipboard into your VB application. You will need to search the

Help file on the Clipboard object to learn the methods and how they are used. Check the GetFormat and GetText methods in particular.

2. Make it so your application does not run unless the screen's width and height do not met your minimum values. Then, set the minimum values higher than the resolution of your monitor. This is often done in commercial applications to enforce the proper monitor setup required by the application. Examine the TwipsPerPixelX and TwipsPerPixelY properties of the Screen object as well as the Screen object's Width and Height properties.

Increasing Your Knowledge

Help Search Strings
- Screen
- Printer
- App
- Debug
- Err
- Clipboard
- Forms Collection
- Controls Collection

Books Online Search Strings
- Screen
- Printer
- App
- Debug
- Err
- Clipboard
- "Forms Collection"
- "Controls Collection"

Mastery Questions

1. To write text data to the Clipboard you would:
   a. use the GetFormat method followed by the SetData method.
   b. use the SetData method followed by the SetText method.

    c. use the SetText method.

    d. None of the above

2. To determine the resolution of the screen, your VB program must:

    a. sample the Width and Height properties of the Screen object.

    b. sample the Width and Height properties of the Screen.Active-Form.

    c. sample the Width and Height properties of the Screen.ActiveControl.

    d. sample pertinent properties of the App object.

3. To force the printer to eject a page, you must:

    a. execute the Printer.Eject method.

    b. execute the Printer.KillPage method.

    c. execute the Printer.NewPage method.

    d. set the Printer.NewPage property to True.

4. The App object:

    a. has a property that indicates how long your application has been running.

    b. has a property that indicates if your application was started as a component instead of a standalone executable.

    c. has a property that contains information about your application's version information.

    d. A and c only

    e. B and c only

5. The Clipboard may contain many data formats. How does your application determine the format of the data currently in the Clipboard?

    a. You must execute the GetText method of the Clipboard object.

    b. You must execute the GetData method of the Clipboard object.

    c. You must execute the GetFormat method of the Clipboard object.

    d. You must execute the GetInfo method of the Clipboard object.

6. The primary purpose of the Debug object is to:

    a. provide a way for your application to trap errors that may occur.

    b. provide a way to set and clear breakpoints in your VB code.

    c. provide an interface between your program and the Immediate window.

    d. None of the above

7. Before retrieving graphical data from the Clipboard object, it is wise to:

    a. execute the GetFormat method of the Clipboard to be sure the data in the Clipboard is the expected format.

    b. execute the GetData method to be sure the data in the Clipboard is the expected format.

    c. execute the Clear method of the Clipboard object.

    d. sample the FormatName property of the Clipboard object to be sure the data in the Clipboard is the expected format.

8. The Screen object's ActiveForm property indicates:

    a. which form has the Windows focus at run time.

    b. which form was last loaded.

    c. which form received the last mouse click.

    d. A and c only

9. If the Screen.MousePointer is set to the value of 99, this indicates:

    a. the Screen has a MousePointer that looks like an hourglass.

    b. the screen has a MousePointer that looks like a pointer.

    c. the screen has a MousePointer that is defined by the user.

    d. None of the above

10. The Raise method of the Err object:

    a. sets the error number and produces a trappable error.

    b. sets the error text and produces a trappable error.

    c. Both a and b

    d. Neither a nor b

Answers to Mastery Questions

    1. c

    2. a

    3. c

    4. b, c

    5. c

    6. c

    7. a

    8. a

    9. c

    10. c

CHAPTER 3

Menus

What You Need to Know First

To understand this section you need to know about the VB form object and how to use the Show method to navigate from form to form. A solid understanding of how to use methods and event procedures will be necessary to understand how to implement right mouse button menu support.

What You Will Learn

In this section of the book you'll learn about the single event and multiple properties associated with menus. You'll also learn about standard user interface (UI) devices that are familiar to users such as accelerators, context menus, and shortcut keys.

- The Menu Editor dialog
- Menu object properties
- Accelerators and shortcuts
- UI standards for menus
- Popup menus
- The common dialog control

Using the Menu Editor Dialog

A menu is a set of drop-down commands that can be installed into a form. In VB, one form may possess only a single menu structure. There is no simple way to install a new menu structure within a form. You can think of each form in your VB project as a kind of container for a menu. A menu in VB can have many submenus, up to four levels deep. (See Figure 3-1.)

If you define a menu for a form using the Menu Editor dialog, it is stored within the form file that VB creates and maintains in your project directory. For example, after you build a simple menu within a form and save the form, you could examine the .FRM file using a word processor and see the menu structure as it is stored in the form file. We will be doing exactly this shortly after we work with the Menu Editor dialog to build a menu system that looks like the menu system of VB itself.

Building a Simple Menu

To begin building a menu structure for display on a form, you start with the Menu Editor. (See Figure 3-2).

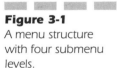

Figure 3-1
A menu structure with four submenu levels.

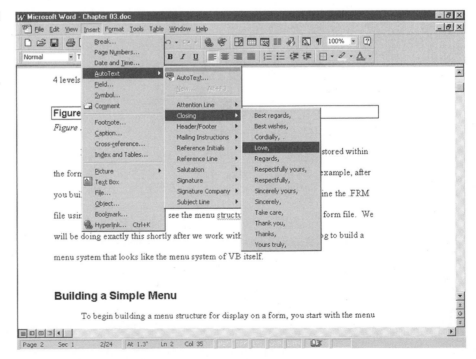

Figure 3-2 The VB Menu Editor.

There are two ways to bring up the Menu Editor. One way is to select the Tools/Menu Editor menu command. The other is to right-click on a form in your project, which brings up a context menu. When you right-click on a form, one option contained on the menu that appears is Menu Editor. Selecting this option brings the Menu Editor into view.

──── ─── ─── ─── ─── ─── ─── ─── ─── ─── ─── ─── ─── ─── ─── ────

Context Menu: *a menu that appears as a result of clicking the right mouse button.*

▨ **Check It Out: Getting to the Menu Editor**

1. Start a new VB project.

2. Right-click on Form1. A context menu appears for the Form object. Select the Menu Editor command from this context menu to bring up the Menu Editor.

3. Click on the Cancel button of the Menu Editor to dismiss it.

4. Click on Form1 to give it the focus. Now display the Menu Editor by selecting the Tools/Menu Editor menu command. After it appears, click on the Cancel button to dismiss it.

5. If the Toolbox is not visible, make it appear by clicking the View/Toolbox menu command. Now click on the Toolbox, which will give it the focus. Next, try to select the Tools/Menu Editor menu command again. It will be disabled, or grayed-out. Why? Because a form does not have the focus-the Toolbox does.

6. Now click on Form1. Try to select the Tools/Menu Editor command one more time. Since Form1 now has the focus, the Menu Editor command will be enabled. Dismiss the menu dialog after it appears by clicking on the Cancel button.

The Menu Editor creates menu objects. Menu objects have several properties, but only a single event, the Click event. After menu objects have been created with the Menu Editor, you can view and change the properties of each menu object from the Properties window of VB itself. The menu objects will appear in the Object box of the Properties window, where they can be selected easily. We will check this out later, after we build a simple menu structure.

Accelerators and Shortcuts

For our model menu, we will use the menu system of VB itself. VB's menu starts with these menu commands:

File   Edit   View   Run

Each of these contains submenus. For example, the File menu item contains these submenus:

```
New Project  Ctl-N
Open Project… Ctl-O
Add Project…
Remove Project
Save Project
Save Project As…
```

There are several items to note regarding these submenu items. First, some menu items have three dots following the menu command text. The three dots are known as an ellipsis. This UI device is used to signal the user that a dialog box will follow, allowing the user a way to back out if necessary. You may not realize it, but this UI device is common to many applications. Users are familiar with it, and because of this you should employ it as well.

Second, each menu command has an underlined character. The underlined character is the accelerator key. This key, when pressed concurrently with the Alt key, selects the menu item with the keyboard instead of the mouse. For example, holding down the Alt key and pressing "f" (for File) and "o" (for Open) selects VB's File/Open Project menu command. This is a keyboard alternative to using the mouse to select menu commands.

Third, some items in the menu structure support shortcuts. A shortcut is a key combination that, when pressed, takes you directly to the menu item-without the need to navigate the menu at all. It makes execution "go to" that menu command. For example, the Open Project menu command can be accessed directly by holding down the Ctl key and pressing "o". The effect of a shortcut on a menu command is to make it simple to select that menu command with one keystroke.

Enabled, Checked, and Visible Properties

Each menu item that you create with the Menu Editor possesses several properties, and many of them are based on the True-False Boolean data type. For example, the Visible property of the menu object, if set False, makes the menu item invisible. Setting any menu item's Enabled property to False makes it appear gray and makes using the menu item impossible. You will want to do this to your own menu items, based on the state of your application. Many applications do this. For example, when you run a VB project, many of VB's menu items will be disabled and appear gray. That's because VB knows you are in Run mode and that certain menu items do not apply in Run mode, so it grays them out. You should do the same in your own applications. In the exercise that follows, you will manipulate these True-False properties of menu objects in Visual Basic code at run time.

UI Issues with Menus

Menus are part of the overall user interface of your application. As such, your menus should adhere to a standard look and feel. For example, in most Microsoft Office applications, you'll always see these standard menu options:

File   Edit   View   Insert

Like it or not, users grow accustomed to this standard ordering of menu choices. Because of this, you should adhere to this familiar look. If you examine the menus of various commercial products, including VB itself, you will see extensive use of menu separators-horizontal lines that delineate the boundary of related menu commands. Visual Basic's File menu command, for example, makes use of no fewer than six menu separators. In the exercise that follows, you'll see how to define menu separators.

Submenus

Submenus are branches that grow from the basic menu you create. For example, if you have defined File, Edit, Run, and Insert as the top-level menu commands in your menu structure, each of these can have subordinate items. For example, File may contain New, Open, and Save menu commands. Each of these basic menu choices could in turn be defined as a submenu. A submenu reveals more indented choices from which you can select a single option.

Check It Out

1. Start a new Standard EXE VB project.

2. Right-click on Form1. Select Menu Editor from the context menu. The Menu Editor appears.

3. In the textbox labeled Caption, enter this text: &File

4. In the textbox labeled Name, enter this text: mnuFile

5. Now you will add two items under the File menu command. Press the Next button that appears in the Menu Editor. This creates a space for a new menu item. Now click on the _ button (the right-arrow button). This indents the currently selected menu item by one level. By clicking on this button, you are defining the next menu item you define as an indented item of the File menu command. You should see this under the File menu: ——. This indicates that the menu item you are defining is indented under the File menu.

6. Now define the Caption and Name for this menu item. In the textbox labeled Caption, enter this text: &New. In the textbox labeled Name, enter this text: mnuNew.

7. Now define one more indented menu item under File. Click on the menu item labeled New and then click on the Next button. This creates a new indented item that will appear under the File menu item. Set the new menu item's Caption to &Open and the Name to mnuOpen.

8. Click on the OK button of the Menu Editor and fiddle with your menu. When you click on the File menu command, the menu items New and Open should drop down.

9. Now click on the File/New menu command-while in VB's design time. A code window will appear, displaying the menu command's only event procedure: the Click event procedure. Add the following code to this event procedure:

```
Msgbox "File/New menu command clicked"
```

10. Now run the project and click on the File/New menu command. Does the Msgbox output appear when you click this menu command?

11. Return to design time. Now you will define the File/Open menu command as a submenu. Bring up the Menu Editor.

12. Click on the menu command labeled Open, and then click on the Next button. This creates a new menu item. You want to indent it under the File/Open menu command, so click on the _ button to indent it now. You will see the item indent under the menu item labeled Open if you did everything right.

13. Next, define the Caption and Name for the menu item. Set the Caption to Thing1 and the Name to mnuThing1.

14. Create one more submenu item by clicking on the Next button. This creates a new menu item at the same level as the previous one. Set the Caption of this menu item to Thing2 and the Name to mnuThing2.

15. You have just defined the File/Open menu command as a submenu. Click on the OK button of the Menu Editor to dismiss it.

16. Now examine the menu. Click on the File/Open submenu. The submenus Thing1 and Thing2 will appear. Clicking on either of these will reveal a code window and the Click event procedure for the menu item.

Manipulating Menu Items

Once you have a menu structure defined in your form, you can access the menu items in many ways during VB's design time. As you know, the Properties window contains something called the object box. This is a drop-down combination box. Clicking on it reveals the name of the form and every item contained on the form-including the menu items. Try it. Clicking on any menu item from the Properties window Object box will display all of the properties of that menu item. From the Properties window you can make changes to the properties of any menu item on a form. Thus, the Menu Editor is where you create menu items and the Properties window is where you edit and/or maintain the properties of menu items.

You can add your code to the Click event procedures of menu items by clicking the menu item at design time; this will display the Click event procedure in a code window for the menu item you click. Once you have a code window displayed, you will see a drop-down combo box to the upper left of the code window. This is the object box of the code window. Clicking on it will reveal all of the menu items defined on the form, and selecting one will bring you to the Click event procedure for that menu item.

Helpful Hints with Menus

Though the menu item is a control and has properties and events, it is not in the Toolbox. It is not available to click and drop onto the form. There is no simple way to install a new menu structure into a VB form without modifying the .FRM file that contains the form definition. Any menu you define for the form is saved as text in the form's .FRM file in your project directory. After you define a menu for a form, it is instructive to examine the .FRM file after you have saved the form safely to disk. When you examine the form file (using a text editor like NotePad), you will see your menu structure saved in the form file as text. The Menu Editor is the only facility in VB that reads and writes this information to the form file.

VB will not let you add a shortcut key to a menu item that is defined as a submenu. Neither will VB allow you to use a menu separator (covered in the exercise) as a top-level menu item. It needs to be indented.

Be careful when editing a menu structure that already has code in the Click event procedures. If you have a top-level menu item with no subordinate menu items, then you probably have code in the Click event procedure. If a user asks you to subdivide the top-level menu command called Tools into three or four subordinate menu items, you may want to simply add the subitems under the Tools menu using the Menu Editor. Then, of course, you will test the menu at run time. Guess what happens? The original code under the Tools menu "short-circuits" your new submenu items. That is, when you click on Tools, the original code under the Click event procedure executes and you never see the new options you defined for your users. The only way to fix this situation is to bring up a code window for the form, click on the Object box dropdown, select the Click event procedure for the Edit menu object, and remove the original code.

Each menu item is a control. It has both a Caption property and a Name property. It's best to avoid duplication of names on more than one menu control. Duplication of the Caption property may cause confusion to your users. But duplication of the Name property may confuse *you* because of the way VB wants to handle that situation. This is because when you name two objects the same, VB assumes you want to create a control array. Control arrays are arrays of similar controls (or menu objects) that share the same name and a common event procedure. Control arrays are covered in a subsequent section. There are beneficial uses of menu control arrays, but for now, be sure to carefully name each menu item with a unique name. If you noticed, we named menu items starting with the prefix "mnu". For example, mnuOpen refers to the menu object used as the Open menu command. Always use prefixes for your Menu object, variable, and control naming. We will have much more to say about naming standards in a subsequent section of this manual.

Popup Menus

With the advent of 32 bit operating systems like Windows 95 and Windows NT workstation, right mouse button support has become mandatory for all applications. Users have grown accustomed to context menus appearing when the right mouse button is clicked. VB itself, for example, has right mouse button support. Try clicking on any VB window to get an idea of how this works. Right mouse clicking on the Toolbox reveals a context menu for the Toolbox, while right mouse clicking on any form gives you a context menu for the form. Try it and see. The context menus that appear list commonly needed menu options in a convenient and natural format.

Context menu support is easy to implement in VB. First define a menu on any form. Next, add code to the MouseDown event procedure to enable right mouse button support. Under the MouseDown event, you call the PopupMenu method of the form object, specifying the menu name you want to display. It's that simple! The exercise that follows explains the techniques.

The CommonDialog Control

Some commonly used menu items can easily be supplied with powerful functionality. File/Open, for example, is a menu command supported by many applications. It typically displays an Open dialog. This functionality is available with the CommonDialog control. It provides your VB app with a simple way to display familiar dialogs such as for selection of File,

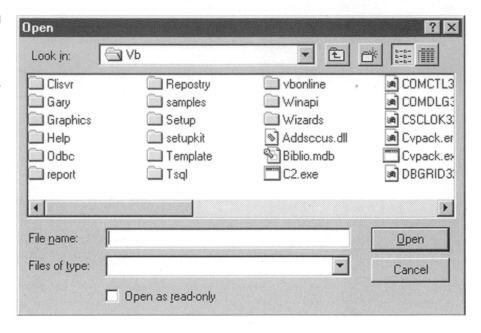

Font, or Color. With the CommonDialog control, it's simple to display the same dialogs that appear in apps like Excel, Word, and Visual Basic.

▇ Check It Out: Showing a Common Dialog Box

1. Start a new Standard EXE VB project.

2. Click on the Project/Components... dialog. Here, you will find a list of all the custom controls that can be added to your project. Select the Microsoft Common Dialog Control 5.0 by clicking on the checkbox and then clicking on OK. VB will add the new control to your Toolbox and dismiss the Project/Components... dialog.

3. The Common Dialog control will appear at the bottom of your Toolbox. Double click on it now to place one of these controls on Form1.

4. Place a button on Form1. Double click on it and place this code under the Click event procedure of the button:

```
Form1.Commondialog1.ShowOpen
```

5. Now run the project (with the Run/Start menu command) and click on the button. What happens? You get a fully functional File Open dialog-all accomplished with one line of code under the click of a button.

6. Return to design time. Try using the ShowColor and ShowFont methods of the CommonDialog control. Search VB Help for more information on how to do this.

The exercise that follows uses the CommonDialog control. To see which file a user selects after the ShowOpen method displays the dialog, you must examine the FileName property. VB fills in this CommonDialog's FileName property with the file name selected by the user. You then use this string-valued property in your code to open the file with calls to database or file input/output commands that VB supports.

SUMMARY

Menus are strongly supported in VB. You get accelerator keys, shortcuts, submenus, and various adornments such as menu separators and checkmarks. You also get the ability to support context menus via the PopUpMenu method and MouseDown event procedure of the form object. Menus are an important part of your overall application UI and should be defined for every application you develop.

Increasing Your Knowledge

Help Search Strings

- Menu Commands
- Menu Editor
- Menu Items
- Menu Shortcut
- Menus, Popup
- PopupMenu
- MouseDown
- CommonDialog Control

Books Online Search Strings

- Menus
- Creating Menus
- Menu Basics
- Displaying Popups

Mastery Questions

1. The PopupMenu method of the Printer object
   a. Provides a way to specify where a popup menu will appear on the form
   b. Provides a way to specify that only a click of the middle mouse-button will result in a menu click
   c. Provides a way to specify that only a click of the right mouse button will result in a menu click
   d. A and C only
   e. B and A only

2. Which property of the Commondialog control will be populated with the name of a file selected by the user after the FileOpen method is executed?
   a. File
   b. Path
   c. Filename
   d. Pathname

3. The Enabled property of a menu item may be accessed via
   a. The Tools/Options/Settings dialog box
   b. The menu Editor dialog
   c. The Properties window
   d. All of the above
   e. B and C only

4. Is it possible to bold a menu item in a Popup menu made visible with the PopupMenu method?
   a. Yes
   b. No

Answers

1. c

2. c

3. e

EXERCISE: MENU CONSTRUCTION AND THE COMMONDIALOG CONTROL

The following exercise is presented in three main parts. They are intended to be done together as a unit.

The Menu Editor

The Menu Editor is what you use to create menus in VB applications. You will start by building a menu structure that simulates the VB menu structure. When you are done, it will look like this on your form:

```
File
New Project  Ctl-N
Open Project…  Ctl-O
Add Project…
Remove Project
Save Project
Save Project As…
_____

Exit
```

This same menu structure, when completely defined by you using the Menu Editor, will be represented like this in the Menu Editor:

```
&File
—&New Project
—&Open Project…
—A&dd Project…
—&Remove Project
—Sa&ve Project
—Sav&e Project As…
—  -
—  E&xit
```

1. Create a new directory named VBCLASS/MENUS. Start a new Standard EXE project in VB with the File/New Project… menu command. Save it immediately to the new directory using the File/Save Project menu command. Be sure to periodically save your work during the exercise.

2. Now you will begin work with the Menu Editor. Right-click on Form1. Select the Menu Editor menu command that appears in the context menu. This displays the Menu Editor.

3. Now you will define some menu items. First, you will define the File menu command, and then you will define all of the subordinate items beneath it. Define the File menu command by specifying the following attributes:

```
Caption: &File
Name: mnuFile
```

4. Now define the subordinate items that will appear when you click on the File menu command. Click on the Next button. The selection will move down one line, from File to a blank line.

5. All of the subordinate menu items need to be indented one level. The indent buttons of the Menu Editor appear together and display arrow shaped icons pointing left, right, up, and down. For the first item you will define in the next step, click the right-arrow indent button. This will indent the menu item under the File menu command.

6. Now, define the Caption and the Name of this menu item as follows:

```
Caption: &New Project
Name: mnuNew
```

7. Click on the OK Button of the Menu Editor, then run your project. When you click on the File menu, you should see the File/New Project menu command. Return to VB design time.

8. Go back into the Menu Editor and add the following items. Be sure each one gets one level of indentation, so they appear under the File menu command. Also, be sure that after you define the Caption and Name for each menu item, you press the Next button. This sets up the Menu Editor to define the next menu item. Notice how the same level of indentation is assumed for the next item.

Define the following items:

```
Caption:   &Open Project… Name:mnuOpen
Caption:   A&dd Project… Name: mnuAdd
Caption:   Sa&ve Project  Name: mnuSave
Caption:   Sav&e Project As… Name: mnuSaveAs
```

Menu Separators, Shortcuts, and Inserting New Menu Items

The next item you will add is a menu separator. A menu separator is a horizontal line that logically groups menu items. You cannot click on a menu separator; its function is cosmetic only. You must, however, give the separator a name, even though it has no functionality. By convention in VB, any menu item whose caption is the hyphen or minus sign or dash (it looks like this: -) is automatically assumed to be a menu separator. You will add a separator to the menu now and add an Exit menu item right under it.

1. While still in the menu editor, begin the definition of a new menu item by pressing the Next button. Now, define a menu separator by specifying the following:

```
Caption: - (this is a single dash or hyphen character)
Name: mnuSeparator1
```

2. Now press the Next button again, to define one more item. Specify the Caption and the Name as follows:

```
Caption: E&xit
Name: mnuExit
```

3. Your menu structure should appear as follows in the Menu Editor:

```
&File
——&New Project
——&Open Project…
——A&dd Project…
——Sa&ve Project
——Sav&e Project As…
—— -
—— E&xit
```

From here, you will insert one final item in the menu. In the previous steps, all you did was add new items one after the other. With production VB applications, you will often find a need to insert new menu items after the initial structure is defined. You may decide, for example, to add new items or regroup existing items, or add one or more menu separators. You'll be editing the menu structure often. VB provides a way to insert new menu items, with the Menu Editor's Insert button, which is located adjacent to the Next button.

4. To insert a new menu item, the rule is you have to click on the menu item *under* the location where you will insert the new one. We want to insert a menu item *before* the Save Project menu item, so click on this and then press the Insert button. This opens up the menu so that you can insert a new item.

5. Define the new menu item as follows:

```
Caption: &Remove Project
Name: mnuRemove
```

Notice what happens if you leave the name blank. Blank out the Name text you just entered and press the Menu Editor's OK button. What happens next? VB complains, with the error message "Menu control must have a name." After you click on OK, VB will place the cursor on the Name textbox. Reenter the text mnuRemove in the Name textbox of the Menu Editor and then click on OK. Everything should be fine. Remember, every menu object must have a Name. This allows VB to display the menu object's Click event procedure in the code window. It also allows VB to display the menu object's properties in the Properties window.

6. A shortcut key allows the user to get to a menu item with a single key combo such as Ctl-O or Ctl-N. VB provides a simple way to provide this functionality. Click on the Open Project menu command, and then click on the drop-down combo box labeled Shortcut within the Menu Editor. (It appears to the upper right of the Menu Editor dialog.) Scroll all the choices. Then, select Ctl-O as the shortcut for this menu item by clicking on it. Next, look at the menu hierarchy. See how the menu shortcut is attached to the menu command?

7. Click on OK, then click on your new menu structure on Form1. Now run the project and click on the menu again. Although the menu structure is defined and will display, it has no functionality. Now we will add some functionality to the menu.

8. At design time with the Menu Editor closed, click on the File/Exit menu command. A code window will appear. Add the following lines to the Click event of the mnuExit menu command:

```
Msgbox "App closing down…"
End  rem this is a special VB reserved word that kills the app
```

9. Now run the project and click on the File/Exit menu command. See how the app just stops executing?

10. Now click on the File/Open menu command; this brings you to a code window displaying the mnuOPEN_Click event procedure. Add the following line of code under that event procedure:

```
Msgbox "You have clicked the File/Open menu command."
```

11. Now run the app. Click on File, then click on Open. This should display the message box you just defined. Dismiss the message box.

12. Now hold down the Alt key and press "f" followed by "n", all the while holding down the Alt key. You are now using the keyboard accelerators you defined for this menu item. Again, dismiss the message box.

13. Now, here is one more way to execute the Click event procedure of the File/Open menu command: Hold down the Ctl key and press "o" (the letter "o"). This is the shortcut for File/Open that you defined earlier. Clicking with the mouse, using the Alt accelerator keys, or

using the shortcuts-these are the three ways to select menu commands from your applications.

The CommonDialog Control

In most commercial applications with menus, the ellipsis (…) characters follow some of the commands. This means a dialog box will appear with at least a set of OK and Cancel buttons, and maybe more functionality and choices that are optional. You might not realize it, but when you see a menu command followed by three dots, you start thinking, "Here comes a dialog box. If I click on this option, I can back out without any problem." Some of our menu commands in this exercise are followed by the ellipsis. Under one of these, we will display a Common Dialog box.

1. Click on the Project/Components… dialog. Here you will find a list of all the custom controls that can be added to your project. Select the Microsoft Common Dialog Control 5.0 by clicking on the checkbox and then clicking on OK. VB will add the new control to your Toolbox and dismiss the Project/Components… dialog.

2. The Common Dialog control will appear at the bottom of your Toolbox. Double click on it now to place one of these controls on Form1.

3. Now click on the File/Open Project menu command of Form1 (not VB's menu, but Form1's menu-in your project!) while still in VB design time. A code window will appear, and here, you will add the following lines line of code to the mnuOPEN_Click event procedure:

```
Form1.Commondialog1.ShowOpen
MsgBox "File selected was : " & Form1.CommonDialog1.FileName
```

4. The Msgbox line of code just listed takes a string ("File selected was : ") and a string-valued property (Form1.CommonDialog1.FileName) and mashes them together to form one string. Computer scientists call this concatenation. In VB, the concatenation operator is the ampersand. This is why the "&" character appears between the two items following the MsgBox command. Be sure to type the line exactly as shown.

5. The CommonDialog control will not appear on the form at run time, it will be invisible. That's because this control simply provides services to your application, and does not need to be displayed. The CommonDialog does, however, display various familiar dialog boxes (known as "common dialogs") when you call one of its methods such as ShowOpen.

6. Run the project and click on the File/New menu command. See the File Open dialog? Use this dialog to specify a file you see in the list. When you select a valid file, a message box will appear, displaying the value of the Filename property of the CommonDialog control. What actually happens is, the ShowOpen method will fill in the File-Name property if the user selects a file via the dialog display.

7. Return to design time by clicking on your functional File/End menu command.

8. Now search the Help file for ShowOpen, ShowSave, ShowPrinter, and ShowColor. Experiment with these methods of the Common Dialog control by calling them under some of the other menu items you have defined. For example, add this line under the Add Project... menu command:

```
Form1.Commondialog1.ShowColor  'Showcolor example
```

EXTRA EFFORT

1. Add a new menu item somewhere under File, named mnuTOGGLE. In the Menu Editor, click on the Checked checkbox. Run the app. See the checkmark? Now you will add code so that the Toggle menu item switches between two states: checked and unchecked. This involves the use of the Not operator, a reserved word in Visual Basic. The Checked property of a menu object is Boolean, meaning it's either on or off. VB uses True and False to signify the two states. You will now write code that toggles the Checked property between True and False.

2. Click on the File/Toggle menu command of Form1, this brings up a code window. Add the following code to the mnuTOGGLE_Click event:

```
Form1.mnuTOGGLE.Checked = Not(Form1.mnuTOGGLE.Checked)
```

3. Run the app. Click on the File/Toggle menu command several times. What happens?

4. Search Help on Not for more info on this operator

CHAPTER 4

Data Types, Scoping Rules, and the Visual Basic Language

This portion of the book deals with topics you already are familiar with from other languages and programming systems you know. Topics such as data types, subroutines, functions, and constants will be familiar to you from other programming systems, so a long-winded definition of these terms will not be provided here. However, since constants, built-in functions, and scope rules are supported in Visual Basic, we provide a brief explanation of the concept followed by the pertinent details regarding the facilities VB provides.

What You Need to Know

If you have not programmed a computer before reading this book, it's unlikely you will have fun with this chapter. You'll need to be familiar with a previous version of the BASIC language, or be familiar with COBOL, Pascal, C, Fortran, or some other complete programming language before diving into this chapter. We assume you understand what iteration is and how a function differs from a subroutine. (As you know, a function returns a value while a subroutine does not.)

What You Will Learn

- Data Types Supported in Visual Basic
- Scoping Rules for Variables
- Special Object-Based Variable Types
- Writing Subs and Functions with Parameter Passing
- Scoping Rules for Procedures (Functions and Subs)
- The Visual Basic Programming Language
- Built-In (Intrinsic) Functions and Keywords
- Intrinsic Visual Basic Constants
- Using the Object Browser

Overview

This chapter covers and demystifies the declaration of variables and procedures as well as the VB language and some of the built-in keywords and constants that make the language rich in functionality. When you are done with this chapter you will know most of the places where VB procedures and variables can be declared. This chapter, therefore, will fill in many gaps in your knowledge, and is required reading for developers new to VB.

Data Types and Scoping Rules

Variable Data Types

VB provides many data types; these are described briefly here. The general form of a variable declaration in a VB project looks like this:

```
{Public|Private|Dim|Static} variablename [As] datatype
```

where datatype can by any of the intrinsic types, or a even a user-defined Type or Class name. For example, the line

```
Private strAppName As String
```

defines a variable named strAppName whose data type is String, meaning it can contain characters. The prefix "str" is a coding convention used by many programmers to denote the data type of the variable within the variable name. This is known as Hungarian notation, in part because a prolific Microsoft programmer named Charles Simonyi, a Hungarian by descent, came up with the prefixing scheme as a way to make code more readable. Consistent naming clarifies code, and prefixing helps in this regard.

Public, Private, Dim, and Static are all VB reserved words. Each one is a declarative, nonexecutable statement that assigns storage for a variable or array. These will be covered in detail in this chapter. First and foremost, we will describe the data types VB provides to your application. Table 4-1 lists the data types for VB variables.

Byte
The Byte data type is used for the representation of UniCode and for a number of other uses. It occupies 1 byte of memory.

TABLE 4-1:
Visual Basic
variable types

Datatype	Size	Value
Byte	1 byte	0 to 255
Boolean	2 bytes	True or False
Integer	2 bytes	-32,768 to 32,767
Long(long integer)	4 bytes	-2,147,483,648 to 2,147,483,647
Single (single-precision floating-point)	4 bytes	-3.402823E38 to -1.401298E-45 for negative values; 1.401298E-45 to 3.402823E38 for positive values
Double (double-precision floating-point)	8 bytes	-1.79769313486232E308 to -4.94065645841247E-324 for negative values; 4.94065645841247E-324 to 1.79769313486232E308 for positive values
Currency (scaled integer)	8 bytes	-922,337,203,685,477.5808 to 922,337,203,685,477.5807

(continued)

TABLE 4-1:
Visual Basic
variable types

Datatype	Size	Value
Decimal	14 bytes	+/79,228,162,514,264, 337,593,543,950,335 with no decimal point; +/-7.92281625142643375 93543950335 with 28 places to the right of the decimal; smallest non-zero number is +/-0.000000000000000000 0000000001
Date	8 bytes	January 1, 100 to December 31, 9999
Object	4 bytes	Any Object reference
String (variable-length)	10 bytes + string length	0 to approximately 2 billion
String(fixed-length)	Length of string	1 to approximately 65,400
Variant(with numbers)	16 bytes	Any numeric value up to the range of a Double
Variant(with characters)	22 bytes + string length	Same range as for variable-length String
User-defined(using Type)	Number required by elements	The range of each element is the same as the range of its data type.

Boolean
A Boolean is used to hold True/False values, and also to declare functions that return True or False.

Integer
The Integer occupies 2 bytes. The fastest VB data type and the most general purpose, it holds values between _32K and +32K, with no fractional part.

Long
The Integer's big brother, the Long, occupies 4 bytes and can hold _2M to +2M as a value.

Single and Double
Single and Double are used to hold very small and very large numbers. The value is stored in scientific notation with mantissa and exponent.

Date
This data type holds a date and time. For example, this code can capture the current date and time in a variable of type Date:

```
Dim dtmCurrDateTime As Date
dtmCurrDateTime = Now
```

Be sure to search help on DateDiff() and DateAdd(), two special date functions that can operate on the Date data type. You can find the difference between two dates and add a number of days to a date with these functions.

Currency

Currency is the data type used for money; it always maintains four digits of decimal precision. This data type is also known as a scaled integer. It's pretty slow compared to most of the other data types, so use it sparingly. It occupies 8 bytes of memory.

User-Defined Types, or UDTs

The user-defined type is something you create by declaring it in the [general][declarations] section of a form or module. For example, if you wanted to define a set of variables associated with an employee, you would start with this Type statement:

```
Type Employee
  LastName As String
  FirstName As String
  EmpNum As Integer
  BirthDate As Date
End Type
```

With this user-defined type declared, you could then declare a variable of that type, like this:

```
Private e As Employee
```

When you do this, you automatically get all the items associated with the type. To assign the employee an employee number, in code you may write

```
e.EmpNum = 23456
```

and to determine the employee's birthday, you may write

```
Dim DOB As Date
DOB = e.Birthdate
```

You must define the user-defined type (also known as a UDT) before attempting to declare a variable of that type.

Variant

The Variant is the default data type in VB, and it is the slowest data type in VB. The Variant can cause you special problems, even though it has many special uses. If you do not specify the data type for a variable when you declare it, you get a Variant. For example, both of these lines declare a variable of type Variant:

```
Dim v As Variant  'Explicit declaration is best, but….
Dim MyVariant  'Here, you get a variant by default!!
```

The Variant data type can have data of any type whatsoever assigned into it. When you assign string data into a Variant, for example, you can perform string operations on the Variant, just as you would if it was actually a string variable. Because of the very flexible nature of Variants, they occupy much memory and can be very slow. They are, in fact, the slowest data type available in VB.

Form and Control

Form and Control are special VB reserved words. For example, you can define variables of type Form or type Control. You can also receive parameters in procedures that are of type Form or Control. You may, for example, write a procedure that sets many properties of a form or manipulates a control on a form. Form and Control are known as object variables, something we will discuss in great detail in the section on Class modules and objects.

Using Data Types

Problems with Variable Declaration

When you refer to a variable for the first time, by default Visual Basic will create a Variant and then allow you both read and write access to the variable. This is known as implicit variable declaration, and for the most part it should be avoided. If a form or module has this code in the [general][declarations], VB will require explicit variable declaration in the entire form or module:

```
Option Explicit
```

You should always use Option Explicit.

■ **Check It Out**

1. Start a new VB project. Show a code window for Form1, and navigate to the [general][declarations] of the Form. Make sure there is no Option Explicit line coded there. Then, add this code to the Form_Load of Form1:

```
V = 10
MsgBox v
```

2. Run the application. What you are doing is referring to a variable that has not been declared *at all*. Visual Basic in this case creates a new Variant and hands it over for you to use in your program. Thus, the code works.

3. Now go back to the [general][declarations] of Form1 and add this line:

```
Option Explicit
```

4. Now run the application again. This time, you get a compiler error in the Form_Load of Form1, because Option Explicit tells VB that undeclared variables are invalid.

The whole point of using Option Explicit is to require variable declaration. You can type the statement into the [general][declarations] of a form or module, or you can have VB do this for you automatically, every time you add a new form or module. The steps are outlined next; Figure 4-1 shows the dialog where you tell VB to Require Variable Declaration.

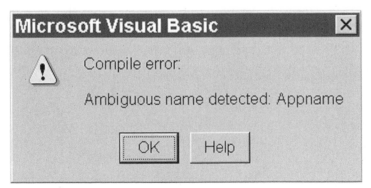

Figure 4-1   Requiring variable declaration.

■   **Check It Out**

1. Start a new project. Click on the Tools/Options menu command and select the Editor tab.

2. To the upper right, you will see a checkbox labeled Require Variable Declaration. Click on this box and then click on the OK button. Now, every time you add a module or a form to your project, VB will automatically add the Option Explicit statement, which will force variable declaration in the module or form.

3. Please keep in mind that changing the Tools/Options/Editor tab setting *does not* have any effect on existing forms or modules. You must type the Option Explicit statement into each existing module or form where you want variable declaration required. If Require Variable Declaration was not checked before, now that you have checked it, the existing Form1 will still need Option Explicit added manually.

SPECIAL NOTES ON VARIABLE DECLARATION

As you already know, this code will get you a Variant:

```
Private MyVar As Variant
```

So will this code, because you are not explicitly specifying the data type:

```
Private MyVar
```

Keep in mind that even if you have Option Explicit set up, as you should, this code still gets you a Variant. So you *can* get a Variant by mistake if you do not explicitly declare it, even with Option Explicit in the [general][declarations].

The next thing to note is that this code is valid also:

```
Private a As Integer, b As Integer, c As Integer
```

So you can declare multiple variables on a single line. You can even mix the types like this:

```
Dim x As Integer, y As String, z As Double
```

We advise against this technique because it can get you into trouble. You might notice that VB will format your valid lines of code, performing capitalization and turning valid keywords to the color specified in the

Tools/Options for that kind of text. Sometimes, what looks good is not. Consider this code:

```
Dim a, b, c, d, e, f As Integer
```

In this sample a, b, c, d, and e will be Variants, since no type is specified for any of those variable names. Only the variable "f" will be an integer! Even worse, VB will format this line with capitalization for the Integer keyword, change the color of the Dim keyword, and so on. If you are coming from C or another language that allows similar syntax, watch out for this "gotcha." The problem can be especially acute if a, b, c, d, or e is used as a loop counter in a For...Next loop. The Variant is about 30% slower than an integer as a loop counter!

Declaring Arrays

Arrays in VB are simple to implement, but there are some special rules. First, arrays can be declared anywhere a variable is valid. This code creates an array of Integers:

```
Public KeyCodes(10) As Integer
```

There will be not 10 but rather 11 items in the array; all arrays in VB start with 0 by default.

To force the array to start with 1 and have 10 items, just use this syntax when you declare the array:

```
Public KeyCodes(1 to 10) As Integer
```

If you want all the arrays in a form or module to start with 1, add this line to the [general][declarations] of that form or module:

```
Option Base 1
```

When you do this, all arrays start with 1. You can also declare multidimensional arrays, like this:

```
Public KeyCodes(10,100) As Integer
```

Assuming there is no Option Base statement in the module, this array declaration sets up an array of 11 rows with 101 array elements per row, since both dimensions start with 0. You can have many dimensions, not just two. Scan the Help file for more information on this.

You can also declare a dynamic array, which can be restructured dynamically at run time. There will be times when the upper bound of the array cannot be known until run time; an example would be when the array size is based on data that is in a database. In such cases, declare a dynamic array, like this:

```
Private MyNames() As String
```

This array does not specify the upper boundary. It may be redimensioned at run time, like this:

```
ReDim MyNames(Max)   'Max is an integer with the value 100 in it.
```

Assuming all 101 elements of this array had data stored within, this next statement would reduce the size of the array to 10 and destroy the data in the first 10 elements:

```
ReDim MyNames(10)
```

If you wanted to preserve the original contents of the array while making it bigger or smaller, use the Preserve clause of the ReDim statement. This statement, again assuming 101 elements of the array with data, adds 50 new elements and preserves the original data in the first 101 array items:

```
ReDim Preserve MyNames(100+50)
```

Note that any valid integer expression is valid as shown here. Also note special functions provided by VB, Ubound() and Lbound(). Ubound() returns the ordinal number of the last element in the array, Lbound() returns the ordinal number of the first element. Thus, to add 50 items to an array, regardless of size, use this notation:

```
ReDim Preserve MyNames(Ubound(MyNames) + 50)
```

Scoping for Data and Procedures

Variables can be defined in many places in a VB program and for many purposes. We will now handle some of the more common uses of variables and use the String and Integer data types for all examples to keep things simple.

Public in Module

Many applications will have a need for variables that are available for reading and writing by the entire application. Such variables are known as global variables. To declare a global variable, you declare with the Public keyword, in the [general][declarations] section of a module. This section is where the variables in a module are declared.

Check It Out

1. Start a new VB project. Click on the Project/Add Module menu command to add a module. Module1 will be the default name. A code window will appear displaying the [general][declarations] section of the Module. Add this line to that section now:

```
Public strAppName As String
```

2. Now insert a new form and drop two CommandButtons on the form. Under the click of Command1, add this code to write a value into the variable:

```
strAppName = "VB BOOTCAMP"
```

3. Under the click of Command2, add this code to read the value of the variable:

```
MsgBox "Value of strAppName = " & strAppName
```

4. Run the application. Click on Command1, then click on Command2. You are writing and reading a global variable (declared Public in Module1) from a form in your application.

Variables declared Public in a module are available to the entire application from the time the application loads until it is terminated. Multiple modules are allowable in your project; each module can have Public variables defined. You can think of all the Public variables in all the modules in your project as being available at all times to your program.

Private in a Form

Variables can also be declared in a form. Variables declared Private in a form are not available to the entire application; they are available only to the procedures that reside in the form itself. Private variables in a form are sometimes referred to as member variables of the form. The typical use of these variables is support for the processing contained in procedures located in the form, such as event procedures of the Form object or event procedures of buttons and list boxes located on the form. The advantage to variables declared this way is their limited lifetime. Variables declared Private in a form come into existence when the form initializes. (The first loading of a form explicitly initializes the form.) The storage associated with these variables is automatically reclaimed after the form terminates. (Unloading a form explicitly causes termination of the form.)

Check It Out: General Declarations on a Form

1. Start a new project and insert a form. Double click on Form1 to reveal a code window for Form1. You will notice that Form_Load is what appears in the code window. This is because the Form object has event procedures and Form_Load is the default event procedure. The author of the Form object designated the Load event procedure as the default event procedure-one that would appear in the code window first.

2. Often new VB developers who do not read the documentation miss the fact that the form object has a place to declare variables. To see the [general][declarations] of Form1, click on the Object: dropdown that appears to the upper left of the code window. The items in the list will be Form and [general]. Click on the [general] item in the dropdown list.

3. The next thing you will see is the [general][declarations] area of the Form object. Here you can define variables for use in the form. Variables declared here as Private can be accessed only by procedures (such as Form_Load) that are located within the form itself.

Dim

The Dim statement in VB5 is used to declare variables that are used at the procedure level. This means that if you need to use a variable inside the Form_Load event procedure, you will be using the Dim keyword. Dim is short for "dimension." Variables declared in a procedure with Dim come into existence when the procedure starts executing and are destroyed when the procedure is terminated with End Sub or Exit Sub. This type of variable is known only to the event procedure that declares it.

Static

The Static keyword is similar to Dim. With the Dim statement, variables are initialized to zero (in the case of numerical data types) or an empty string (in the case of string variables) each and every time the procedure is called. Not so with the variables declared with the Static keyword. Variables declared with the Static keyword retain the value from the procedure's previous execution. This is useful for applications where you need to preserve the state of variables in the procedure.

Check It Out: Dim Versus Static

1. Start a new Standard EXE project. On Form1, drop a command button. Under the Command1 click event procedure, add this code:

```
Dim countA As integer  'Dim an integer
Static countB As integer  'Statically declare an integer
countA = countA + 1  'Increment countA by 1
countB = countB + 1  'Increment countB by 1
'Report the results of incrementing both variables:
MsgBox "countA = " & countA & " ...countB = " & countB
```

2. You can see that the variable declared Dim is reset to zero each time the Click procedure executes. The variable declared as Static retains its last value, and each execution of the Click procedure increments it by one. Static variables are useful for counting the number of executions of a procedure and a variety of other tasks.

Refinements: Private in a Module, Public in a Form

Variables can be declared Private within a module. Similar to declaring a variable Private in a form, a "Private in a module" variable declaration limits the scope of the variable to only the procedures in the module. This type of variable will not be declared very much in a module, since a variable private to a module has limited uses. However, when we discuss Class modules a little later, the use of Private in a Class module becomes very important. Although the Class module, like the so-called "standard module" also has a [general][declarations] section, we defer the Class module discussion for a subsequent section of the book.

You may be asking yourself, "If I can declare a variable Private in a module, can I do the opposite; that is, declare a variable Public in a form?" The answer is Yes. When a variable is declared Public in a form, it is available to the entire application by using a special syntax:

```
[Formname].[variablename]
```

For example, if I have two forms named Form1 and Form2, and Form1 has a public integer variable named NumberOfClicks, then the value of this variable is accessible from anywhere in the application. For example, if Form2 has a CommandButton on it, this code would report the value of the variable NumberOfClicks contained in Form1:

```
MsgBox "Form1 has been clicked " & Form1.NumberOfClicks & " times!"
```

This fully qualified reference to the variable is required to access Form1's public variables. The exception to the rule is that if the code referring to the variable resides within the form where the variable is declared, qualification is not required, and just specifying the variable by name is OK.

You might be asking, "Why not just use a global variable, declaring it Public in a module?" One reason is this: Since this variable contains data that pertains to the form, it should reside in the form to make things simple to follow if you are maintaining the program at a later date. Second and more important, Public variables in a form can be thought of as custom properties of the form. Even the syntax is the same. For example, to access the Caption of a form, you may write:

```
Dim strWindowText As String
strWindowText = Form1.Caption
```

Accessing a Public variable declared in a form looks very similar. In fact, it looks just as if you are sampling a property of the form! Here is the code:

```
Dim intNumClicks As Integer
intNumClicks = Form1.NumberOfClicks
```

This idea of a custom property is something that you will use extensively when you begin to work with Class modules.

Duplicate Variable Names

You can run into unexpected problems and error messages when you use the same name for two or more variables. This is possible in Visual Basic and can cause unpredictable results. Let's look at a couple of examples.

SAME VARIABLE NAME, DECLARED PUBLIC IN TWO MODULES

In VB it is possible to have multiple standard modules in your project. Each can have Public variables. If you have two modules named Module1 and Module2, and each module has a variable declared in this way in its respective [general][declarations] section:

```
Public strAppName As String
```

the application has a single form-Form1. If this code appears in the Form_Load event procedure for Form1, which variable does VB write to?

```
Private Sub Form_Load()
strAppName = "VB BOOTCAMP"
End Sub
```

The answer is, VB gets really confused, and issues the message shown in Figure 4-2. VB is saying "I'm confused, and it looks like you are, as well!"

The way to fix this problem is to use a fully qualified reference to strAppName. If you want to alter the contents of the variable found in Module1, use this code:

```
Private Sub Form_Load()
   'FULL QUALIFICATION REMOVES THE AMBIGUITY!!!
   Module1.AppName = "VB BOOTCAMP"
End Sub
```

Normally, you would never actually name two variables at the same level of scope with the same name, but just in case, you'll know what will happen if you do this by mistake. Often, professionally written VB applications will contain a module named GLOBALS.BAS, which contains just the Public variables. By using one module to contain all global variables declared with the Public keyword at the module level, you can cut down on irritating error dialogs such as the one shown in Figure 4-2.

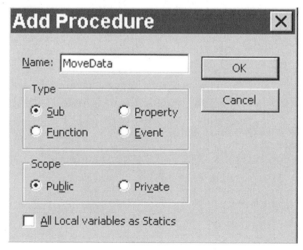

Figure 4-2 Two global variables, same name. The VB error message.

SAME VARIABLE NAME DECLARED AT
TWO OR MORE LEVELS OF SCOPE

Another possible scenario you could run into is called shadowing variables. It is possible to declare two variables with the same name at two differing levels of scope. For example, you could declare one variable named intCount as Public in a module and another with the same name as a variable local to a form (Private in a form). Let's look at an example.

Check It Out

1. Start a new Standard EXE VB project. Click on the Project/Add Module menu command to add a new module. Declare this variable in the [general][declarations] of Module1:

```
Public iMyInt As Integer
```

2. Now navigate to the [general][declaration] of Form1. Add this declaration there:

```
Private iMyInt As Integer
```

3. You now have two variables with the same exact name. One is global because it is Public in a module. The other is local to Form1 because it is declared Private in the form named Form1.

4. Now drop a button on Form1 and double click on the button. This gets you a code window displaying the Command1_Click event procedure. Add this code to the event procedure:

```
Dim iMyInt As Integer
```

5. The big question now is this: When you reference iMyInt, which variable gets the assignment? The global one, the form-local one, or the one we just defined at the procedure level? The answer is this: The lowest level of scope gets the assignment.

6. Add this line to the Command1_Click procedure to alter the variable defined in the procedure:

```
'The local one gets the assignment
iMyInt = 10
```

7. To access the form-level variable, use a qualified reference. Append this line to the Click event procedure of Command1:

```
'This one is declared in the Form1 [general][declarations]
Form1.iMyInt = 20
```

8. Why the qualification? Since the variable is declared Private to the form, what is the need for the qualified reference Form1.iMyInt? The reason is that the Click procedure also has a variable with this same name, so qualification is necessary to make sure the intended variable gets the assignment.

9. In this scenario, it is even possible to reference the global variable. You simply qualify the reference with the module name. Append this line of code to the list of lines that now reside in the Command1_Click event procedure:

```
'This one is declared in the Module1 [general][declarations]
Module1.iMyInt = 30
```

10. The entire Command1_Click should now look like this:

```
Private Sub Command1_Click()

Dim iMyInt As Integer

'The local one gets the assignment
iMyInt = 10

'This is the one declared in the Form1 [general][declarations]
Form1.iMyInt = 20

'This is the one declared in the Module1 [general][declarations]
Module1.iMyInt = 30

End Sub
```

Prefixing the name with "i" as in iMyInt helps to make it clear that the variable holds an integer. The prefix, however, does not indicate the level of scope. Many teams will add this information into the standard prefix for naming variables. For example, all globals may start with a "g", followed by characters indicating the data type. Thus declarations that look like this in a module:

```
Public iMyInt As Integer
```

would be altered as follows:

```
Public giMyInt As Integer
```

This additional scope information in the prefix can make the variable scope and data type immediately obvious to maintenance programmers who examine the code later. Another immediate benefit is that the duplication of variable names at multiple levels of scope is impossible by virtue of your naming standard. However, another school of thought says that long prefixes are bad because they render variable names unreadable. We disagree. We are in favor of prefixes up to four characters long. Maintainability is always a central concern. Adopt a naming standard for variables and stick to it.

Functions and Subs

Functions and subroutines are procedure types you can create and use in VB. Both types of procedures can receive zero or more parameters, which are data variables passed to the procedure. Both procedures return to the calling program when the processing in the procedure completes. In the case of a procedure coded as a function, a value will be returned. Both subs and functions can receive parameters, but only functions return a value.

Creating functions and subs in VB is quite simple. For example, if you want to create a function that is accessible from the entire application, you will define it as a Public function in a module. The same scope rules apply to procedures that apply to variables; the rules are consistent across data variables and procedures you create. With respect to creating a procedure, you have two options. You can type the function declaration directly into the [general][declarations] of a module, or you can use the Add Procedure dialog. We will explore both techniques shortly.

Creating Functions

The first and most pressing concern is how to write a function your application can access on a global basis. We will now create a globally accessible function that accepts a single parameter consisting of text in a string variable. The function will convert the text to uppercase and return the converted text via the function's return value. Here is how the function would look when completed:

```
Public Function fCapitalize(strTEXT As String) As String
'UPPERCASE WHATEVER IS PASSED
'FUNCTION NAME = RETURN VALUE
fCapitalize = Ucase$(strTEXT)
End Function
```

There are a couple of things to note here. First, the function name is fCapitalize. The prefix "f" is our standard for denoting that this was a function created in-house. As you will see, a subroutine or function can originate from a variety of places. It could be created in-house or be a built-in VB function or an API call, or it could come from a third party Dynamic Link Library, or even from a reusable component. Using the prefix "f" for custom functions we write makes it simple to see in code where the function comes from. (We use "s" for subroutines.) You can decide how far you want to go with prefixing and naming schemes. We know it aids maintenance programmers who must examine our code.

Second, functions take parameters. This line:

```
Public Function fCapitalize(strTEXT As String) As String
```

specifies that the function fCapitalize takes one parameter of type String. Also, the As String portion of the declaration indicates that this function shall return a string to the calling program.

Once properly defined, this function may be called as follows:

```
Dim r As String
R = fCapitalize("vb is fun indeed!")
MsgBox "The return value is " & r
```

Functions can be called in interesting ways. For example, a more compact way to accomplish the same task would be to write

```
MsgBox "The return value is " & fCapitalize("VB is fun indeed!")
```

This works because the entire function call resolves to a string, and a string is expected after the ampersand. It works fine, try it and see for yourself. This is valid usage and you will see this type of usage often in programs written by more experienced developers. Now let's write this function and a couple of subroutines.

Check It Out

1. Start a new Standard EXE project, and click on the Project/Add Module menu command to add a module named Module1. When the code window is displayed, click the Tools/Add Procedure menu command. (Please note that this menu item will be disabled unless a code window is active.) The Add Procedure dialog is depicted in Figure 4-3.

2. Click on the option button labeled Function in the Add Procedure dialog and type fCapitalize for the procedure name. Click on OK and VB will automatically insert the function declaration into the code window.

Figure 4-3
The Add Procedure
dialog.

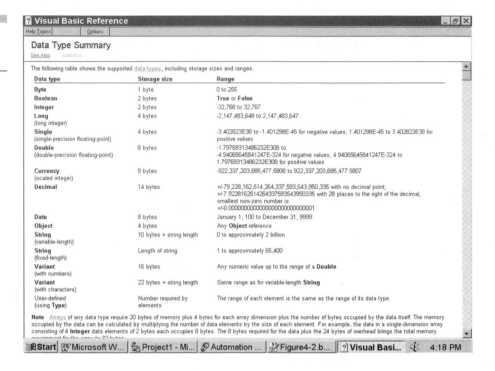

Figure 4-3
The Add Procedure
dialog.

3. When VB creates a function with the Add Procedure dialog, you have to edit the code it creates for you. VB does not define the parameters (if any) passed in, and it does not specify the data type of the returned value. Edit the function to (a) accept a string and (b) return a string. When you are done, it should look like this:

```
Public Function fCapitalize(strTEXT As String) As String
'UPPERCASE WHATEVER IS PASSED
'FUNCTION NAME = RETURN VALUE
fCapitalize = UCase$(strTEXT)
End Function
```

4. Now go to the form and drop a CommandButton; it will be named Command1. Add this code to the click of the button. The code will alter the Caption of Form1.

```
Form1.Caption = fCapitalize("VB Bootcamp!!")
```

5. Run the application and click the button. The window title should change from Form1 to VB BOOTCAMP!! when you click the button.

6. End the application. Display the code window for Module1. Now bring up the Add Procedure dialog again by clicking on the Tools/Add Procedure menu command.

7. Define a subroutine named MakeRecord by (a) clicking on the option button labeled Sub, (b) specifying MakeRecord for the subroutine name, and (c) clicking on the OK button.

8. So far, you have defined a function and a subroutine using the Add Procedure dialog. You can also type the function or subroutine declarations directly into the [general][declarations] section of a module or form. With the code window up for Module1, navigate to the [general][declarations] section and enter this line:

```
Public Sub MySub
```

9. Press Enter, and watch VB complete the subroutine declaration by adding End Sub automatically. You can also enter functions in this manner.

10. Now click on the Procedure dropdown located at the upper right of the code window. The dropdown will list all of the procedures (both functions and subroutines) defined inside Module1. As you add procedures, this dropdown allows you to navigate from procedure to procedure.

Scoping Rules for Procedures

The rules regarding the scope of a procedure are exactly the same as the rules for variables. We have not mentioned this yet, but procedures can be defined within forms as well as within modules. These procedures in forms can be declared as Public or Private. If they are Private, they are callable only from another procedure (such as an event procedure) inside that same form. If a procedure defined in a form is declared Public, you can call it from anywhere in your application, with qualification, like this:

```
[formname].[procedurename(parameter list)]
```

If you are calling the procedure from inside the form where it is defined, the qualification is not needed when specifying the procedure name. Note that these rules for referencing procedures are completely consistent with what you have learned regarding the scoping rules for data variables declared Public and Private in forms and modules.

Procedures can also be defined Private to a module. This is rarely used but in the occasional case when a procedure is called only from within the module, it makes sense to limit such procedures in scope by declaring the procedure Private in the containing module.

Parameter Passing

VB offers a variety of ways to call procedures and pass parameters to procedures.

USING THE CALL STATEMENT WITH SUBS... AND FUNCTIONS

You can call a subroutine using a Call statement, with or without the Call keyword provided by VB. We encourage the use of the Call keyword because it makes it clear in the code that this subroutine exists in your own source code. Consider this subroutine, which simply echoes back the three passed parameters Name, Age, and IDNUM:

```
Public Sub sEchoParms(Name As String, Age As Integer, IDNUM As
String)
  MsgBox "Name is " & Name
  MsgBox "Age is " & Age
  MsgBox "IDNUM " & IDNUM
End Sub
```

This subroutine can be called in VB code with a Call statement, without the optional Call keyword, as follows:

```
sEchoParms "Dan", 29, "333448888"
```

You can also call the same routine and get the same outcome by using the Call keyword. Using the Call keyword, however, requires the use of parentheses if you are passing any parameters; you'll get an error otherwise. You can use the Call statement like this:

```
Call sEchoParms("Dan", 29, "3334448888")
```

Using the Call statement has the advantage of clearly indicating that EchoParms is not a built-in VB subroutine or reserved word. For this reason, we encourage the practice of using the Call keyword. Prefixing the same of the subroutine also speeds the process of understanding the code.

One use of the Call statement concerns the execution of a function that returns a value. Normally, a function would have to be used in an assignment statement, like this:

```
Dim r As Boolean
r = fMakeNewDatabase("C:\MYDB.MDB")
```

But perhaps you just want the effect of the processing that the function performs, and you do not care about the return value. In such cases, you can use Call to execute the function and effectively disregard the return value, like this:

```
Call fMakeNewDatabase("C:\MYDB.MDB")
```

When the function is called it will process normally, but the return value from the function will not be available to the calling program.

DIFFERENT WAYS TO CALL FUNCTIONS

Functions can be called in many interesting ways because, by definition, all functions return a value. This means that any call to a function is interpreted by VB as an expression value in the data type of the return value.

Consider a function that returns a Boolean data type. Say the function simply indicates whether a passed-in integer is negative or not in value. Here is the declaration:

```
Public Function fIsNegative(intParm As Integer) As Boolean
  If intParm < 0 Then
    fIsNegative = True
  Else
    fIsNegative = False
  End If
End Function
```

The function returns a Boolean value of either True or False. This means the function can be used anywhere a Boolean expression may be used. For example, this If statement works great and is very easy to read:

```
Dim intMyInt As Integer
intMyInt = -99
If fIsNegative(intMyInt) Then MsgBox "intMyInt is negative"
```

You can also use the Not operator to invert a Boolean value. Not True is False, and Not False is True. Thus, you can express a test for a positive integer with the preceding function as follows:

```
If Not fIsNegative(intMyInt) Then MsgBox "intMyInt is positive"
```

You need to work this through on your own if you have questions about the use of the Not operator or you have questions about using a Boolean-valued function as the condition of the If statement.

By the way, in these examples we use a form of the If statement that can express the condition and action on a single line. If you have just a single action, you can code the If...Then conditional statement in the manner just depicted.

You can also use a function as part of an expression. Say a function named fMultiply() returns the product of two integers. This function could be used in an expression as follows:

```
Dim intResult As Integer
intResult = 10 + fMultiply(5, 25) + 100
MsgBox intResult
```

This code is a valid use of the fMultiply function, and it will return the value 225.

A function can also call another function in-line. Say a function named fAdd() accepts two parameters and adds them, returning the result. This statement is valid:

```
Dim intResult As Integer
intResult = fAdd(100, fMultiply(5,25))
```

This code evokes the function fMultiply by coding it as the second parameter to the fAdd() function. Since fMultiply() evaluates to an integer expression, using it as the second parameter is valid. This is referred to as a nested procedure call. You can write compact code with this technique, but be careful you do not nest too deeply. Statements like this can be hard to read and maintain:

```
R = fBigHonkingResult(fAdd(intAGE,fMultiply(intYOURAGE, intMYAGE)))
```

so avoid deeply nested procedure calls.

BYVAL VERSUS BYREF PARAMETER PASSING

VB passes parameters by reference unless you specify otherwise with the ByVal keyword. When you pass by reference, the procedure that receives the passed parameter can alter the contents of your variable in a permanent manner, producing some very bad effects if you are not careful. To ensure that the variables used to pass data to a procedure are protected from alteration by the procedure, use the ByVal keyword when defining the procedure. For example, this definition of the sEchoParms() procedure specifies "by value" parameter passing, by using the ByVal keyword in front of each parameter:

```
Public Sub sEchoParms( _
  ByVal Name As String, _
  ByVal Age As Integer, _
  ByVal IDNUM As String)
  MsgBox "Name is " & Name
  MsgBox "Age is " & Age
  MsgBox "IDNUM " & IDNUM
End Sub
```

By specifying ByVal, this procedure makes private copies of the contents of the parameters passed and uses the copy of each passed parameter to perform any processing. If the value of a parameter is changed, only the local copy gets affected. When the procedure is complete, these private variables are destroyed. Any variables used to call the procedure are preserved when ByVal is used.

In many languages, most notably C and C++, parameters are automatically passed by value, meaning the procedure gets a copy of the parameter data, not the memory address where the data is stored, known as "by reference" parameter passing. Visual Basic's default parameter passing scheme is by reference, while most other computing languages take the by value approach. Programmers familiar with C, Pascal, and other systems should be very careful to specify ByVal in the procedure declaration if that is what you intend.

PASSING NAMED PARAMETERS

Passing named parameters is an option when calling any procedure. Returning to our sEchoParms() example, here is the declaration:

```
Public Sub sEchoParms( _
  ByVal Name As String, _
  ByVal Age As Integer, _
  ByVal IDNUM As String)
```

This procedure can be called by naming the parameters. The parameters must be specified using this syntax:

```
[keyword]:=[value]
```

The cool thing about named parameters is that the parameter list can be specified in any order. This example specifies the parameters defined for sEchoParms() in reverse order of their definition in the sub declaration:

```
Call sEchoParms(IDNUM:="232324", Age:=29, Name:="Scot")
```

As an alternative, you may also write:

```
sEchoParms IDNUM:="232324", Age:=29, Name:="Scot"
```

You do not have to do anything special to begin using named parameters except you must comply with the syntax. Be sure to note the colon, equals sign (:=), which must be used; a simple = will not work, but instead will produce an error.

Named parameters make code more readable and easier to maintain. This is because the actual parameter names are used in keyword:=value format to evoke the procedure. This notation automatically makes the intent more evident. If the function writer knows that the procedure will be called by name, the author is more likely to use meaningful names for the parameters defined in the procedure declaration. Named parameters, combined with the Call statement and procedure name prefixing, make code easier to debug and maintain.

PASSING OPTIONAL PARAMETERS

VB supports the definition of optional parameters. Optional parameters are, as the name implies, not required, but may be optionally passed into the procedure. For example, if the sub\\ sEchoParms was coded such that the IDNUM parameter was optional, it would be coded like this:

```
Public Sub sEchoParms( _
  ByVal Name As String, _
  ByVal Age As Integer, _
  Optional ByVal IDNUM As Variant)

  MsgBox "Name is " & Name
  MsgBox "Age is " & Age
  MsgBox "IDNUM " & IDNUM

End Sub
```

Now the IDNUM is optional and not required. Please note that if you are using ByVal, Optional comes before the ByVal keyword (VB will enforce this). If you use an Optional parameter, you can also set the default value for the argument to be used when the parameter is not passed. This prevents errors in the body of the procedure. For example, the following routine has an Optional parameter with a default value:

```
Public Sub sMyProc(Optional ByVal intTest As Integer = 1)
```

What if the parameter is not passed in and not set by default? The answer is, any reference to the Optional parameter will get you an error. For example, this line in sEchoParms() will get you an error if the optional IDNUM is not passed in:

```
MsgBox "IDNUM " & IDNUM
```

The way around this problem is to test for the parameter and see if it is missing. VB provides a function to do this. The IsMissing() function returns True when the parameter is missing and False otherwise. Here is how it is used in a procedure to test for the existence of a parameter that is declared Optional:

```
If IsMissing(IDNUM) Then
  'BIG PROBLEM
  MsgBox "IDNUM is MISSING!!!"
Else
  'NO PROBLEM, PROCEED
  MsgBox "IDNUM " & IDNUM
End If
```

THE PARAMETER ARRAY

A parameter array is a single parameter with a name that can hold several individual data items. A parameter declared as a ParamArray is actually an optional array of Variants. Any data type can go in the array, and there are functions (search Help for "VarType") to determine the type of each item from inside the procedure, if need be.

When using ParamArray, you must use specify the parameter with a set of empty parentheses to signify that it is an array. For example, this declaration causes a precompile error:

```
Public Sub PA_Sample(ParamArray Items As Variant)
```

By including the required parentheses, the error is avoided. The procedure in the next exercise allows numerous individual items to be passed to the procedure in a single ParamArray. Each item comes in as a Variant. This procedure reports the subtype of each Variant-valued item in the parameter array:

▨ Check It Out

1. Start a new Standard EXE VB project. Double click on Form1, to get a code window. Click on the Tools/Add Procedure dialog, and add a procedure named sPA_Sample. Make it a subroutine.

2. Edit sPA_Sample to look like this:

```
Public Sub sPA_Sample(ParamArray Items() As Variant)
'NOTE THE PARENTHESES AFTER "ITEMS"
  Dim v As Variant
  For Each v In Items()
    MsgBox "This item has a variant subtype of " & VarType(v)
  Next
End Sub
```

3. Now make Form1 visible and drop a button on Form1. We will test out the procedure with a call to sPA_Sample under this button. Add this code under the Click of Command1:

```
Call sPA_Sample("Bill",10)
```

4. Run the application; click on Command1. You should see the sub-types for the two parameters passed in as Variants to sPA_Sample via the ParamArray. The procedure should report an 8 for the first parameter (indicating the String subtype) and a 2 for the second parameter (indicating the Integer subtype.)

5. Search Help now on VarType and get familiar with what this function does.

6. Modify the code found in the Click of Command1 so that it has another parameter, this time a number with a fractional, decimal portion:

```
Call sPA_Sample("Bill",10, 99.99)
```

7. Run the application again. See how the procedure accepts the new item without any additional supporting code required? This is the major advantage to the ParamArray.

ParamArrays will become more significant when you learn about building ActiveX servers.

The Language: Control and Functions

The Visual Basic language looks a lot like previous versions of Basic. You will find the language to be very rich and contain dozens of intrinsic functions and subroutines. There is also a rich set of flow-of-control statements, which is what this section focuses on. We finish with a brief description of the built-in intrinsic functions and the always-available predefined constants supported by VB. This part of the book absolutely assumes you are a programmer, and for that reason it is relatively short. The looping constructs are well documented in the VB Help file. You can expect the Help file to be your primary resource in understanding this part of the language.

For...Next

The primary purpose of the For...Next construct is to iterate over a list of known size. For example, this For...Next loop displays the contents of an entire array filled with information. Please note the use of the UBound() function, from which you can derive the array size:

```
Dim intLastOne As Integer
Dim c As Integer
intLastOne = UBound(strADDRESSES())
For c = 1 to intLastOne
  Debug.Print "Item " & c & " contains this: " & strADDRESSES(c)
Next c
```

For...Next loops can go backward through a list. The Step clause specifics how many to advance with each iteration, and also the direction. For example, to report the contents of strADDRESSES() in reverse order, you would write:

```
Dim intLastOne As Integer
Dim c As Integer
intLastOne = UBound(strADDRESSES())
For c = intLastOne To 0 Step -1
  Debug.Print "Item " & c & " contains this: " & strADDRESSES(c)
Next c
```

Do...Loop

The Do...Loop construct is the most flexible of all the available loops in VB. Figure 4-4 shows the Help file topic that is displayed when you search Help on "Do".

This code gets you an endless loop:

```
Do
Debug.Print "Looping…"
Loop
```

The following code iterates 100 times, using an If statement to test a variable that is incremented by one each time through the loop. It uses the Exit Do statement to terminate the loop based on the condition:

Figure 4-4
The Do…Loop con-
struct as explained by
the VB Help file.

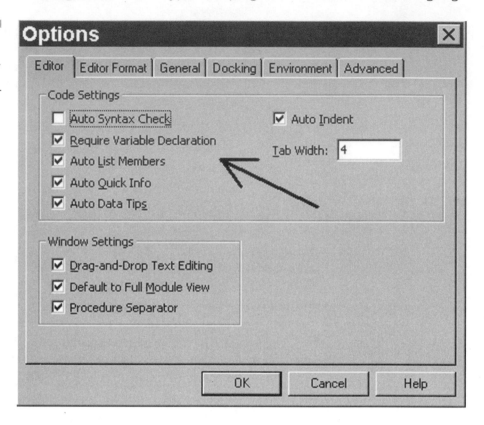

```
Dim c As Integer
Do
  c = c + 1
  If c = 101 Then Exit Do
Loop
```

You can Do While a condition is true or Do Until a condition is true. You can also perform the While or Until test at the bottom of the loop by specifying the While or Until clause with the Loop keyword (at the bottom of the loop) instead of with the Do keyword (at the top of the loop). VB provides the option for testing at the bottom of the loop in case you want to guarantee at least one iteration. Testing at the bottom makes this guarantee possible.

The following code guarantees one iteration and ends when the variable "c" exceeds the value 10:

```
Dim c As Integer
Do
  c = c + 1
Loop Until c > 10
```

Select Case

The Select Case statement is used to implement a multiway branch. For example, if you wish to test a variable for one or more possible values and perform processing based on the value, you can use this statement. It works like this:

```
Select Case intMyVar
Case Is = 10
  MsgBox "That's a 10"
Case Is = 20
  MsgBox "That's a 20"
Case Is = 30
  MsgBox "That's a 30"
Case Else
  MsgBox "What is it?"
End Select
```

Once one of the cases evaluates to True, processing under that Case statement executes. When the processing is done for that Case, the rest of the construct is skipped and execution goes to the first line under the End Select. Note the Case Else, which is the default case. If none of the cases evaluates to True, the default case is executed.

USING SELECT CASE WITH MSGBOX

One interesting use of the Select Case construct involves the use of the MsgBox function. You know the MsgBox statement, but you probably do not know that it takes two optional parameters. One specifies the overall style of the MsgBox window, including the buttons and icon that will appear. The other specifies the window title that will appear in the MsgBox window. For example, this statement specifies a MsgBox that will have Yes, No, and Cancel buttons and a window title that says VB BOOTCAMP:

```
MsgBox "Your message here", vbYesNoCancel, "VB BOOTCAMP"
```

The second parameter specifies vbYesNoCancel. This is a predefined constant that VB provides. A constant is a symbol that evaluates to a value; the symbol is easier to remember and recognize, and is therefore superior to hard-coding the value. VB supplies a good many constants that are used for a variety of purposes. You will learn more about constants in a section that follows. The second parameter, as stated previously, specifies the style of the window that will appear. If you search Help on "MsgBox function" you will find documentation on most of the values that are valid as part of the second parameter to the MsgBox function. For example, this code provides an "exclamation" icon and also makes the second button the default button, in addition to defining Yes, No, and Cancel buttons and specifying the window caption:

```
MsgBox _
"Your message here", _
vbYesNoCancel + vbExclamation + vbDefaultButton2, _
"VB BOOTCAMP"
```

The constants vbYesNoCancel, vbExclamation, and vbDefaultButton2 all evaluate to numbers. These numbers are added together to form the second parameter to the MsgBox function, which determines the style and look of the MsgBox that appears. One interesting thing to note is that vbDefaultButton2 makes the No button the one that VB thinks was clicked on if the user simply presses the Enter key.

The MsgBox function returns an integer indicating the button pressed. But just how will your program test the return value and take appropriate action? The Select Case construct is best for this task.

Check It Out

1. Start a new VB project, and drop a button on Form1. Add this code to the Click event of Command1:

```
Dim intRetValue As Integer
intRetValue = MsgBox( _
  "Your message here", _
  vbYesNoCancel + vbExclamation + vbDefaultButton2, _
  "VB BOOTCAMP")
MsgBox "That was a " & intRetValue
```

2. Run the application. Click on Form1 Command1, and the MsgBox will appear with Yes, No, and Cancel buttons. Click on Yes, No, and Cancel in turn. You will discover the values returned by the MsgBox function for Yes, No, and Cancel. Yes gets you a 6, No gets you a 7, and Cancel gets you a 2 as the return value from the MsgBox function.

3. Now append this additional code to the code that appears already under Form1's Command1_Click:

```
Select Case intRetValue
Case Is = 6  'or you can say vbYes rather than 6
  MsgBox "That's a YES"
Case Is = 7  'or you can say vbNo rather than 7
  MsgBox "That's a NO"
Case Is = 2  'or you can say vbCancel rather than 2
```

```
    MsgBox "That's a CANCEL"
End Select
```

4. Run the application and observe the behavior. The code under Command1's Click event procedure should now look like this:

```
Dim intRetValue As Integer
intRetValue = MsgBox( _
  "Your message here", _
  vbYesNoCancel + vbExclamation + vbDefaultButton2, _
  "VB BOOTCAMP")
MsgBox "That was a " & intRetValue
Select Case intRetValue
Case Is = 6  'or you can say vbYes rather than 6
  MsgBox "That's a YES"
Case Is = 7  'or you can say vbNo rather than 7
  MsgBox "That's a NO"
Case Is = 2  'or you can say vbCancel rather than 2
  MsgBox "That's a CANCEL"
End Select
```

5. Try this last technique: Since Select Case needs an expression or a variable, you can evoke the MsgBox function right in-line with the Select Case, like this:

```
Select Case MsgBox("Your message here",
vbYesNoCancel + vbExclamation + vbDefaultButton2, _
"VB BOOTCAMP")
Case Is = 6  'or you can say vbYes rather than 6
  MsgBox "That's a YES"
Case Is = 7  'or you can say vbNo rather than 7
  MsgBox "That's a NO"
Case Is = 2  'or you can say vbCancel rather than 2
  MsgBox "That's a CANCEL"
End Select
```

Try this and observe how it works. Why does this work?

With Statement

This construct is used to provide a concise notation for the setting of object properties. The With statement makes it simple to refer to an object once, and then refer to properties of the object.

For example, here is a way to change several properties of a ListBox object on Form1:

```
With Form1.List1
  .Height = 500
  .Width = 400
```

```
  .ListIndex = 1
  .Visible = True
End With
```

In previous versions of VB, the With statement did not exist, and you would set consecutive properties of an object like this:

```
Form1.List1.Height = 500
Form1.List1.Width = 400
Form1.List1.ListIndex = 1
Form1.List1.Visible = True
```

The With statement makes this kind of thing easier to do by referring to the object once.

For Each...Next

The For Each...Next statement is used to iterate over arrays and collections. The key to using this construct is to remember that you must define a variable that will be used to refer to the current item in the list or collection when the loop iterates. The variable must be a Variant if you iterate over an array, even if the array is declared to hold just integers or strings. When you iterate over a collection, you can use a Variant, or the Object data type, or a more specific object type such as Control or Form.

This example displays all of the controls located on Form1:

```
Dim ctl as Control
For Each ctl in Form1.Controls()
  Debug.Print "The Control named : " & ctl.Name & " exists in
Form1"
Next
```

This next example lists all of the strings in the array strMyNames():

```
Dim strMyNames(3) as String
strMyNames (0) = "Attila the Hun"
strMyNames (1) = "Hannibal"
strMyNames (2) = "Napolean"
strMyNames (3) = "McArthur"
Dim v as Variant
For Each v in strMyNames
  Debug.Print "strMyNames() contains " & v
Next
```

If...Then...Else Statement

The basic form of the statement for taking a single action based on a condition is

```
If [condition] Then [Action]
```

If you contemplate multiple actions, you will probably use multiple program lines. In that case, you should use the If with the End If, like this:

```
If [condition] Then
   [Action Line 1]
   [Action Line 2]
   [Action Line3]
 End If
```

The condition can be any expression, variable, or literal that evaluates to True or False. What if the expression evaluates to False? In such cases, the Else is provided so that you can force processing for when the condition is False. This example uses the If…Then…Else statement to test the return value from a MsgBox():

```
Dim r As Integer
r = MsgBox("Do you want to continue?", vbYesNo, "Confirm")
If r = 6 Then
  MsgBox "You pressed YES"
Else  'Not a Yes, must be a NO
  MsgBox "You pressed NO"
End If
```

Leftovers from the Cretaceous Period of the BASIC Language

There are several leftover portions of the BASIC language that were useful in years gone by. In the present day, we have little use for them. One such example is the use of the word "Let." These two statements are equivalent:

```
MyVar = 10
'OR, YOU CAN SAY IT THIS WAY:
Let MyVar = 10
'You will have little use for the Let keyword
```

Visual Basic also uses symbols to define the return value of functions and the data types of variables. These two statements are equivalent:

```
Dim Zipcode$
'THIS IS THE PREFERRED WAY TO DO IT:
Dim Zipcode As String
```

In the olden days, there was no As [datatype] syntax, and symbolic declaration was your only option. There are symbols for string, integer, long, and so on, but you will not want to be using them. However, some code you may review from others may date back to VB3, and you may see the symbols used for procedure declarations. This function will accept a string and return a string:

```
Public Function fCapitalize$(UserText$)
```

This more modern function declaration is equivalent:

```
Public Function fCapitalize(ByVal UserText As String) As String
```

That's enough on the symbolic declaration of data and procedure typing. There is more old stuff you will want to avoid, yet be familiar with.

More Ugly Language Stuff You Should Not Use

WHILE...WEND

While...Wend is a looping construct. It works like this:

```
While Form1.Visible = True
  Call sBigProcedure
  'THE BIG PROCEDURE MAY SET Form1.Visible to False
Wend  'While End
```

GOTO

You can GoTo anywhere in a procedure; but this should be avoided. The only legitimate use of GoTo is to branch to a single exit point in a procedure or to branch to an error trapping portion of your code. GoTo remains a valid and dangerous reserved word. Avoid!

GOSUB...RETURN

GoSub also remains a valid reserved word. GoSub branches execution to a code block within the same procedure and returns to the calling statement when the Return statement is encountered. The Help file provides more detail. GoSub has been largely superseded by the Call statement.

Built-in (Intrinsic) Functions

VB provides an impressive set of built-in functions; you can think of these functions as part of the language. For example, UCase$() is a function that will uppercase whatever text you pass in as a parameter:

```
Dim strMyText As String
strMyText = "daniel joseph mezick"
strMyText = Ucase$(strMyText)
```

There are many other built-in functions. See the Help search strings later in this chapter for more detail on the many functions available.

STRING FUNCTIONS

The String functions manipulate strings. To find a string within a string, use the InStr$() function. To get the leftmost or rightmost text in a string, use Left$() and Right$(). To extract some text from the middle of a string, use Mid$().

CONVERSION FUNCTIONS

Visual Basic provides many useful conversion functions to convert one data type to another. Most of these functions start with C. You can guess the names in most cases. For example to convert a string to an integer, use the CInt() function. To convert a Variant to a Double, use CDbl(), and so on.

THE IIF() FUNCTION

Visual Basic provides some very interesting functions that appear in other languages. IIf() is one such function. It evaluates any expression that can be evaluated to True or False, and then it allows you to specify what will be returned from IIf() in each case. For example, if you wanted to return "Dan" if the Integer variable MyVar is greater than 10 expression is True, and "Scot" otherwise, you would code IIf() like this:

```
Dim r As String
r = IIf(MyVar > 10, "Dan", "Scot")
```

IIf() is known as the Immediate If. It is useful as a compact and lightweight way to evaluate an expression and return a value.

FINANCIAL FUNCTIONS

Visual Basic provides a set of financial functions, which allow you to obtain the present value of an annuity, calculate interest, and so on. Functions such as IPmt(), IRR(), and Rate() are part of the financial library. Search the Help file as described later in this chapter to learn more about these functions.

Constants

CONSTANTS AND THE OBJECT BROWSER

You might be wondering about VB's support for values with symbolic names, so-called constants, which make your code easier to read. Does VB support them? The answer is Yes. You can learn about VB's support for constants by examining VB's Object Browser, which is accessible from the View/Object Browser menu command.

Check It Out

1. Start a new standard VB project and click on the View/Object Browser... menu command. The Object Browser will appear.

2. You will see a list of items in the list box to the left of the Object Browser; it will be labeled Classes. In this box you will see many listed items, among these will be the item Constants. (*Hint:* the list

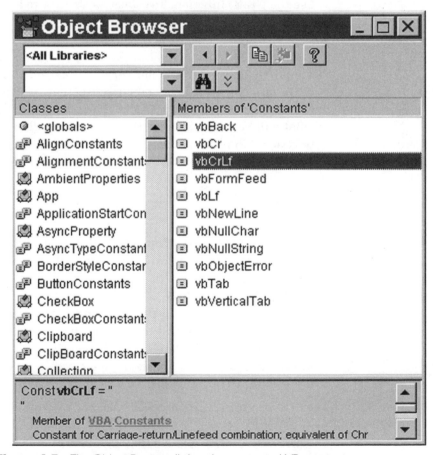

Figure 4-5   *The Object Browser listing the supported VB constants.*

is in a group order and then an alphabetical suborder.) Click on Constants. It is a part of the VBA type library.

3. The Object Browser will now list all of the members of the class named Constants. One of these will be vbCrLf. Most of the built-in VB constants start with the prefix "vb". After you click on the constant vbCrLf, look to the bottom of the Object Browser, scrolling if necessary. At the bottom of the Object Browser you will find some useful information, telling you that vbCrLf is the equivalent of Chr$(13)+Chr$(10).

4. There is a button with a yellow question mark on the Object Browser. Click on this button now to reveal more detailed help on the vbCrLf constant.

The Object Browser is actually providing a lens into the many objects available to your VB programs. At the top of the object browser is a dropdown combo. If you hover the mouse over this dropdown, you will see a yellow ToolTip with the text Project/Library, and dropping down this list provides you access to objects such as VB, VBA, and VBRUN. Even your project is listed in this dropdown box. Each item in this dropdown has members accessible to your program. Each item in this dropdown may

If you click on <All Libraries> in the Project/Library dropdown, you can see various classes, and some will contain the string "constant" in the name. Click on any of these classes to explore the many constants available to your programs.

Searching Help for a given function or keyword will often provide you with some useful information on constants that apply to that particular function or procedure.

Check It Out

1. Start a new VB project; double click on Form1 to reveal a code window.

2. Type MsgBox in the Form_Load. You are doing this just to index into the Help. Click on the keyword MsgBox with the mouse and press the F1 key to index into the Help for the MsgBox keyword.

3. Examine the Help, which describes the constants that can be used in conjunction with the MsgBox function. Be sure to scroll all the way to the bottom. The Help file display should be the same as what is depicted in Figure 4-6.

Figure 4-6
Help on the MsgBox
function, describing
various constants.

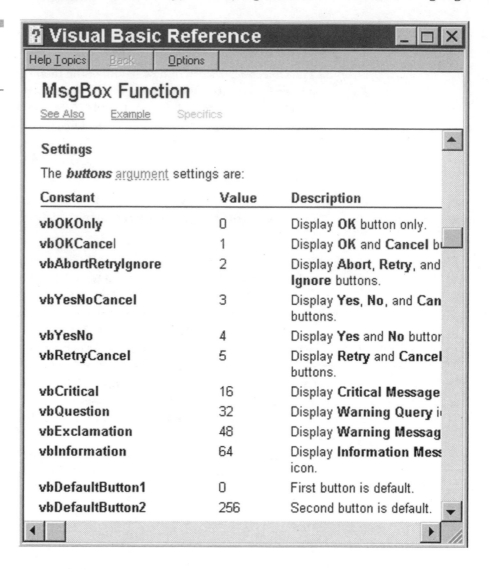

Figure 4-6
Help on the MsgBox
function, describing
various constants.

Defining Your Own Constants

You can define your own constants in VB. It is simple to do. To make an application-wide global constant, define the constant in the [general][declarations] section of a module, like this:

```
Public Const gcAppName As String = "My Little Application"
```

Const is the keyword used to define a constant. Constants differ from variables in that they cannot be changed. The scoping rules concerning the use of Public, Private, and Dim are the same as for variables. You can think of constants as read-only variables.

SUMMARY

This was a large chapter that covered an enormous amount of important material you will need to actually program in VB. We covered all of the data types, wrote procedures, and learned about the VB language. If you have a good grip on this material, you are now in great shape to really dive in and get into the more advanced topics. As of right now, you know all of the VB data types and know all of the places where VB code can reside within your project. With this fundamental knowledge, you are ready to really do some interesting things with VB.

Increasing Your Knowledge

Help Search Strings

- Intrinsic Data Types
- Keywords, Conversion (many of what VB Help lists as "keywords" are actually intrinsic functions.)
- Keywords, Financial
- Keywords, String
- Keywords, Math
- Visual Basic Constants
- Optional Keyword
- Arrays, Overview

Books Online Search Strings

- Intrinsic
- Variables
- Procedures
- "Constant and Variable Naming Conventions"
- "Introduction to Variables, Constants and Data Types"
- "Introduction to Procedures"
- "Introduction to Control Structures"

Mastery Questions

1. A function that you create in a Class module by declaring with the Friend keyword is visible
   a. inside the Class module, only.
   b. to the entire project.
   c. outside of the project.
   d. None of the above

2. A function that you create in a standard module by declaring with the Friend keyword is visible
   a. inside the Class module, only.
   b. to the entire project.
   c. outside of the project.
   d. None of the above

3. Which of the following statements are true about the decimal data type?
   a. You can declare a variable to be of type Decimal.
   b. The Decimal data type is a subtype of the Variant data type.
   c. The Decimal data type is a derived from the Currency data type.
   d. You can create a Variant data type whose subtype is Decimal using the CDec function.

4. Which statements are true for a subroutine?
   a. It remains idle until called upon to respond to events caused by the user or triggered by the system.
   b. It must be explicitly invoked by the application.
   c. It can be either Public or Private. It is Private by default.
   d. It may return a value to the calling procedure.

5. Which statements are true for a function?
   a. It must be explicitly invoked by the application.
   b. It may return a value to the calling procedure.
   c. It remains idle until called upon to respond to events caused by the user or triggered by the system.
   d. It can be either Public or Private. It is Public by default.

6. Which statements are true of the ByRef keyword?
   a. If neither ByVal nor ByRef is specified, this is the default.
   b. When used, it passes the address of the argument to the procedure.
   c. The Option Explicit statement forces this to be the default method of passing parameters.
   d. It can be used with Call whenever you need to force the procedure being called to accept parameters by this method.

7. The Static statement is used at which level?
   a. Procedure level
   b. Module level
   c. Form level
   d. Property procedures

8. In VB5, once you have declared a user-defined type using the Type statement, you can declare a variable of that type using which of the following?
   a. ReDim
   b. Public
   c. Friend
   d. Private

9. Which is used at module level to declare the default lower bound for array subscripts?
   a. Option Bounds
   b. Option Base
   c. Option Compare
   d. Option Explicit

10. In VB5, which of the following are valid variable declarations?
    a. Dim a, b, c As Integers
    b. Dim Static a As Integer
    c. Static Dim a As Integer
    d. Dim a, b, c as Single

11. A Variant data type may contain which of the following?
    a. String arrays
    b. Integer arrays
    c. Variant arrays
    d. User-defined data types

12. If an arithmetic operation is performed on a Variant data type and the result exceeds the normal range for the original data type:
    a. the result is truncated to the maximum value that the data type may hold.
    b. the result is promoted within the Variant to the next larger data type, if it exists.
    c. it will result in an Arithmetic Overflow runtime error.
    d. None of the above

13. What keyword indicates that the Variant variable intentionally contains no valid data?
    a. New

b. Empty

c. Nothing

d. Null

14. What keyword denotes a Variant variable that hasn't been initialized

    a. New

    b. Empty

    c. Nothing

    d. Null

15. According to the largest positive values allowed in the range each data type, which groupings are arranged from smallest to largest?

    a. Boolean, Byte, Integer, Long

    b. Currency, Single, Double, Long

    c. Integer, Long, Single, Double

    d. Byte, Integer, Currency, Variant

16. A variable of Boolean data type is True when it contains which of the following values?

    a. _1

    b. 0

    c. 1

    d. 32767

17. Which set of statements repeats a block of statements while a condition is True or until a condition becomes True?

    a. If...Then...Else

    b. Select Case

    c. Do...Loop

    d. While...Wend

18. A variable of what data type may be used to iterate through the elements of an array using a For Each...Next statement?

    a. String

    b. Variant

    c. Decimal

    d. Long

19. Which of the following statements are proper uses of the Exit statement?

    a. Exit Loop

    b. Exit For

    c. Exit Sub

    d. Exit Next

20. An uninitialized Boolean variable contains what value?
   a. Null
   b. Empty
   c. Nothing
   d. 0

21. Which of the following may be valid assignment statements?
   a. Let X = Y + Z
   b. Let X = Y And Z
   c. Let X > Y If X = Z
   d. X = Z And Y = Z

Answers to Mastery Questions

   1. b

   2. d

   3. b, d

   4. b

   5. a, b, d

   6. a, b

   7. a

   8. a, b, d

   9. b

  10. d

  11. a, b, c, d

  12. b

  13. d

  14. b

  15. c, d

  16. a, c, d

  17. c

  18. b

  19. b, c

  20. d

  21. a, b, d

EXERCISE

Hardware and Software Requirements
Visual Basic Version 5.0 Pro or Enterprise

Required Files
None

Assumed Knowledge
Basic understanding of data types, scoping rules, and procedure writing

What You Will Learn
In this exercise, you will explore the scope of data variables and procedures, and learn to exploit much of the VB language. The application includes global variables, variables local to a form, and some procedures.

Summary Overview of Exercise

Exploring the Scoping of Variables
This section will get you up to speed on variables generally as you explore the three levels of scope for variables.

Exploring the Scoping of Procedures and Procedure Writing
This section will get you started with procedure writing and show you how to pass some interesting things to procedures, such as arrays, forms, and controls.

Programming a Selection List Box
This last part of the exercise explores the Visual Basic language and provides the processing to move items from one list box to another as the user clicks the source list box.

Exploring the Scoping of Variables

Step 1: Setting Up
Start a new VB Standard EXE project and add a module. In the module you will add your own custom constants that are accessible from the entire application. Add the following code to Module1:

```
Public Const gcAppName As String = "My Little Application"
Public Const gcVersion As String = "1.0"
```

Now add this code to the Form_Load of Form1:

```
Form1.Caption = gcAppName & " Version " & gcVersion
```

Run the application. You have access to the constants, so the values appear in Form1's window title.

Now return to Module1 and make the constants Private instead of Public. Run the application. What happens? Why?

Step 2: Experimenting with Constants
Change the constant declarations in Module1 back to Public declarations. Next, try this line as the first line under the Form_Load of Form1:

```
gcAppName = "Your error will occur here. Constants cannot be
changed."
```

Run the application and see what happens. Can you change the constant? You should see the message box in Figure 4-7 when you try to compile.

Step 3: Form-Level Variables
The next step is to work with a variable that is local to a form. One common use of a variable that is Private to a form is to use it for communi-

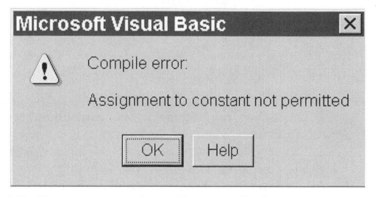

Figure 4-7   The error you get when writing to a defined constant.

cating between procedures. You will explore this use of a Private, form-level variable now.

Add this line to the Form_Resize of Form1:

```
MsgBox "FORM RESIZE"
```

Step 4

Now run the application. You will see that when the form loads, the Resize event procedure executes. This can be suppressed, but only if the Resize event procedure can be sure that the Form_Load just executed. This is fairly simple. You need to set up a variable on the form and make sure that the Form_Load sets it to a value. Later, the Form_Resize can examine the variable and take action based on the value.

Step 5: Putting the Form Level Variable to Work

Declare a Boolean variable Private to Form1. Name it FromFormLoad, like this:

```
Private FromFormLoad As Boolean
```

Next, set this variable to True in the Form_Load of Form1, like this:

```
FromFormLoad = True
```

So far, you have set up this Boolean local to Form1, and Form_Load sets it to True. Now you can test the value in Form_Resize and exit the resize procedure if the Form_Load just executed. Place this code under the Form_Resize of Form1:

```
If FromFormLoad = True Then
  FromFormLoad = False
  Exit Sub
End If
MsgBox "FORM RESIZE"
```

Now run the application. You will not see the Resize event procedure execute when the form loads, but if you resize Form1, the Resize event procedure will execute, as evidenced by the MsgBox that appears when you do so. The effect was accomplished by using a variable so that the Load event procedure could communicate to the Resize event procedure. Since both of these procedures reside in Form1, it made perfect sense to use a variable private to Form1. There was no need for the entire application to know about this variable.

Exploring Procedures and Procedure Writing

Step 6: A Procedure of Your Own

Now you will write a procedure that uses the intrinsic string functions of VB. The procedure will accept a string of text and return that same text with each word capitalized. This part of the exercise shows you how to use some of the intrinsic functions, and also shows you how to write your own procedure that can be called from the entire application.

Bring up Module1 in a code window, and then click on the Tools/Add Procedure dialog. Define a Public Function named UpperAllWords(). Set it up so it will accept a string and return a string. The function declaration should look like this when you are done:

```
Public Function UpperAllWords(TEXTIN As String) As String
```

Here is the complete function. Note, it does not handle a single word in a string (it only handles multiple words in a string), and it does not capitalize the first letter of the last word. We expect you to add this functionality if you have time.

```
Public Function UpperAllWords(TEXTIN As String) As String
Dim strWORD As String
Dim strFINAL As String
Do While InStr(TEXTIN, " ") <> 0
  'Isolate the first word in TEXTIN
  strWORD = Left(TEXTIN, InStr(TEXTIN, " "))
  'Chop the first word off the TEXTIN
  TEXTIN = Right(TEXTIN, Len(TEXTIN) - InStr(TEXTIN, " "))
  'Uppercase the 1st letter
  strWORD = UCase(Left(strWORD, 1)) + Right(strWORD, Len(strWORD) - 1)
  'Append results to final string
  strFINAL = strFINAL & strWORD
Loop
'Return Results Capitalized
UpperAllWords = strFINAL & " " & TEXTIN
'WARNING: This function can't handle a single word, and
'does not uppercase the last word. You need to
'consider fixing this.
End Function
```

Step 7: Testing the Procedure

The procedure UpperAllWords() accepts a stringful of words and returns those words capitalized. Your next step is to test out the procedure. Drop a textbox on Form1, and then drop a CommandButton. Under the CommandButton, add this code:

```
Form1.Caption = UpperAllWords(Text1.Text)
```

Run the application, and enter a sentence using all lowercase letters in Text1, and click on Command1. The uppercase text should appear in the window title of Form1.

Step 8: Fixing the Procedure
Alter the code in the procedure to handle (a) an input string with only one word inside, and (b) setting the first letter of the last word to an uppercase. Good luck!

A Simple Selection List Box

Step 9: Setting up the Controls
One of the more common things you may have seen in installation and setup applications is the selection list box arrangement. Here, two list boxes are provided. The one to the left is the source box, and is filled with items. The one on the right is populated dynamically, as you click on items you desire for inclusion, from the source list box. Typically, these boxes have a set of buttons between them as depicted in Figure 4-8.

The > button moves a single item from left to right.

The < button moves a single selected item from right to left.

The >> button moves everything from left to right.

The << button moves everything from right to left.

Drop two list boxes on Form1; these will be named List1 and List2 by default. Also, drop four buttons and label them as just described. Next, add this code to the Form_Load of Form1:

```
Form1.List1.AddItem "Thing1"
Form1.List1.AddItem "Thing2"
Form1.List1.AddItem "Thing3"
Form1.List1.AddItem "Thing4"
Form1.List1.AddItem "Thing5"
Form1.List1.AddItem "Thing6"
```

Run the application to be sure that List1 fills with information.

Step 10: Making It Work
Now you need to add functionality to the buttons. Add this code to the Click event procedure of Command2:

```
'BE SURE SOMETHING IS SELECTED, Exit Sub if not
If Form1.List1.ListIndex = -1 Then Exit Sub
```

Figure 4-8
The Selection list box
arrangement.

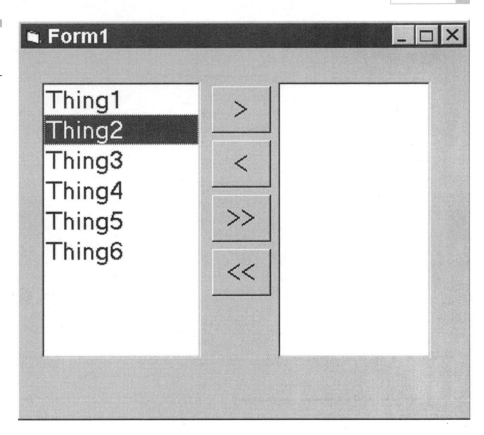

```
`COPY SELECTED TEXT FROM SOURCE TO DESTINATION
Form1.List2.AddItem Form1.List1.List(Form1.List1.ListIndex)

`DELETE TEXT FROM SOURCE
Form1.List1.RemoveItem Form1.List1.ListIndex
```

Be sure to examine the Help file for the ListBox properties and methods used here. They include List(), ListIndex, AddItem, and RemoveItem.

Step 11: Making More of It Work

Test the application. Click on an item in List1 and then click on the Command2 button labeled > and see if the selected data "moves" from List1 to List2. It should! As you might imagine, moving data in the opposite direction is just a matter of moving this code under another button (Command3) and renaming some items such that the selected item in List2 gets moved to List1. Select the code with the mouse, and click on Edit/Copy. Then, open a code window to the Command3_Click, and click on Edit/Paste. Now alter the code so that the data movement is from List2 to List1. The completed code under Command3 should look like this:

```
'BE SURE SOMETHING IS SELECTED, Exit Sub if not
If Form1.List2.ListIndex = -1 Then Exit Sub

'COPY SELECTED TEXT FROM SOURCE TO DESTINATION
Form1.List1.AddItem Form1.List2.List(Form1.List2.ListIndex)

'DELETE TEXT FROM SOURCE
Form1.List2.RemoveItem Form1.List2.ListIndex
```

Step 12: Making It Work Better

Test the application. The data should move in both directions when you click on data. You completed some functionality that works, but you did essentially duplicate code, and this was done needlessly. The processing under Command2 and Command3 is very similar. In fact, wouldn't it be nice if we could call a procedure to move the data? You can do exactly that, by passing in the ListBoxes as parameters. In this last part, we will write a procedure as a subroutine that moves the data from a source list box to a destination list box. The key to this is that you can pass controls around just like variables. Create a new procedure by clicking on the Tools/Add Procedure dialog. (It can be set up as Public in Module1 or as Public or Private in Form1, your choice.) Create a subroutine and name the procedure MoveData. The completed procedure should look like this:

```
Public Sub MoveData(SOURCE As ListBox, DEST As ListBox)
'BE SURE SOMETHING IS SELECTED, Exit Sub if not
If SOURCE.ListIndex = -1 Then Exit Sub

'COPY SELECTED TEXT FROM SOURCE TO DESTINATION
DEST.AddItem SOURCE.List(SOURCE.ListIndex)

'DELETE TEXT FROM SOURCE
SOURCE.RemoveItem SOURCE.ListIndex

End Sub
```

This procedure moves data between any two arbitrary ListBoxes! Now, call the procedure by replacing all that code under Command2 with this single line of code:

```
Call MoveData(Form1.List1,Form1.List2)
```

Test it out. It works, right? Now replace all the lines of code under Command3 with this single line. All you do is reverse the order, making List2 the source ListBox:

```
Call MoveData(Form1.List2, Form1.List1)
```

What do you think?

Exercise Summary

Here you have examined the scoping rules for variables in a module and form, and you have also written a pretty interesting procedure while learning more about VB parameter passing.

Exercise Extra Effort

Finish the last two buttons, which move everything in one box to another. There are many correct solutions to this problem. Did you know that when you remove the first item from a ListBox, all the other items move up?

CHAPTER 5

Tricks, Techniques, and Introduction to Data Access

What You Need to Know

This part of the book assumes you understand the basics of VB-multiple forms, menus, properties, events, and methods-in addition to the language itself. If you are jumping into this section, it's assumed you've worked for a few months with VB. You should have a solid grasp of the scoping rules for data and procedures before starting this section.

What You Will Learn

Here you will learn about coding standards, building simple applications, menu and control arrays, the Multiple Document Interface (MDI), report writing, and the concepts and facilities of data access using the Data Control and Data Access Objects (DAO). This portion of the book covers a set of topics that are not easily classifiable, such as MDI and control arrays. It also foreshadows the more comprehensive data access coverage later in the book by introducing the basics of data access in this chapter.

Overview

Naming Standards for Variables, Procedures, and Controls

Consistent naming of program objects clarifies code. We explain the basics of employing standard naming practices across your entire application to enhance maintainability.

Tricks and Techniques

Now that you understand how event procedures work and the scoping of variables, we can show you how things really work.

Multiple Document Interface

A special kind of VB form can be designated the container of other forms. In this part of the book we introduce MDI and show you the techniques to put it to work in your applications.

Control Arrays and Arrays of Dynamic Menu Items

A control array is a group of related controls with the same class name and Name property setting. A common set of event procedures addresses the entire group, with each item identified by a unique index number. Arrays of both controls and menu items are possible. Control and menu arrays have many uses in production applications.

Data Access Concepts and Facilities

Data access is a huge topic. Here we introduce the basic tools and techniques to access data in VB and get you ready for more advanced data access topics that follow later in the book.

Report Writing

VB ships with the Crystal Reports report writer. Mentioned only briefly in this book, it is worth knowing about and checking out.

Naming Standards for Variables, Procedures, and Controls

Consistent naming clarifies code. The basic technique of prefixing variables with a short lowercase code that denotes the data type is a powerful, self-documenting way to clarify your code. This technique can also be effective when applied to procedures and controls.

Standard Naming for Variables

Variables always have a data type. As such, this type can be denoted in the variable name. For example, a variable that will hold the sum of a person's annual salary with all bonuses could be declared like this:

```
Public sngTOTALPAY As Single
```

The "sng" in sngTOTALPAY denotes the data type. If this same variable had global scope because it was declared in a module, you could write:

```
Public g_sngTOTALPAY As Single
```

Here, the "g_" denotes that the variable has global scope. Thus, the prefix can convey both datatype and scoping information to a maintenance programmer. All of this is accomplished without the use of comments in the code.

There are many religiously held arguments both for and against the use of prefixing (also known as Hungarian notation). Some folks like only three-letter prefixes; others argue for a postfix notation. Pick a standard for naming and stick to it, because it clarifies code. For example, here a prefixed variable is used in an expression, making it easy to see the data type and the level of scope where the variable was declared:

```
dblResult = (g_sngTOTALPAY * .10)/12)
```

Prefixing makes it easy to see the data type of variables used in expressions and easy to see what the programmer intended regarding the data type of the return value. This speeds comprehension of the code and greatly aids maintainability, which is something we strongly favor. Gaining a nanosecond of processing speed may seem sexy, but often that blazing code is impenetrable in terms of comprehension. We favor maintainability over incremental performance enhancements that render the code unreadable. Table 5-1 lists a set of suggested prefixes for the intrinsic VB variable types:

Standard Naming for Procedures

Like variables, procedures have scope. Procedures are either subroutines or functions. In the case of functions, the procedure returns a value. A prefixing scheme for procedures can be formulated and enforced, thereby making code more clear and readable. Table 5-2 lists the suggested prefixes for each type of procedure:

TABLE 5-1:

Suggested intrinsic datatype prefixes

Byte	by
Boolean	bl
Integer	int
Long	lng
Single	sng
Double	dbl
Currency	curr
Date	dat
Object	obj
String	str
String	fstr (fixed length)
Variant	var

TABLE 5-2:

Suggested prefixes for procedures

Scope Level	Procedure Type	Prefix
Global	Function	gf
Global	Sub	gs
Form, Public	Functions	fpf
Form, Public	Sub	fps
Form, Private	Function	flf
Form, Private	Sub	fls

Standard Naming for Controls and Forms
Every control has a class name. You can see the class name in the object box of the Properties window. The class name should be denoted in the name of the control. For example, you could use "txt" for TextBox, "btn" for CommandButtons, and so on. This you should do. Table 5-3 lists suggested prefix names for controls.

ORPHAN CODE
Here is a situation that arises as you work with VB. If you drop a CommandButton, the default name will be Command followed by the first available integer. For example if the CommandButton is the first one on the form, VB will name it Command1. You might not know this, but if you have a form with buttons named Command1, Command3, and Command4, the next time you drop a CommandButton, guess what the name will be? The name will be Command2, since that is the first available

	Class Name	Prefix	Sample Use
TABLE 5-3:	CommandButton	cmd	cmdOK
A naming standard	Listbox	lb	lbNAMES
for selected intrin-	OptionButton	opt	optYES
sic controls	HscrollBar	hscr	hscrAge
	Vscrollbar	vscr	vscrHeight
	Textbox	txt	txtLocation
	Data Control	dta	dtaPUBLISHERS
	Label	lbl	lblAddress
	Frame	fr	frUsageOptions
	Checkbox	cb	cbOK
	Timer	tim	timCheckFile
	ComboBox	cmb	cmdMembers

integer. After that, the next CommandButton drop will get the name Command5. This is how VB behaves when you drop controls.

A more interesting and potentially troublesome situation arises when you rename a control. Let's say you have a button named Command1, and you have attached code to the Command1_Click event already. If you rename it to, say, btnCANCEL, the code becomes "orphaned." That is, the event procedure Command1_Click remains in the form, but unattached to any control, since there is no longer a control with the name Command1. (This bloats your code if not removed before compilation.)

But wait. There's more! If, after you rename the control, you decide to drop a new CommandButton, it will get the name Command1. And your previously "orphaned" code will attach itself to that new button. Isn't VB great?

▮ Check It Out: Orphan Code

1. Start a new Standard VB project. On Form1 drop four Command-Buttons. These will be named Command1, Command2, Command3, and Command4 by default.

2. Now delete Command2 and drop another button. What does VB name it?

3. Now add this code to Command1_Click:

```
MsgBox "ISNT THIS SPECIAL?"
```

4. Run the application, click on Command1, and see the MsgBox output. Command1 works.

5. Now go to the Properties window and set the Name property of Command1 to btnSAMPLE.

6. Now run your application again and click on that button you just renamed. Shouldn't you be seeing MsgBox output?

7. End the application and double click on btnSAMPLE, which is still labeled Command1. There is no code under the Click event procedure. Where is it?

8. Navigate to the [general][declarations] of Form1. Click on the Proc dropdown. See the procedure Command1_Click? This is orphan code.

9. Now drop a button on Form1. The name will be Command1 by default. Double click on it to show a code window. See the orphan code? Command1 has just "adopted" it because it has the name Command1.

As you can see, the first thing you should do when you drop a control is to name it properly. This is because if you attach code to a control and subsequently rename (or delete) it, the code becomes orphaned in the [general][declarations]. Orphan code can and will bloat the size of your EXE if it is not removed prior to compilation.

Avoiding "Career Secure" Coding Techniques

Throughout this manual, we have been suggesting that maintainability is a virtue, and specific techniques make your code more readable. The summary statement of fact here is that code that is explicit is good, and code that uses implicit knowledge is bad. Consider this line of code:

```
AppName = Label1
```

This is career secure code. Career secure code is code no one else can understand, not even you after about 4 weeks or so. However, since no one else can read the code, people come to you when a maintenance fix is needed. With career secure code, everyone else has to do detective work to find out what is going on, because everything is implied. What is the data type of AppName? It's probably a string, but it could be variant and still work. What about the scope of AppName? We have to go find it, adding time to the maintenance task. Now, consider Label1. It sits on the form in which this code resides. But what is going on? Implicit coding techniques, that's what. Taken to extremes, career secure code that uses implicit techniques at every turn can be downright impossible to maintain. You'll need the "decoder ring" to understand the code. The preceding example is a very simple example of what can become very complex very fast.

You see, the Caption property of the Label object is the default property of that object. This is the property that is used if no property is specified. So actually, this is going on:

```
AppName = Label1.Caption
```

This code can be improved with prefixing and some simple explicit coding techniques. How about this?

```
gstrAppName = Me.Label1.Caption
```

Here I can understand what is going on and what is intended by the original coder. The "gstr" in gstrAppName tells me that this is a string declared to be global in scope, and known to the entire application. I also know that the Caption property of Label1 is intended to be assigned into gstrAppName. It seems so simple, it's just so hard to actually do. Multiple programmers, contract programmers, multiple versions-all of these things work against the application and enforcement of a standard for the naming of variables, procedures and controls. Ditto for forms, which should be prefixed with "frm".

Develop a standard set of naming prefixes and stick to it! Your code will be more readable and require few, if any, comments. Comments are rarely revised, and therefore become misleading when a maintenance change is made to the code. Keep your comments in a single block at the start of a procedure, and use consistent naming to write self-documenting code instead. If you use consistent naming practices from the start of a project through to completion of the code, you get the resulting code clarity for free.

One final note: This is a book that teaches you VB such that you get to know substantially the entire product if you go through the entire book. We use all of the default names (Form1, Command1, etc.) to keep things simple. If we were writing an application instead of a book, we would definitely be sure to practice what we are preaching.

Application Building Techniques and Tricks

Visual Basic has many trick parts to it, and they are easy to understand once you have the basic knowledge you need to try to "break" things. Here we offer some interesting knowledge related to forms, variables, and the ListBox control.

Trick: Calling an Event Procedure from Another Form

Event procedures are nothing more than procedures that are Private to the form in which they reside. You can call them anytime you like, just like any other procedure. This raises many interesting possibilities. You could, for example, have some code under a button that performs a useful task. You may then execute that code by calling the Click event procedure, perhaps from a menu item. This has the effect of enabling the same functionality in two places on your form. A little imagination applied here will present all sorts of possibilities for creative ways to reduce the code size and increase the functionality of your applications.

Check It Out

1. Start a new VB Standard EXE application. On Form1, drop two CommandButtons. Under Command1_Click, add this code:

```
MsgBox "Command1 CLICK event fired!!!!"
```

2. Now run the application and click on Command1 with the mouse. It works, right?

3. Return to design time, and add this code to Command2:

```
MsgBox "Here I am clicking Command2, which 'clicks' Command1!!!"
'Command1 will now think someone 'clicked' it!!!
Call Command1_Click
```

4. Run the application and click on Command2. (Make sure you click on Command2, not Command1.) What happens? Command1's Click event procedure gets called from VB code.

5. Return to design time.

You can even click on a button on another form! To do this, you must change the Click event procedure from Private to Public so that it will be callable from outside the form. After you do this, you can click the button, but remember, the rule for Public procedures in a form applies. The rule is, you must qualify the procedure name with the name of the containing form.

Let's say you're somewhere in Form1 and you want to click on the button named Command1 on Form99. Assuming you modified the Click so that it was Public, the code to click on the button would go like this:

```
Form99.Command1_Click
```

That's all there is to it. This technique has many applications.

You might be asking, "Can I click a button on another form, even if that form is not loaded?" The answer is Yes. The next section explains the difference between initializing a form and loading a form. The difference is important.

Trick: Accessing the Procedures and Data on an Unloaded Form

Say a form with procedures you want to call is not loaded yet into memory. When you attempt to access an event procedure that is modified to be Public, or you attempt to call a Public procedure in the form that you authored, it works OK. Why? Because the procedures and variables that are Public in the form are available immediately after the Form_Initialize event of the form executes. When you access a Public procedure or a Public variable on a form from outside the form, the Initialize event procedure executes. This makes the Public procedures and data in the form available to your program. The form does not have to actually load.

Check It Out: Using Procedures and Data in an Unloaded Form

1. Start a new VB Standard EXE project. Add a new form. It will be named Form2. Now add this MsgBox to the Form_Load:

```
MsgBox "FORM2 LOADING"
```

2. Also add this code to the Form_Initialize of Form2:

```
MsgBox "FORM2 INITIALIZING"
```

3. Click on the Tools/Add procedure, and add this procedure to Form2:

```
Public Sub MyTest()
  MsgBox "MyTest Procedure executing"
End Sub
```

4. Now switch to Form1. Drop a button on Form1. It will be named Command1. Add this code to Command1_Click, which will execute the Form2.MyTest procedure:

```
Call Form2.MyTest
```

5. Run the application and Form1 will appear, with Command1 on it. Click on Command1. What happens? The Initialize event procedure of Form2 executes. The Load event procedure of Form2 does not. This is the default behavior when you access Public procedures or Public data on a form that has not yet been initialized.

6. Click on Command1 again. This time, the Initialize event procedure does not execute, because this event only executes once.

So now you know what happens if you access Public procedures and data on a form that has not yet been loaded. You may want to experiment further to examine the behavior of the Initialize and the Load events of the form object.

Tricks: Sorting with a ListBox, and the Click of the ListBox with ListIndex

You might need to sort some strings from time to time. One way to accomplish this is to utilize the Sorted property of a list box. The basic idea is, you set this read-only at run time property to True during design time. Then, you load it with the text you'd like to sort. Finally, you just pull the text out, after the list box sorts it for you. Let's assume the strings are in an array named MyText(). Here is the code to load the list box:

```
'LOAD THE LISTBOX
Dim i As Integer
Dim max As Integer
max = UBound(MyText): MsgBox max
For i = 0 To max
  Me.List1.AddItem MyText(i)
Next
```

Remember, the list box has a Sorted property set to True. This means all of the items in the list box will be in sorted order. Here is the code to reload the array with the sorted data:

```
'UNLOAD THE LISTBOX
Dim i As Integer
Dim max As Integer
max = Me.List1.ListCount
For i = 0 To max - 1
  MyText(i) = Me.List1.List(i)
  Debug.Print MyText(i)
Next
```

The key to the effective use of this technique is to set the Visible property of the list box to False, so your users cannot see the list box at all. You just use it as a cheap but effective sorting machine for text.

Multiple Document Interface

Concepts and Facilities

The notion of a Multiple Document Interface is easy to understand. You have seen the Multiple Document Interface (MDI) style of user interface in popular applications like Word and Excel. In applications that use the MDI interface, there is one dominant window and one or more sibling (or child) windows. The child windows are always contained in the containing (or parent) window. The parent window is a special kind of VB form called an MDIForm.

Defining the Parent Window

To create an MDI application, you must first add a single MDIForm to your application. Selecting Project/Add MDIForm from the VB menu does this. You can only add one MDIForm per application, so the menu item will be disabled after you add an MDIForm to your application.

Defining Child Windows

The child windows that will inhabit the MDIForm are just regular VB forms with their MDIChild property set to True. If the MDIChild property is set to True, then this form will always display as a subordinate, fully contained window of the MDIForm in the project.

Implementing MDI

Even after you have added an MDIForm to your project and set the MDIChild of other forms in the project to True, you are not yet done. There are a couple more steps to perform. First, go to Project/Project Properties and set the Startup object to the name of your MDIForm. This makes the MDIForm the startup form for the project, so that it will load and show to start things off. If you use the Show method to show a child window before the MDIForm, the child window will force the MDIForm to load also. However, any additional child windows will not automatically load.

You cannot drop controls on an MDIForm unless they have an Align property, which forces the control to align itself along the top, bottom, left, or right of the parent window. Search Help on "Align Property" and check the controls in the "Applies To" link.

Special Menu Options for the MDIForm

When building MDI applications, you can ask for the "Window List" to appear in the menu in the MDIForm. You accomplish this by specifying that you want the window list in the menu editor. Figure 5-1 shows the

Menu Editor and shows the WindowList checkbox, Figure 5-2 displays a menu within an MDIForm with the Window list displayed.

The Window list displays the window title of each child window as a menu item. With the Window list feature, you can shift the focus from one child window to another by clicking an item in the Window list.

Behavior Peculiar to MDI Applications

One thing that happens in an MDI application is the display of a child window's menu within the MDIForm parent when the child has the focus. The best way to describe this is with a small demo. The parent MDIForm will have a menu and two child windows, Form1 and Form2. Form1 will have a menu, Form2 will not. When Form1 gets the focus, its menu will supersede the parent MDIForm menu until focus is moved to a child window without menus. This is demonstrated in the following example:

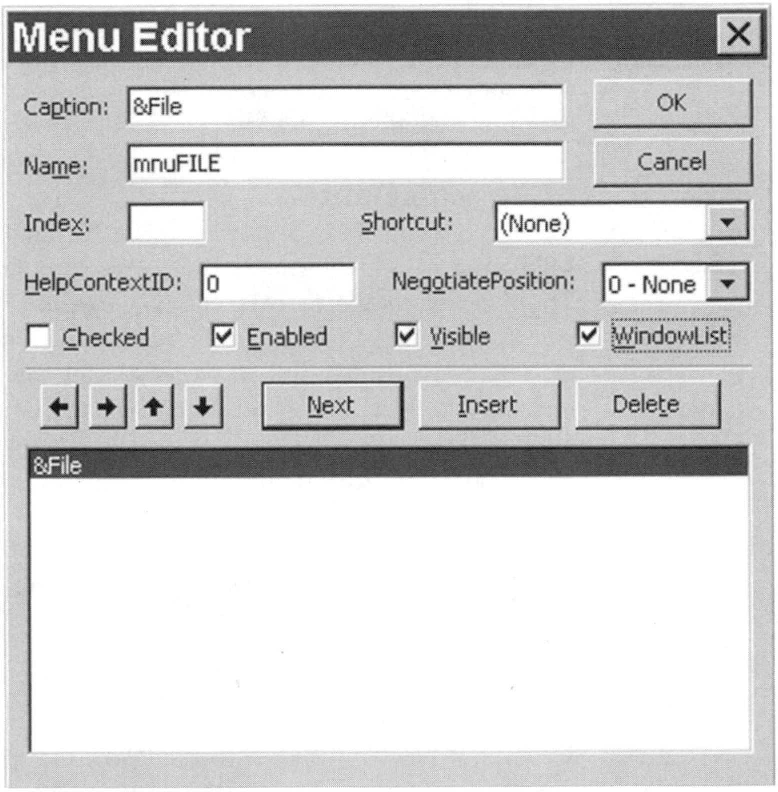

Figure 5-1   The Mwnu Editor.

Figure 5-2

Child Forms in an
MDI Window.

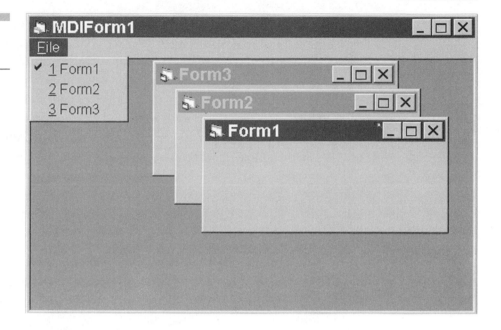

Check It Out: Multiple Document Interface

1. Start a new VB Standard EXE project. Add an MDIForm by clicking on Project/Add MDIForm.

2. The MDIForm does not become the Startup object by default. You have to specify that from Project/Project1 Properties. Click on this menu item now and set Startup Object to MDIForm1.

3. Start the application. You will see MDIForm1 appear, but it will not have any child windows. You will add them now.

4. Return to design time. Add a form to the project. It will be named Form2.

5. Set the MDIChild property of Form2 to True. Also set the MDIChild property of Form1 to True. This makes both forms subordinate to MDIForm1. In the Project window, you will see the icon of both Form1 and Form2 change slightly to reflect the new status of these forms.

6. Don't run the application yet, because the children will not automatically appear in the parent. You have to load them first. Add this code to the MDIForm_Load of MDIForm1:

```
Load Form1
Load Form2
```

7. Now run the application and you will see Form1 and Form2 appear as child windows within MDIForm1. Fiddle with each window. Try normalizing, maximizing, and minimizing Form1 and Form2.

8. Return to design time. Now set the AutoShowChildren property on MDIForm1 form True to False. Run the application. Where are the children?

9. Now back in design time, comment out the Load statements, and add the following statements that use the form's Show method.

```
Form1.Show
Form2.Show
```

10. Run the application and see the children again. Return to design time. Change the AutoShowChildren property back to its default value, and replace the Show methods with the Load statements again.

11. Now you will explore menu behavior under MDI. Add a single menu item captioned MDIForm1 MainMenu to MDIForm1 using the Menu Editor. Also, check the WindowList option for this menu item of MDIForm1. Finally, add a single menu item captioned as Form1 Child Menu.

12. Run the application. Click on the menu of MDIForm1. The window list will appear, displaying Form1 and Form2. Notice the menu that currently appears; it belongs to MDIForm1. Now, click on Form1 or select it from the Window list. What happens? Form1's menu appears until you give Form2 the focus by clicking on it or selecting Form2 from the window list.

13. Return to design time. Set Form1 as the StartUp object in the Project/Project1 Properties dialog. Run the application. When you do, MDIForm1 will appear with only Form1 contained as a child window. This is because when you load a child window, it drags its parent along.

MDI applications are more structured than independent, form-based applications that do not impose the structure of Multiple Document Interface. Just remember the simple rules and you are all set: (a) Set the MDI-Form as the Startup Object, (b) You must set the MDIChild property to True for each child window, and (c) Set the MDIForm's AutoShowChildren property so that you get the desired results.

Control Arrays and Arrays of Dynamic Menu Items

A control array is a set of controls of the same class that share a common set of event procedures and a common name. The Index property of each control array element differentiates it from all others and is passed in to the common event procedures they all share. The control array has many applications.

Control arrays can be dealt with at design time or run time. During Run mode, you can dynamically add new items to the control array. You can also add new items to a menu dynamically at run time. This is also very useful for many applications. You may be familiar with menus displayed in Word and Excel and even VB that display the last four documents, or last four spreadsheets, or last four projects you have worked on. These menus change dynamically at run time because Windows allows you to dynamically alter the menu. VB provides support for menu arrays and control arrays.

Control Arrays

A control array is a group of controls that are all of the same class (for example, all CommandButtons, all ListBoxes, or all Labels) that share a common name. Because they share a common name, they all share the same event procedures. Each item in a control array has a unique value for the Index property. The first item is always gets an Index value of zero. Let's look at an example:

Check It Out: Creating Control Arrays

1. Start a new VB Standard EXE project. Drop a CommandButton on Form1. It will be named Command1.

2. Drop another button on Form1. It will be named Command2. Go to the Properties window and change the name to Command1. You will be prompted with the message "You already have a control named `Command1`. Do you want to create a control array?" Click on Yes in response to this dialog. The dialog appears in Figure 5-3.

3. Now double click on the first button and look at the Click event. You will see the parameter Index As Integer passed in by the system. This parameter has the value of the Index property of the CommandButton that was clicked by the user. Enter this code in the Command1_Click event procedure:

Figure 5-3

The prompt about creating a control array.

```
MsgBox "Index of button clicked is " & Index
```

4. Run the application. Click on the first button, then click on the second button. What happens? The first button has an Index of zero, while the second button has an Index of one.

5. Return to design time. Open up the Properties window so you can see both the Name property and the Index property. Drop yet another CommandButton on Form1. Name it Command1 from inside the Properties window. Notice that you are *not* prompted about the control array, because you have already created it in the previous step. Also, notice that as soon as you name it Command1, VB automatically sets Index to the value 2.

6. Run the application. Click on the third button. What happens now?

7. Control arrays can be created in many ways. Try this: Start a new project, and drop a CommandButton on Form1. Click on it so it gets the design time focus.

8. Now go to the VB menu and click on the Edit/Copy menu command. You have just copied the control to the clipboard.

9. Now click on Edit/Paste. This will copy the contents of the clipboard to the form. When you do this, you'll be trying to paste another Command1 onto Form1. The result is you will get the prompt shown in Figure 5-3. Specify Yes to create a control array, and the new control will have the value 1 for its Index, while the original control it was based on will now have an Index of 0. This is another way to create a control array at design time.

Control Array Processing at Run Time

There is a way to add new controls to a form dynamically at run time, which involves a control array. The basic idea is this: You place a control with an Index of zero on the form, and make it invisible. This item is the base element of a control array that can grow dynamically at run time. In VB code, you add new controls to the control array and make them appear dynamically, at run time. You will need to use the Load command to do this. The Load command is not just used for loading forms.

Check It Out

1. Start a new VB project. Drop a TextBox control on Form1. Set the Index property to zero. This makes it the base element in a control array. You'll add new elements to the control array at run time.

2. Set the Visible property of Text1 to False, and set both the Left and Top properties to zero.

3. Drop a button on Form1. When this button is pressed, code in the Click event will create new TextBoxes. Add this code under the Command1_Click event:

```
Static c As Integer
'THIS WILL BE USED AS THE INDEX OF THE NEW ELEMENT:
c = c + 1
'THIS LINE CREATES THE NEW CONTROL ARRAY ELEMENT:
Load Text1(c)
'THIS LINE MAKES THE TEXT BOX VISIBLE:
Text1(c).Visible = True
'THIS LINE POSITIONS THE TEXT BOX ON THE FORM:
Text1(c).Top = Text1(c - 1).Top + 500
```

Since all control array elements share a common name, the notation you use in code to refer to an individual element in the array must include the array index, such as [ArrayName] (ArrayIndex). [property | method]. The code just shown uses a Static variable that is incremented by one each time the button is clicked. This variable's value is used as the index for the new item created with the Load command.

Dynamic Menu Items

The same way you can add new control to a form, you can add new menu items to a menu. The basic idea is the same: You create a menu item that has an Index of zero, making it a base element in an array of menu items. Then in code you use the Load command to add a new menu item to the menu, change the Caption of the new item, and set the Visible property of the new item to True. All of the menu items will share a common Click event. The system will pass in the Index value to the Click event procedure so that you know which menu item was clicked.

Check It Out: Dynamic Menus

1. Start a new VB Standard EXE project. Add this menu to Form1 using the menu editor:

```
Caption  Name
File  mnuFILE
Open  mnuOPEN
Menu Array  mnuARRAY
```

2. For mnuARRAY, set the Visible property to False and set the Index property to zero, all from the Menu Editor.

3. Drop a Textbox and a CommandButton on the Form. The way this is going to work is as follows: You will enter text into the Textbox and then click the button. When you do, the text in Text1 will be inserted as a menu item. Add this code to the Command1_Click event procedure:

```
Static c As Integer
c = c + 1
Load mnuARRAY(c)
mnuARRAY(c).Caption = Text1.Text
mnuARRAY(c).Visible = True
```

4. Run the app. Enter your name into Text1 and click on the Command1 button, then click on the menu. Your name should appear in the menu. Type in some other names and check the menu after each click on Command1.

5. Each menu item shares a common event procedure for the Click event. Add this code under the mnuARRAY_Click(Index as Integer) event procedure:

```
MsgBox "That was menu array item number " & Index
```

6. Run the application add some dynamic manu items as done previously, and then click those dynamic menu items. You will see how the system passes in the Index of the menu item you clicked. This allows you to perform conditional processing based on the Index.

With a little imagination, you can see the applications afforded by menu arrays and control arrays. In both cases, the concepts are identical.

Data Access

Concepts and Facilities

With all of this background on the basics of using VB, you are ready to explore data access using Visual Basic. VB offers several levels of data access support.

DATA CONTROL

Data Control is one of Visual Basic's intrinsic controls. It provides a simple properties, events, and methods interface to a variety of data formats. It sits on top of and relies on Data Access Objects to perform processing.

DATA ACCESS OBJECTS (DAO)

Underlying the Data Control are the Data Access Objects. The DAO provide a more comprehensive interface to data using VB. For example, with the DAO you can add a table to a database, something that is impossible to do with the Data Control. When you use the DAO, you take on more responsibility for processing the data when compared to using the Data Control. What you get in return is more power to control the database under program control

OPEN DATABASE CONNECTIVITY OVERVIEW

The Data Control and DAO require the Access runtime database Engine, also known as Joint Engine Technology or JET. JET is collectively a set of Dynamic Link Libraries that contain the callable procedures that read and write databases. While you can access ODBC (Open Database Connectivity) data sources using JET, the DAO, and the Data Control, this is not recommended because the processing overhead is prohibitive in most cases. This is because JET is optimized for Microsoft Access databases, not ODBC data sources.

The Remote Data Control and the Remote Data objects do not rely upon JET to perform database processing. Instead, these facilities rely on ODBC only. ODBC is a Microsoft technology that seeks to make all databases look the same to your application. This is accomplished by accessing the data via an ODBC database driver. For example, if your application was written using ODBC to access an ORACLE database, at a later time migrating to SQL Server would be rather simple, since your application does not access ORACLE directly. Instead, your application communicates with ODBC, and ODBC communicates with the ORACLE database engine. Since SQLServer also has ODBC access as an option for this database format, migrating to SQLServer typically will not require a rewrite of the application. ODBC technology uses "drivers" to make most data formats look uniform to your application in much the same way as printer drivers make all printers look similar to your desktop.

THE REMOTE DATA OBJECTS

Existing as a thin layer atop ODBC, the Remote Data Objects are a set of programmable objects that provide properties, events and methods access to ODBC data. Objects with names like rdoConnection and rdoResultset exist in the RDO ActiveX server, a reusable component that ships with the Enterprise Edition of VB. The RDOs provide a easy-to-program alternative to using the ODBC API functions to access data from VB. To the programmer, the RDOs looks much like the Data Access Objects. For client server applications, the RDOs are superior to the DAO.

THE REMOTE DATA CONTROL

The Remote Data Control is similar to the Data Control. It can be dropped on a form and provides a reduces set of properties, events and methods for accessing ODBC data. The Remote Data Control "wraps" the programmable RDOs, allowing ODBC data access with almost zero code.

In this section we cover only the basics of the Data Control and the Data Access Objects as an introduction to data access in Visual Basic. This section gives you the basics of data access using VB and also introduces you to the world of reusable components, also known as ActiveX servers. The Data Access Objects form a kind of reusable component, something that plays a major role in rapid application development using Visual Basic.

Using the Data Control

The Data Control allows simple data access with almost no coding. Using it is very easy to do. Visual Basic ships with a database called BIBLIO.MDB, an Access database that contains information on technical books, listing publishers, authors, and titles. We will use this database to explore the Data Control. The Data Control is an intrinsic control of VB and is depicted in Figure 5-4.

The primary properties of the Data Control are as follows:

Databasename: Specifies the database to connect to.

Recordsource: Specifies the table in the database (or the SQL) for the data.

Once these properties are set, other controls known as data bound controls can access the data pointed to by the Data Control. There are a set of properties associated with data bound controls that are important:

Figure 5-4   The Data Control in the Toolbox.

DataSource: This design time only property specifies the Data Control to connect to.

DataField: This property of a bound control, if present, specifies which field to display in the control.

You can give the Data Control a try in the following example.

Check It Out: Using the Data Control

1. Start a new VB Standard EXE project. Drop a Data Control on Form1. It will be named Data1.

2. Drop a Textbox control on Form1. It will be named Text1. It is a data bound control. In this control you will display some data.

3. Click on Data1. Set the Databasename property so that it points to BIBLIO.MDB in your VB installation directory. To do this, click on the Databasename property in the Properties window and click on the button that appears with the 3 dots (known as an ellipsis). When you click, you will get a dialog box similar to that shown in Figure 5-5. Navigate to your VB installation directory and click on the file BIBLIO.MDB to specify this Access database as the one that will appear in the DatabaseName property of Data1. After you click on the BIBLIO.MDB, click on OK and the full pathname of the BIBLIO database will be written into the Databasename property of Data1 in the Properties window.

4. The next property to set on Data1 is the Recordsource. Click the Recordsource property of Data1 in the Properties window; then click on the drop-down button that appears. When you do this, all of the tables that are in BIBLIO.MDB will appear in the drop-down list. Select Publishers and you will see this table indicated for the Recordsource in the Properties window.

Figure 5-5
Specifying the Databasename property of the Data Control.

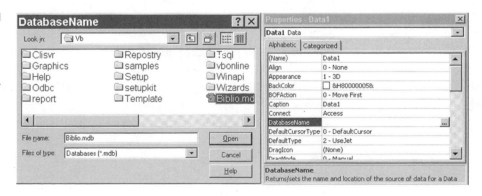

5. Now you have set up Data1 to point to a table in BIBLIO.MDB. The next step is to display some of the data. Click on Text1 and set the DataSource property to Data1 and the DataField property to Company Name. (The Datafield will expose a list of available fields. Just drop down the list and select one.) You are now done setting up the data access with the Data Control.

6. Run the application and click some of the Data Control buttons. You can navigate to the first, last, previous and next records using the buttons on the Data Control. When you click on a button, the user interface updates automatically. If you edit the data and navigate to another record, that edit is saved to the database.

7. End the application and return to design time.

The major advantages of the Data Control are the automatic update of the user interface as you scroll the data and the automatic update of the database when you make edits. You can specify a SQL statement in the Recordsource instead of a table name, and the SQL can express a complicated query.

The Data Control contains a ReadOnly property and an Exclusive property. The ReadOnly property makes the data read-only; the user cannot make updates when this property is True. The Exclusive property gives your application exclusive use of the data and locks out other users until your application breaks the connection.

The Data Control also has events. The Validate event fires as you navigate from record to record. It executes before a different record becomes the current record, allowing you to examine the record under program control before it is written back to the database with changes. The Reposition event fires after a record has become the current record, allowing you to examine it before display.

The Data Control maintains a kind of container called a Recordset; the Recordset object can contain a Table, a Snapshot, or a Dynaset. You specify which in the RecordsetType property of the Data Control. Table is the default. The table-type Recordset is read-write. The Snapshot-type Recordset gives a copy of the database records as of the time the query executed; Snapshots are fast, but read-only. The Dynaset type of Recordset is a collection of indexes that point to the data, not the data itself. This means that if other users change the underlying data, you see the changes if you rescroll those records. Dynasets can be slow and should therefore be avoided if possible for performance reasons. As you will see later, properties of the Data Control map to functionality found in the Data Access Objects. This is because the Data Control is a "wrapper" of

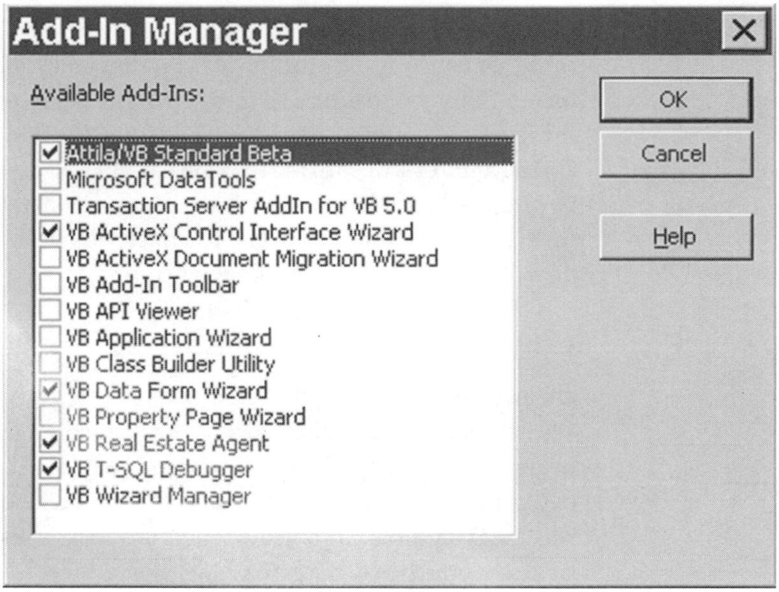

Figure 5-6 The Addin Manager dialog.

the DAO functionality. When you use the Data Control, you must ship all of the Dynamic Link Libraries that comprise the Access runtime ("Jet") database engine.

The Data Control makes it simple to access data without programming, but this access is not without a cost. The Data Control does not allow you full access to the database; you cannot add a new table to the database or add a new field to a table using the Data Control. That type of processing will require the Data Access Objects, covered later in this chapter.

The Data Form Wizard

The Data Form Wizard is a tool that ships with VB. This wizard builds a form that displays data, and it employs the Data Control on the form it generates. You can learn a lot about how to use the Data Control from examining the output of this wizard.

You can select the Data Form Wizard in the Addins Manager, selectable by clicking on the Addins/Addin Manager menu command. The dialog that appears is depicted in Figure 5-6. When you click on the VB Data Form Wizard item, the wizard will appear in your addins under the Addins menu of Visual Basic. The wizard walks you through a series of steps in the process of building a form. The form that is created by the wizard provides a complete add/change/delete facility for maintaining the table you specify. It eliminates the drudgery of building file maintenance forms. This wizard has several options, some of which produce some very interesting source code. For example the Data Form Wizard uses the Data Bound Grid control to create a master/detail form for data display of two related tables in a database. Sample output from the wizard appears in Figure 5-7. You will want to check out both what the wizard does and how it does it by giving this wizard a try.

Using the Data Access Objects

OVERVIEW

The Data Access Objects provide properties, events, and methods access to data, such that VB programmers can leverage existing knowledge to exploit the power of the Access runtime database engine. You can think of the DAO as a bag of objects, each object possessing its own set of properties and methods. The DAO sit below the Data Control and above the Access runtime engine. The objects contained within the DAO make data access very easy in Visual Basic. Keep in mind that with the DAO you do not get the automatic UI (user interface) update and the automatic data-

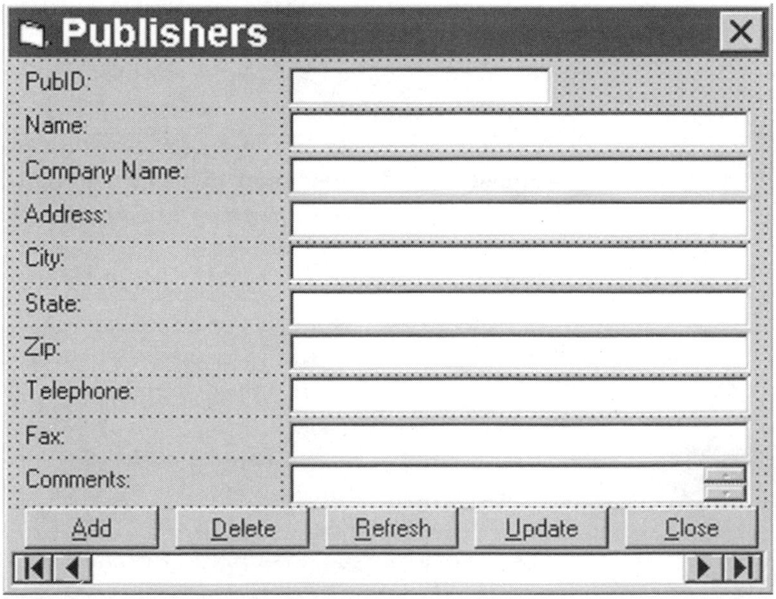

Figure 5-7 A form for maintaining the Publishers table in the BIBLIO database, as produced by the Data Form Wizard.

base update that is provided by the Data Control. You are responsible for both the UI and the database updates when you use the DAO.

THE HELP FILE AND THE OBJECT MODEL

The DAO are a set of related objects. You can see the relationship via the Help file.

Check It Out: Examining the DAO Help File

1. Start a new VB Standard EXE project.
2. Click on the Help/Microsoft Visual Basic Help Topics menu command. From the Index tab, specify object models, DAO in the text box to access the DAO object model help.
3. Note the containment of the Field object within the Recordset object, and also the containment of the Recordset object within the Database object.

The DAO are a set of related objects that make it easy to do data access in Visual Basic. Each object has properties and methods. A partial view of the object model as depicted by the Help file appears in Figure 5-8.

The basic idea with an object model is that objects are related. The relation is typically implemented via the methods of the objects. In the DAO, the Database object contains a collection of Recordsets. Each Recordset contains a Fields collection. Each Field object has properties,

Figure 5-8
The DAO Object
Data

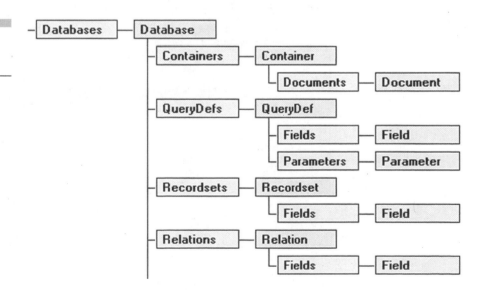

just like any other VB object. (This is also true of the Database and Recordset object). The containment relationships between the Database, Recordset and Field objects are depicted in Figure 5-8.

Note the one-to-many relationship between Database and Recordsets. A single database contains, potentially, many recordsets. This one-to-many relationship is implemented with collections, which are similar to arrays. You can iterate over a collection just as you can over an array to gain access to the members of the collection.

REFERENCES AND THE OBJECT BROWSER

Before you can begin using DAO, you must establish a reference to the DAO ActiveX server. This reference makes all the objects in the DAO available to your program. This is done by clicking on the Project/References menu command. When you do this, you will see the References dialog. This dialog displays all of the reusable components available on your computer. When VB installs, it also installs the DAO reusable component. For this reason, the references dialog displays the DAO as an available item for use by your project. Figure 5-9 shows the References Dialog.

Figure 5-9
The References Dialog.

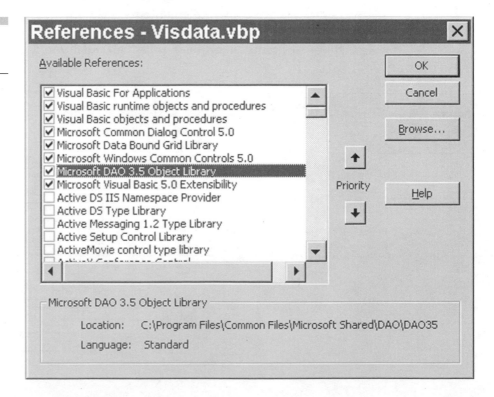

Click the item titled Microsoft DAO 3.5 Object Library and make sure it is checked off before you start working with DAO. This makes the objects in the DAO available to your program.

The following example shows how to determine how many rows of data are in a database table by using simple data programming. @3ah:Check It Out: Using DAO

1. From the Windows Explorer, copy the BIBLIO.MDB that is found in your VB installation directory to the root directory of your computer. By placing the file in the root or topmost directory, we are simplifying the task of opening it under program control, because it will not be preceded by a long pathname.

2. Start a new VB Standard EXE project. Make sure the Microsoft DAO 3.5 Object Library entry in the VB References is checked off before proceeding to the next step. You get to the references by clicking on the Project/References menu command.

3. Now drop a button on Form1. Clicking on this button will tell you how many rows of data are in the Publishers table of the BIBLIO.MDB you copied to your root directory. You will experiment with DAO programmability under the click of this button.

4. Add this code to the Command1_Click event procedure:

```
Dim db As DAO.Database
Dim rs As DAO.Recordset
Set db = OpenDatabase("C:\BIBLIO.MDB")
Set rs = db.OpenRecordset("Publishers")
MsgBox "Publishers table contains " & rs.RecordCount & " rows of
data."
Set db = Nothing
Set rs = Nothing
```

5. Run the application and see what happens. Click on Command1. You should get a MsgBox telling you how many rows of data are in the Publishers table.

There were two new statements used in this example: Set and Nothing. You saw how we dimensioned variables of type DAO.Database and DAO.Recordset. These are known as object variables and are used when accessing the objects in reusable components such as DAO. Next, notice how we used a function named OpenDataBase() to get the whole process started. This function returns a database object. The Set statement is used here because "db" is declared as an object variable. Object variables always require the use of the word "Set" when initializing them to a value or when using them in assignment.

The next thing that happens is we use a method of the Database object to return a Recordset. We specify that the Recordset will contain the data from the Publishers table. We could have specified the Recordset type with the call to OpenRecordSet, but since it returns a Table-type Recordset by default, there is no need. You may wish to check the Help on this method to learn more about it now. With the Recordset in hand, it is a simple matter to report the value of the RecordCount property, which is what we do.

These two lines:

```
Set db = Nothing
Set rs = Nothing
```

are not actually required, but are important to understand. The Nothing keyword disconnects the variable from whatever it was previously referencing, making it available for use again. This code is not needed because the variable has procedure-level scope and will be destroyed when the procedure terminates. We show the Nothing keyword for your information. It is worthwhile to search Help soon on the Set and Nothing keywords.

Here is another example that iterates over the TableDefs collection of the Database object and populates a list box with the Table names contained within BIBLIO.MDB:

Check It Out: The TableDefs Collection

1. Start a new Standard EXE project. GO to the references and make sure that Microsoft DAO3.5 Object Library is checked off.

2. Drop a Listbox and a Commandbutton on Form1. Add this code under the Command1_Click event procedure:

```
Dim db As Database
Dim td As TableDef
Set db = DBEngine.Workspaces(0).OpenDatabase("C:\BIBLIO.MDB")
'ITERATE OVER THE TABLEDEFS COLLECTION
For Each td In db.TableDefs
  Me.List1.AddItem td.Name  'Add the .Name to List1
Next
```

3. Run the application and click Command1. List1 should populate with the name of every table definition in BIBLIO.MDB.

The whole point is that the database contents have been reduced to a set of objects with properties and methods. Groups of related items are contained in collections, and accessing the members is simple with the For Each...Next statement. DAO is a pretty large topic. Exercises following this chapter make it easy for you to gain mastery of DAO programmability basics.

Other Data Access Resources:
Visdata Sample Project and the Data Explorer
Visual Basic ships with an addin called the Data Explorer. This item is accessible from the Addins/Visual Data Manager menu command. This application allows you to create new Access databases and modify existing databases. With the Data Explorer you can for example browse the structure of the BIBLIO database and add new tables or delete existing tables if you so desire. The tool is pretty good for creating test tables or browsing existing tables to verify your results when using the Data Control or DAO.

Perhaps more important is the code in the SAMPLES subdirectory of your VB installation. Here you will find a subdirectory named VISDATA. This directory contains the source code for the Data Explorer utility. Think about this. The Data Explorer allows you to create new databases and populate those databases with tables, queries, and indexes. Since the VISDATA project in the SAMPLES directory contains the source code, you have examples there that will demonstrate just about any task you may have in mind when you use DAO. The dirty little secret is that many "consultants" simply swipe source code from DAO to get tasks completed. While you are new to VB and DAO, you can do the same. VISDATA is a great study tool to learn how DAO really works.

Report Writing

Visual Basic ships with the Crystal Reports Report Writer version 4.6. You can access the report writer by clicking on the Addins/Report Designer menu command.

The basic idea with Crystal Reports is to create a report file that is distributed with your application. The report file contains the details of your report. You must ship the Crystal Reports runtime DLLs with your executable to make everything work.

The product makes it simple to issue on-screen and paper output reports on database data such as the contents of BIBLIO.MDB. Since the report writer is well documented, we choose not to explain it in detail here.

SUMMARY

Upon conclusion of this portion of the book, it should be pretty obvious that you are in a position now to actually use VB to build medium to large applications. Typically, a production VB application has dozens of forms and perhaps a great many modules containing data and procedures that are used in the application. Now that you know the basics of MDI, control arrays, and data access, you can start to consider building that application you had in mind.

Increasing Your Knowledge

Help Search Strings

- Control Arrays
- Load
- MDI
- Windowlist
- DAO
- Object Model
- Data Access
- Data Control
- Naming Conventions

Books Online Search Strings

- "MDI Child Forms"
- "MDI Applications"
- "Control Arrays"
- "Naming Conventions"

Mastery Questions

1. To create a control array at design time, you can:
   a. copy a control and paste the copy into the form, answering Yes to the prompt about creating a control array.
   b. drop a control with the same class name as an existing control, set the Name property to the same name as the existing control, and answer Yes to the prompt about creating a control array.
   c. set the Index property of a control to zero, and then drop another control of the same class and give it the same name as the control with the zero Index property setting.
   d. All of the above

2. To get a child MDI window to show up inside the parent MDI form with all other siblings, you must:
a. set the MDI Child property to True.
b. be sure the child form is loaded before the MDI Parent is visible.
c. load the child form as the startup form for the application.
d. Both a and b

3. Which method must be executed when the Data Control has had any property changes at runtime?
a. The Update method
b. The AddNew method
c. The Refresh method
d. The FindLast method

4. Which DAO collection contains all of the SQL statements that are stored in the database?
a. The Recordset collection
b. The QueryDefs collection
c. The TableDefs collection
d. None of the above

5. Assume five CommandButtons are members of a control array. When one of the buttons is clicked, how can you determine which one was clicked, in VB code?
a. Examine the CurrentItem property of the CommandButton.
b. Examine the Index parameter passed into the Click event procedure by VB.
c. Check the Item property of the control array.
d. None of the above

6. To add a new record to a recordset maintained by a Data Control when the user scrolls to the end of the recordset, you must:
a. execute the AddNew method.
b. execute the NewRow method.
c. set the EOFAction property of the Data Control to 2-AddNew.
d. None of the above

7. When using an MDIForm with all child forms containing menus, when you run the application you can expect that:
a. the child window with the focus will be checked off in the parent MDIForm's window list portion of its menu.
b. the child window's menu will be displayed when the child has the focus.
c. the child windows will allow group operations such as tiling and cascading display.
d. B and c only

8. The default names provided by VB for controls you drop on forms are based on:
   a. the Caption of the control.
   b. the next available integer appended to the default name for the control as defined by the control itself.
   c. Your preferences as defined in the Tools/Options dialog.
   d. None of the above

9. An instance of the DAO recordset object can be obtained via the:
   a. OpenRecordSet method of the Database object.
   b. OpenResultSet method of the Database object.
   c. CreateRecordset method of the Database object.
   d. CreateResultSet method of the Database object.

10. The Load command of the VB language can be used in which of the following ways?
   a. To load a new member of an array of menu items
   b. To load a new member of an array of controls
   c. To load a VB form
   d. All of the above

Answers

1. d
2. d
3. c
4. b
5. b
6. c
7. d
8. b
9. a
10. d

EXERCISE: DAO BASICS

Hardware and Software Requirements
Visual Basic Professional Edition

Required Files
BIBLIO.MDB (Ships with Visual Basic)

Assumed Knowledge
Understanding of DAO concepts and facilities

What You Will Learn

- Database object
- Recordset object
- Field object
- MoveFirst, MoveLast, MoveNext, MovePrevious methods

Summary Overview of Exercise

This exercise builds a small form that allows you to scroll forward and backward through the Authors table of BIBLIO.MDB. You must be sure the DAO is checked off in the References before you begin.

Data Acess Objects

Step 1
Create a directory called \DATAOBJ\. Create a new VB project. Save it immediately to the directory created in step 1 by selecting the File/Project menu command and specifying the \VBCLASS\DATAOBJ directory.

This application will access a file using *not* the Data Control, but instead a database object variable. Without the Data Control, the programmer is responsible for (a) updating the database and (b) updating the user interface. In return for doing these things, you get more power and control.

Step 2
Drop two text boxes on the form. Name them txtNumber and txtName. Label them (with labels) "Author Number" and "Author Name". The form should look like this:

Author Number [   ]
Author Name [   ]

Step 3
Drop four buttons on the form. Name them cmdFirst, cmdLast, cmdPrevious, and cmdNext. Arrange them in a row along the bottom of the form.

Add the following code to the [general][declarations] section of Form1:

```
Option Explicit
' The purpose of this example is to simulate the VB datacontrol
' using VB's data object variables and command buttons
'These data types are available to Pro and Enterprise VB users

Dim datBIBLIO As Database
Dim tblAuthors As Recordset
```

Step 4

Add this Sub to the [general][declarations] of Form1:

```
Sub DataToUI
  txtNUMBER.Text = tblAuthors.Fields("AU_ID")
  txtNAME.Text = tblAuthors.Fields("AUTHOR")
End Sub
```

You need this code because the text fields are not bound to a Data Control. You will be using a database object variable instead. Since there is no Data Control, the program must do the work of filling in the data fields on the user interface. This routine will be called each time your application scrolls a record. The overall effect will be to display the new record.

Note that we have fully qualified object called a .Fields object. Also note that tblName was defined as a Table Object variable in a previous step; we are using it now in this code. Object variables have methods and properties, just as controls do. The basic idea is this: You define a database variable (datBIBLIO), and after you set it up the right way, it has properties and methods available to use. Some methods can be used to set up new, subordinate objects such as, for example, a Table object (tblAuthors). In turn, tblAuthors has properties and methods (.Fields for example) that you can use to get at stuff in the table. That is what this exercise is all about: using the hierarchy of DAO to get some work done.

Step 5

Now add this code to Form_Load()

```
'This line initalizes the database object variable
'NOTE: BE SURE TO SPECIFY THE RIGHT PATH TO "BIBLIO.MDB"

Set datBIBLIO = OpenDatabase("[your path here]\BIBLIO.MDB")

Set tblAuthors = datBIBLIO.OpenRecordset("Authors")

Call DataToUI
```

Here we are utilizing a database variable. The .OpenDataBase method assigns it a value. After it has a value, we can use the .OpenTable method of the newly initialized database object variable named datBIBLIO. .OpenTable provides access to a selected table in the database. In this code the table object variable named tblAuthors gains the value returned from datBIBLIO.OpenTable().

To summarize: the preceding code initalizes a database object variable (datBIBLIO) and a table object variable (tbl). You see, a method of the database object *returned* a table object, in effect initializing it. Now we can have some *fun.*

Step 6
Add the following code to cmdFirst_Click():

```
tblAuthors.MoveFirst
Call DataToUI
```

Add the following code to cmdPrevious_Click:

```
tblAuthors.MovePrevious

If tblAuthors.BOF Then
  Beep
  tblAuthors.MoveNext   ' GO BACK TO WHERE WE WERE
Else
  Call DataToUI
End If
```

Step 7
Add the following code to cmdLast_Click:

```
tblAuthors.MoveLast
Call DataToUI
```

Now add the following code to cmdNext_Click:

```
tblAuthors.MoveNext

If tblAuthors.EOF Then
  Beep
  tblAuthors.MovePrevious   ' GO BACK TO WHERE WE WERE
Else
  Call DataToUI
End If
```

Run the app. All buttons should work.

Step 8
Review the code and examine everything closely. (*Note:* When you use the Data Control, it is *automatically* maintaining Recordsets, tables, and Field objects for you.) In reality the Data Control is a higher-level interface to VB's database objects. Note that the database objects used here are much faster than the Data Control, but the Data Control is more convenient. Data Access Objects also allow you to create indexes, tables, fields and more under program control, something the Data Control cannot do.

Extra Effort

1. Search Help for "Object Models" and select the "DAO Object Model" subtopic. Check it out.

EXERCISE: DATA ACCESS OBJECTS QUERY BUILDER

Hardware and Software Requirements
Visual Basic Professional Edition

Required Files
BIBLIO.MDB (ships with Visual Basic)

Assumed Knowledge
Understanding of DAO concepts and facilities, plus an understanding of DAO collections

What You Will Learn

■ Tabledef object

■ QueryDef object

■ DAO Tabledefs, Querydefs Collections

Exercise Overview

In this exercise you will build an application that uses DAO extensively to explore .MDB databases. The application will allow selection of an .MDB for examination, browsing of available tables and fields within any selected table; the execution of freeform SQL queries, and the saving of

SQL queries by name into the database. Additionally, this database browser will be able to browse existing queries stored in the database, and even examine the SQL that makes the query work.

Selecting a Database to View

Step 1

Make a directory named DAOQUERY. Fire up VB and immediately save the new project to this directory. Don't forget to save often during development. The first thing you need to do is set up a menu that looks like this:

```
File
Open…
————-
Exit
```

Build this menu. Name the menu objects mnuFILE, mnuOPEN, and mnuEXIT.

Step 2

Drop a Common Dialog control on Form1. Name it dlgDATABASE. (You may have to select it from the Tools/Custom Controls dialog to make it available to your project.)

Examine the Help for the common dialog control, press F1 with focus on it to get this Help page. Check out the methods available. Check out .ShowOpen. Also check the .Filename property. You'll be using these.

You'll display a common dialog to select an .MDB file for browsing. Add this code under the mnuOPEN_Click() event:

```
Private Sub mnuOPEN_Click()

dlgDATABASE.filename = "*.mdb"
dlgDATABASE.Filter = "Access DBs (*.mdb)|*.mdb"
dlgDATABASE.ShowOpen
Data1.DatabaseName = dlgDATABASE.Filename
```

Building the Graphic User Interface (GUI)

Step 3

Drop a DBGrid control on the form. This is the Microsoft Data Bound Grid found under the Components tab of the Components dialog. Be sure it is checked off within the Components dialog. This will make the control available in your Toolbox. (You can reach this dialog via the Project/Components menu command.) Also drop a text box and a Data Control on the form. Use the default names DBGrid1, Text1, and Data1.

Set properties as follows:

```
Text1
Multiline = True
Scrollbars = 3 (both vertical and horizontal)
DBGrid
Datasource = Data1
Font.Name = SmallFonts
Font.Size = 6
```

Step 4

Now add a button next to the Text1 textbox. Keep the default name Command1. Set the .Caption to "Run Query". Command1 will be used to execute the SQL statement that you type into Text1. To make a query entered into Text1 actually work, you'll assign it to the .Recordsource property of the Data Control Data1, like this:

```
Private Sub Command1_Click()

Data1.Recordsource = Text1.Text
Data1.Refresh

End Sub
```

Step 5

Search the Help on the Refresh method to learn how it works with bound controls. Next, under the Exit menu's mnuExit_Click event, enter the following line of code:

```
End
```

Start the app. Select the FILE-OPEN menu command and select VB\BIBLIO.MDB or VB32\BIBLIO.MDB. Then, type this SQL into the Text box:

```
SELECT * FROM PUBLISHERS
```

Click on Command1 to make it go. The grid named DBGrid1 should fill up with the results from the query.

Building the Database Explorer

This project so far allows the user to open an .MDB database and run some queries. OK, now let's do some more interesting things. For example, for any database I open, what are the table names? That would be useful information in constructing a query. The next step accomplishes this.

Step 6

You can iterate through all the tables using DAO syntax. But first, you need a database object and a table object. Add this code to the [general][declarations] of Form1:

```
Private db As DAO.Database
Private td As DAO.TableDef
Private fld As DAO.Field
```

Step 7

These variables will be used extensively in this exercise. Right now, you want to display all the tables in a list box. Before you can, you'll need a valid database object. So add this code to the File/Open menu command:

```
If UCase(Right(dlgDATABASE.filename, 3)) <> "MDB" Then
MsgBox "File name selected is not an Access database."
Else
  On Error Resume Next
  db.Close
  On Error GoTo 0
  Screen.MousePointer = vbHourglass

  'Open the Database using a method of the Workspace Object
  Set db = DBEngine.Workspaces(0).OpenDatabase(dlgDATABASE.file-
name)
  Form1.Caption = "DAOQUERY: Browsing " & dlgDATABASE.filename
  Screen.MousePointer = vbDefault
  Data1.DatabaseName = dlgDATABASE.filename
End If
```

Step 8

Search the Help for each item to understand what is going on in this code. The result is, we have a database object named "db" that can now be used. You will use it to fill a list box with table names in the user-selected database.

Drop a list box on the form. It will be named List1. Under the File-Open Menu command, add the following code to fill the list box. *All of this code must be placed just before the End If statement:*

```
'Fill List Box with names
```

```
List1.Clear
For Each td In db.TableDefs
  List1.AddItem td.Name
Next
```

Run the app! The list box should fill with table names that are in the selected database.

Step 9
Drop another list box on the form. It will be named List2. Now go to the click event of List1 and add this code:

```
Private List1_Click()

'FIND THE RIGHT TABLEDEF IN THE COLLECTION IN THE DB
Set td = New TableDef
For Each td In db.TableDefs
  If td.Name = Me.List1.List(Me.List1.ListIndex) Then
    Exit For
  End If
Next

'NOW DISPLAY THE FIELDS IN THAT TABLE's DEFINITION
List2.Clear
'LOAD LIST2 WITH FIELDS
For Each fld In td.Fields
  List2.AddItem fld.Name
Next

End Sub
```

Step 10
If you have everything working, you should be able to open a database, which will fill List1 with table names. Clicking any one table name in the list should fill List2 with field names. This information is useful for constructing queries. In the next section, we will list any queries stored in the selected database. (*Note:* Tables that start with the letter "M" are system or "catalog" tables and are not part of this exercise. Clicking on these tables *will not* display any fields in List2.)

Drop another list box on the form. It will be named List3 by default. Then, add this code to the File/Open menu command just before the End If statement:

```
Dim qd As QueryDef
For Each qd In db.QueryDefs
List3.AddItem qd.Name
Next
```

Step 11

This code iterates through the collection of QUERYDEFS that are part of each .MDB database. Now, run the application select BIBLIO.MDB, and notice what shows up in List3. You should get query definition names that are defined inside BIBLIO.MDB.

Add this code under the click of List3:

```
'FILL TEXT1.TEXT WITH QUERY SQL
Dim qd As QueryDef
For Each qd In db.QueryDefs
  If qd.Name = List3.List(List3.ListIndex) Then
    Text1.Text = qd.SQL
  End If
Next
```

Run the application, open BIBLIO.MDB, and click on List3. What happens?

Step 12

For the last part of this exercise, you will save a query to the selected database. The basic idea is, you test some SQL, and when you get it right, you save it by name to the database as a QueryDef.

Drop a new button on the form, and name it Command2. Set the .Caption to "SAVE QUERY". Add this code under the Click event of this button:

```
Private Sub Command2_Click()

'PLACE THE SQL INTO A QUERY OBJECT
Dim q As QueryDef
Set q = New QueryDef
q.SQL = Trim$(Text1.Text)
MsgBox "Query to Save is " & q.SQL

'SAVE THE QUERY WITH A NAME
q.Name = InputBox("Specify a QUERY NAME: ")
db.QueryDefs.Append q

End Sub
```

Step 13

Run your app. You should be able to select BIBLIO.MDB, specify a query, run it, and then press this button to save the query to a specific name as a Querydef in the database. To be sure it worked, reopen the database. You should see your saved query in List3.

EXTRA EFFORT

1. Update List3 with the new query name every time you add a query to the database.

2. Add the ability to *delete* a named query definition from the database. (*Hint:* Search the object browser and help for tips on how to do this.)

CHAPTER 6

Error Handling and Debugging

What You Need to Know

To effectively learn from this portion of the book, you need a good understanding of the structure of a Visual Basic project. You should be familiar with modules, forms, procedure writing, Sub Main, and all of the scoping rules and variable types. You should also be strong in the language keywords and control structures.

What You Will Learn

Up to this point in the book, you have constructed small applications without the benefit of understanding error handling or debugging. After you digest this part of the book, you will know how to perform detailed debugging at design time. You'll also understand how to set up error handling in your applications so that your users will not face a terminated application due to an unexpected, untrapped error from your program.

Overview

Error handling in VB has not changed much since the days of GWBASIC and BASICA. The facility remains one of the weaker elements of the VB language. Debugging, on the other hand, has steadily improved with each new release of VB. As a result debugging in VB is very powerful and simple to practice.

Basic Error Handling

Error handling in VB is designed to trap and handle runtime errors so that the user will not be faced with the unpleasant prospect of a terminated program. If for any reason your application causes an untrapped error, you users will get a VB message and then be returned to the desktop after the program aborts. Here is an example:

Check It Out: Unhandled Errors.

1. Start a new VB Standard EXE project. Add these lines to the Form_Load of Form1:

```
Dim names(3) As String
Names(99) = "This causes a VB error #9"
```

2. Run the app and observe what happens.

The attempt to address the 99th element in the names() array causes an error #9, "Subscript out of range." This produces an error message that comes from VB. When your users encounter this error, they get the same message. Then, VB terminates the application. This is a situation you seek to avoid. This is done via the definition of an error trapping procedure.

Defining an Error Trap

To handle an error, you must first define an error trapping procedure. It starts with this line:

```
On Error GoTo Form1_ErrorTrap
```

The On Error statement tells VB to jump to the specified label when an error is encountered. The label identifies the error trapping code. Here is how the error trap would look:

```
'THIS LINE STOPS EXECUTION FROM 'FALLING IN'
'FROM ABOVE:
Exit Sub
'THE ONLY WAY INTO THIS TRAP IS IF YOU
'JUMP BECAUSE OF AN ERROR:
Form1_ErrTrap:
MsgBox "Error # " & Err.Number & " Description: " & Err.Description
Resume Next
```

The Exit Sub line (or Exit Function) should be the first line in any error trap. This prevents execution from falling into the error trapping routine. Error traps always start with a label, which is a VB program line that is used as a place your application can jump to with the GoTo statement. As we said before, you will want to avoid the GoTo. However, the error trapping On Error GoTo statement requires the label, so we define one. All labels end with a colon. They must be unique within a procedure. Our label is named Form1_ErrorTrap.

The next thing we do in the error trapping procedure is (a) report the value of some properties of the Err object via a MsgBox statement, and (b) execute the Resume Next statement. The Err object's Number property contains the last error number; Err.Description contains the error text. We report both and then Resume Next. This statement forces a branch in execution to the line immediately following the line where the error occurred. Resume without the Next clause will cause execution to branch back to the line where the error occurred. This usage of Resume without the Next clause assumes you have performed processing (in the error trap), which will allow the line to execute without error. The more typical setup is to Resume Next, effectively skipping over the problem line of code.

Check It Out: Trapping a Runtime Error

1. Start a new Standard EXE VB project and add this code to Form1's Form_Load event procedure:

```
On Error GoTo Form1_ErrTrap
Dim names(3) As String

'THE FOLLOWING LINE CAUSES AN ERROR
'AND JUMPS TO THE DEFINED ERROR TRAP:
names(72) = "Error #9 encountered here!"

'RESUME NEXT IN THE TRAP TAKES EXECUTION TO HERE:
names(3) = "No Error here!"

'THIS LINE STOPS EXECUTION FROM 'FALLING IN'
'FROM ABOVE:
```

```
Exit Sub
'THE ONLY WAY INTO THIS TRAP IS IF YOU
'JUMP BECAUSE OF AN ERROR:
Form1_ErrTrap:
MsgBox "Error # " & Err.Number & " Description: " & Err.Description
Resume Next
```

2. Run the app and watch how the application issues a message and recovers from the error to display Form1.

3. Now stop the application and comment out this line:

```
'COMMENT OUT THIS LINE:
On Error GoTo Form1_ErrTrap
```

4. Run the app now, without the benefit of error handling, to see what VB does when an error is encountered without the benefit of an error trap.

Shutting Error Trapping Off

There will be times you will want to shut off the error trap for certain program lines. One example would be when you are opening a text file: If the file does not exist, let's say your program will create it later using VB code. In such a case as this, you do not care if you get the "File not found" error when you try to open the file. To shut error handling off for a group of program lines, use the On Error Resume Next statement. The code snippet that follows shows how your program might shut error handling off momentarily, and then continue processing and trapping any errors you may encounter.

```
Close #1
'SHUT ERROR HANDLING OFF TEMPORARILY,
On Error Resume Next
Close #1
Open App.Path & "\MYAPP.TXT" For Input As #1
Line Input #1, strCOMPANY  'get file data
Line Input #1, strADDRESS  'get file data
Line Input #1, strZIP  'get file data
Close #1
'RESTORE ERROR HANDLING FOR THIS PROCEDURE:
On Error GoTo MYAPP_ErrorTrap
```

If you want to shut off all error trapping, use On Error GoTo 0. Why is this VB line so ugly looking? Because in the early days of the BASIC language, On Error GoTo could branch only to program line numbers, not labels. So, the convention was if you specified line zero, that meant, "shut error handling off." The syntax survives to this day. When you use this line, you will only get the default VB error processing when your application encounters an error. There will be times during debugging error traps, for example, when you will want this kind of setup.

Errors in User-Written Procedures

Every procedure you write needs an error trap. If the procedure with the error has no caller, VB ends the search and reports the error, dumping you into "break" mode if you are testing, or terminating the program in the case of an executable. If there is a caller to your the procedure with the error, VB looks to that caller to see if it contains an active error handler.

If the calling procedure has an error handler active, VB will use it. If the calling procedure has no error trap, VB will look to the caller of *that* procedure, if any, for an error handler. This process is depicted in Figure 6-1 as a flowchart. VB traverses the list of procedures in the Call list, checking each one for an error handling procedure. If the procedure has

Figure 6-1
Error Handling
Control Flow.

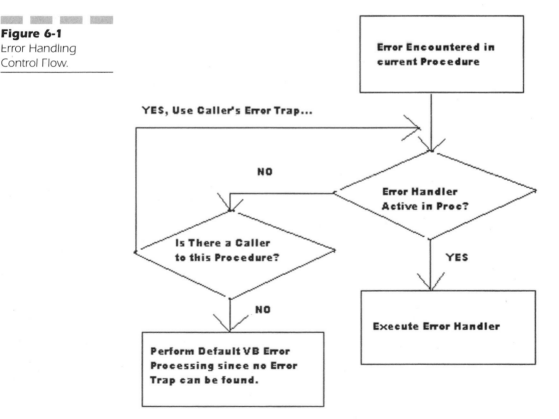

no callers or none of the calling procedures has an error handler, then VB does the default handling with all of the dire consequences associated with an untrapped error.

Errors in Event Procedures

For every event procedure where an unanticipated error may terminate your program, you should code an error trap. This means every event procedure for which you write code should have a corresponding error trap somewhere in the code. (An error trapping code block can occur anywhere within a procedure.) It is customary to code the error trap at or near the bottom of a procedure and to precede it with an Exit (Sub or Function) statement.

Since event procedures are typically executed by VB in response to events, there is no calling procedure. For this reason, unhandled errors in event procedures usually terminate your application.

Typical Setup for Error Handling in a VB Application

Most of the time, you will be coding error handling as a natural part of each event procedure you add code to and for each procedure that you write. Since the trapping of errors in a procedure is the same every time, this is a simple and straightforward task. Each error trap will usually issue a message to the user and try to recover from the error condition. These tasks are perfect for packaging into an error handling subroutine.

It is customary to pass the error number, description, and procedure name (where the error occurred) to an error handling subroutine, which is called from each error trap in your application. This error handling routine usually issues a message to the user in a standard format and might write a log to a text file, detailing the particulars of the error encountered.

Advanced Error Handling; Raising Errors

One aspect of VB error handling that is quite interesting is called raising errors. VB provides a set of preformatted error numbers that have associated text descriptions. You can see some of these by executing the following code:

```
Dim i as Integer
For i = 1 to 100
  Err = I
  Debug.Print Err, Error$
Next i
```

Err and Error$ are the VB3 reserved words that correspond to Err.Number and Err.Description, respectively, in VB5. We are using these now so you can see the VB error codes. These are rarely used and are supported in VB5 only for backward compatibility.

Regardless of Err and Error$, as you can see that some of these numbers are not reserved by VB, and you can use them yourself. For example, VB error 53 is the error "File Not Found." When this error occurs in your Sub Main routine, it may be that the file named MYAPP.TXT file could not be opened. You know this within the context of your app. You can use an unreserved error number and pass it to the Err.Raise method. Err.Raise creates a trappable error that can be handled in your error-trapping routines. The primary use of Err.Raise is to create your own custom error numbers.

This method requires at least the error number. This value is written into the Err.Number property. You can use *any* number but it is wise to stick with error numbers not currently in use by Visual Basic. You can also optionally specify the Source and Description properties. The Source property specifies the name of the application that caused the error. It is also possible to specify a Help file that the user may access as the error is reported.

In the following example, we raise errors from the click of a button. Each button has an error trap that calls the same error handling subroutine. The error handling subroutine contains a Select Case statement that recognizes the two errors raised with the Raise method from each CommandButton:

▧ Check It Out: Defining an Error Handling Routine

1. Start a new Standard EXE VB application. Add a module to the project and add this routine, which will be called from all error traps:

```
Public Sub MyErrHandler(errno As Integer, errdesc As String)
'DONT TRAP ERRORS HERE
On Error GoTo 0
Select Case errno
Case 64
  MsgBox errno & " :" & errdesc & ". This error can only come from
Command1"
Case 65
  MsgBox errno & " :" & errdesc & ".This error can only come from
Command2"
Case Else
  MsgBox "Unrecognized error."
End Select

End Sub
```

2. Note that the first thing you should do in an error handling subroutine is shut all error handling off so that you do not get into a recursive loop by mistake if you get an error while in the error handling routine. This routine gets the error number passed in and then checks against all the codes it recognizes; 64 and 65 are unreserved by VB and used by this application.

3. Now drop two buttons on Form1. Place this code under Command1:

```
On Error GoTo Command1_Err

'MAKE AN ERROR HAPPEN WHEN CLICKED:
Err.Raise Number:=64, Description:="Custom error from Command1"

'ERROR TRAP:
Exit Sub
Command1_Err:
Call MyErrHandler(Err.Number, Err.Decription)
Resume Next
```

4. Now, place this code under Command2:

```
On Error GoTo Command2_Err
'MAKE AN ERROR HAPPEN WHEN CLICKED:
Err.Raise Number:=65, Description:="Custom error from Command1"
'ERROR TRAP:
Exit Sub
Command1_Err:
Call MyErrHandler(Err.Number, Err.Decription)
Resume Next
```

5. Now run the app and click each button. The click of Command1 will produce a trappable error 64, which will be handled by Command1's error trap. The click of Command2 will produce a trappable error 65, which will be handled by Command2's error trap. Both traps call MyErrHandler, which is set up to recognize both custom error codes 64 and 65.

There is more to error trapping, which you will explore in the exercise that follows. The error handling in VB is not great, but it is adequate. It is important to understand that each routine needs its own error trapping code block. This is a tedious but necessary chore.

Basic Debugging

Debugging in VB is very rich. You can set multiple breakpoints, set bookmarks, and walk code in a variety of fashions. You can interrogate the value of variables and execute code immediately by using the Immediate window, also known as the Debug window. A brief demo will acquaint you with the basics of debugging in VB.

Check It Out: Debugging

1. Start a new VB Standard EXE project. Enter this code into the Form_Load event procedure of Form1:

```
Dim i, j As Integer
For i = 1 To 100
  j = j + 1
  Call MySub
Next
```

2. Now you need to define the sub named MySub. Add a subroutine procedure to form1 that looks like this:

```
Public Sub MySub()
Dim names(2) As String
names(0) = "Thing1"
names(1) = "Thing2"
names(2) = "Thing3"
msgbox "MySub Called"
End Sub
```

This sub does nothing except execute; we will use it to demo basic debugging techniques.

3. Now set a breakpoint; the line with the breakpoint, when executed, will drop your app into VB's Break mode. From inside the Form_Load of Form1, click on the line "j = j + 1" and then click on the Debug/Toggle Breakpoint menu command. This will set a breakpoint on that line of code.

4. Run the application. Now you will be in VB's Break mode and you can watch the execution of the application, line by line. To go line by line, click on the Debug/Step Into menu command. You must click this menu command once to advance the execution point one line. F8 is the shortcut key for this menu command, so just press F8 to single step the code.

5. Note that when you press F8, the execution point moves. When you hit the call to MySub, the execution point drops down into the subroutine. What if MySub() is already debugged? What if MySub calls SomeOtherSub()? For these situations, VB provides the Debug/Step Over menu command.

6. Keep pressing F8 until the next line to execute is the call to My-Sub(). When you get to that line of code, select the Debug/Step Over menu command. This will execute MySub(), but will not trace the code. The code display will stay at the caller. This is great when you do not want to drop into debugged procedures when debugging at a higher level. Shift-F8 is the shortcut key combination for the De-bug/Step Over menu command.

7. Do not reinitialize the project as you will be using this project again shortly. Return to design time.

Breakpoints are not saved with your code, so if you have spent much time determining where the breakpoints should be set, it's a good idea to comment the lines with a code such as MYBP or something similar so that you can find the breakpoint lines later to reestablish the break-points. You may want to churn some output while debugging to the Imme-diate window; this is accomplished with the Debug.Print statement. Debug.Print lines do not effect your executable, so you can use as many as you like without removing them prior to compilation.

You can force VB into Break mode with the Stop statement, if you wish. You can, for example, test a variable for a value and then conditionally go to Break mode with the Stop statement. Experimentation is strongly encouraged. Please note that the Stop statement, if encountered by your user, will terminate your application. For this reason you should use the Stop statement with great care, to avoid this situation.

Intermediate Debugging

VB's debugging provides some nice features. You can for example exam-ine the call stack, which is a list of all the subroutines called to get to where you are right now in terms of execution.

▨ Check It Out: Viewing the Call Stack.

1. Using the last example you worked with, get the For…Next loop go-ing until you hit the subroutine call to MySub. Press F8 to step into the procedure MySub().

2. Click on the View/Call Stack menu command. You will see the dialog depicted in Figure 6-2.

3. Try setting a breakpoint by clicking inside the left margin of the code editor. Clicking sets a breakpoint; clicking again clears the breakpoint.

Figure 6-2
The Call stack.

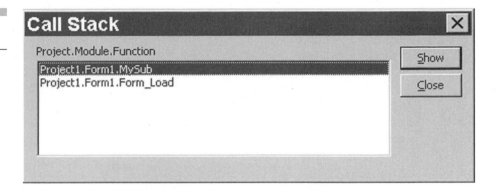

4. Try dragging and dropping the execution point from one place in the code to another using your mouse. Visual Basic supports this. The default color for the line execution indicator is yellow. Try clicking on it while the app is in Break mode and moving it to another location in the loop.

5. Return to design time

USING THE IMMEDIATE WINDOW

The Immediate window can be used to find the values of variables and also to alter variables. Executable lines such as this one are valid:

```
J = J + 99
```

However, lines such as the following are invalid in the Immediate window because the statement is declarative, not executable:

```
Dim foo As Integer
```

Check It Out: Exploring the Immediate Window

1. Return to the project you were checking out before. Set a breakpoint in the code that is found under the Form_Load, and when the app pauses, click on the View/Immediate Window menu command.

2. The Immediate (or Debug) window appears. You can find the values of variables in your program from this window. Try typing the following lines into the Immediate window:

```
Print "I = " & I
Print " J = " & j
? I
```

```
? j
I = I + 90
```

All of these lines are valid in the Immediate window. So is this call to MySub():

```
Call MySub()
```

3. The following line, when you type it into the Immediate window, will cause an error since declarative statements are invalid in the Immediate window:

```
Dim myvar as String
```

Try typing this in. What message from VB do you receive? You should get "Compiler Error. Invalid in Immediate Pane."

4. Click the View/Locals menu command. This window shows you the name and current value of all the variables local to the current procedure.

5. Return to design time.

BOOKMARKS

Bookmarks are placeholders you can set in the code window. Often you will want to skip around all of the bookmarked portions of your code, to pick up some information, view the source, and so on. While not strictly a debugging feature, bookmarks are a very natural way to keep tabs on your code.

Check It Out: Bookmarks

1. Using the current project, display the code from Form1's Form_Load in a code window. Click any line inside the For...Next loop.

2. Click the View/Toolbars menu command and select the Edit sub-menu. This will make a floating toolbar appear. Click the leftmost button to set a bookmark in your code.

3. Dock the Bookmarks toolbar under VB's menu.

Bookmarks provide an easy way to set placeholders in your code, and the Bookmarks toolbar makes it easy to keep track of your bookmarks, with the ability to scroll forward, scroll back, and clear all bookmarks. You may decide to make bookmarks a part of how you work with VB code.

SUMMARY

Debugging and error handling are essential skills as you scale your basic knowledge up to building larger and more robust applications. This chapter has introduced you to the most important features of error handling and debugging. You now need to apply this knowledge with a search of the Help and Books Online as well as some hands-on experimentation.

Increasing Your Knowledge

Help Search Strings and Books Online Search Strings

- On
- "Error Handling"
- Debugging
- "Immediate Window"

Mastery Questions

1. How is can all error handling be shut off when running your VB application?
   a. Use the On Error Off statement.
   b. Use the On Error Resume Next statement.
   c. Use the On Error GoTo 0 statement.
   d. Error handling cannot be disabled.

2. When trapping errors with the On Error GoTo statement, you must:
   a. be sure no other error handlers are active.
   b. specify a properly formatted label to jump to within the On Error GoTo statement.
   c. make sure that your error trap can only be entered via an On Error GoTo jump, and not executed as part of the normal program flow.
   d. Both b and c

3. Bookmarks in the code editor are used primarily for:
   a. locating placeholders that you can jump to or scroll in serial order.
   b. establishing breakpoints.
   c. applying annotations to the code that the bookmark denotes.
   d. None of the above

4. The main reason for trapping errors in your code is to:
   a. write a log of program exceptions to a file using the App.WriteLog method.
   b. examine the state of the program's error status by checking the Err.LastErrNumber property, and take action.
   c. prevent users from experiencing a program termination.
   d. execute your global error handling subroutine.

5. When debugging, you may press the Shift-F8 key sequence to:
   a. dump variables to the Immediate window.
   b. execute but not view code from called procedures.
   c. single step in-line code in the same manner as pressing F8.
   d. Both b and c

6. You can use the Stop statement to:
   a. shut error handling off.
   b. skip over error lines and ignore any trappable errors.
   c. send the program into Break mode when the statement is encountered.
   d. None of the above

7. You can use the On Error Resume Next statement to:
   a. shut error handling off.
   b. skip over error lines and ignore any trappable errors.
   c. send the program into break mode when the statement is encountered.
   d. None of the above

8. The Debug/Step Out menu command performs what action?
   a. It skips out of the current loop or the current procedure and continues.
   b. It concludes the debug session.
   c. It moves the execution point to the cursor location.
   d. None of the above

9. To set a breakpoint, you must:
   a. select the Debug/Toggle Breakpoint menu command.
   b. select the Debug/Step Into menu command.
   c. click on the left margin of the code window at the program line where you want the breakpoint set.
   d. Both a and c

10. The Call Stack informational dialog shows which procedures are active during execution. You can view the Call Stack by:

> **a.** clicking on the View/Call Stack menu command at run time or clicking on the Toolbars/Debug menu command and clicking the Call Stack button from the Debug toolbar.
> **b.** pressing the Ctrl-L shortcut key.
> **c.** clicking on the View/Call Stack menu command at design time or run time.
> **d.** Both a and b

Answers

1. c
2. d
3. a
4. c
5. d
6. c
7. b
8. a
9. d
10. d

EXERCISE: ERROR HANDLING

In this example, we are going to demonstrate the types of runtime error situations that you can get in your project and how to effectively rig your project for error handling.

Hardware and Software Requirements
Visual Basic 5.0 Pro or Enterprise Edition

Required Files
None

Assumed Knowledge
Basic knowledge of error handling concepts and facilities

What You Will Learn

- On Error GoTo, Resume
- Error Handling Options

Summary Overview of Exercise

Part 1: Setting Up an Error Handler

This first part of the exercise gets you going with error handling basics.

Part 2: Error Handlers for Each Routine

The next part of the exercise shows you how to set up an error handler for each routine in your project and how to disable error handlers.

Part 1: Setting Up an Error Handler

Step 1

Start a new project in Visual Basic. Using VB's Tools menu, add a new Public Sub called CalledSub into Form1. In this new procedure, add code so that the procedure appears as here:

```
Public Sub CalledSub( )
Dim MyVar As Integer

MsgBox "Attempting to assign a value to MyVar."
MyVar = "This Integer Var will not accept a string."
MsgBox "MyVar = " & Format$(MyVar)

End Sub
```

Step 2

In Form1's Form_Load event, add the following line of code:

```
'The error will occur inside CalledSub
MsgBox "Starting Sub CalledSub."
Call CalledSub
MsgBox "Sub CalledSub is finished."
```

This is a good place to save your project. Create a directory and save the following files into your new directory:

Save Form1 as FORM1.FRM

Save Project as ERRORH.VBP

Step 3

Now run your project. Visual Basic gives you a message box that gives you the opportunity to go debug the problem. *Do not* try to solve the problem at this time.

Make an executable called DEBUG.EXE by selecting the File/Make Project1.EXE menu command. You will learn about making and distributing applications later. For now think of the EXE as simply a final version of the application that can be given to others to run.

Run the EXE by locating it in the File Manager and double clicking on it. (You will find the file PROJECT1.EXE in your working directory if you saved the project.) Notice that when you try to run your executable, you get a message box and then the application simply goes away. This is exactly what you *don't* want to happen when you distribute your application.

Before we go any further, let's review what happened when you ran the application. First, the Form_Load event fired, which caused Sub Called-Sub to be called. You got a message just before the variable was to be assigned a value. Then, you got the standard VB error message for an unhandled error when you actually tried to assign the variable. Finally, you got VB's default action when it runs into an unhandled error-the application just terminates. Gone. Notice that the lines after your variable assignment didn't even get a chance to execute. If they had, you would have gotten the message, "MyVar = 0" and "Sub CalledSub is finished."

Step 4

Now we are going to add error handling to the Form_Load event. Add code so that the Form_Load event appears as below:

```
Private Sub Form_Load ( )
On Error GoTo MyFirstHandler
  'The error will occur inside CalledSub
  MsgBox "Starting Sub CalledSub"
  Call CalledSub
  MsgBox "Sub CalledSub is finished."

ExitRoutine:
  Exit Sub

MyFirstHandler:
  MsgBox "Form_Load - This error occurred: " & Err.Description
  Resume ExitRoutine

End Sub
```

Step 5

Now save and run the project. Notice that instead of VB giving you a message, you get your own message. This is what happens when an error is "handled." Also, notice that the error is handled in the Form_Load procedure. This is because Sub CalledSub inherits the error handler from the Form_Load event because Sub Form_Load is calling Sub CalledSub.

Step 6

Remake the DEBUG.EXE executable using FILE-MAKE EXE, and run it. This time, the application continues in spite of the error. Now your application is more robust.

Part 2: Error Handlers for Each Routine

Step 1

Currently, everything called by the Form_Load event has its errors handled by the Sub Form_Load's error handler. But what if you want Sub CalledSub to have its own error handler? Simply add an error handler to Sub CalledSub. Sub CalledSub's error handler will override the error handler in Sub Form_Load until CalledSub is finished. When Sub Called-Sub is finished, error handling control reverts back to Form_Load. As long as Sub CalledSub does not call any other procedures *or* any procedures called by Sub CalledSub have their own error handler, you can ensure that Sub CalledSub's error handler will be active *only* when that procedure is actually running.

Change Sub CalledSub so that it appears as here:

```
Private Sub CalledSub( )

  Dim MyVar As Integer

On Error GoTo CalledSubHandler

  MsgBox "Attempting to assign a value to MyVar."
  MyVar = "This Integer Var will not accept a string."
  MsgBox "MyVar = " & Format$(MyVar)

CalledSubExit:
  Exit Sub

CalledSubHandler:
  MsgBox "CalledSub ( ) - This error occurred: " & Err.Description
  Resume CalledSubExit

End Sub
```

Step 7

Save and run the project. Replace Resume CalledSubExit with Resume and run the project. This will cause the execution to go back to the error line and retry. This will get you into an endless loop of execution as the error line pings control back to the error handling routine. When you get in an endless loop situation, press the Ctrl and the Break keys at the same time to get control back. That will get you out of the loop. Remember, the Resume statement returns control to the error line.

Now replace Resume with Resume Next and run again. No more endless loop. This is because Resume Next resumes execution at the next available line after the error line, not the error line itself. You will notice that for the first time in this entire exercise, you are getting the message, "MyVar =0"?

Step 8

When you build error handlers, they should be created while you develop your application. Do not wait until the app is nearly complete. However, when error handlers are placed in the app before it is fully debugged, they can cause unnecessary headaches by trapping errors that you should actually be debugging. Fortunately, Visual Basic allows us to turn off all error handlers in the app.

Click on the Tools/Options menu command and then click on the General tab. In the General tab, select the Break on All Errors option from the Error Trapping section of the tab. Run your project again. All of the error handlers should be disabled, and you should get the VB default error processing.

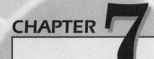

CHAPTER 7

Polishing and Distributing Your Application

After you have developed and debugged your application, it's time to compile, package, and distribute your completed work. This chapter explains several topics: the basics of understanding VB executables, how to package and distribute your VB application, and how to add online Help features to your application. This segment of the book also introduces you to using the Windows Application Programming Interface (API) from Visual Basic. The functions in the Windows API reside within Dynamic Link Libraries, files that typically are already installed on the user's machine. The Windows API can add significant functionality to your VB applications.

What You Need to Know

To get the most from this chapter you need to be familiar with the entire development cycle: designing, developing, testing, and debugging your application. Every part of the book that precedes this chapter should be understood before you begin here.

What You Will Learn

This segment introduces two new features you can add to your apps: the inclusion of online, context-sensitive Help topics, and the use of the Windows API. This chapter also details the facilities provided by Windows for compiling and distributing your application to end users.

Including Context-Sensitive Help

Help Files

Visual Basic does not supply a utility for creating Help files. In most cases, you will not be creating the Help file but instead incorporating it into your VB app. Help files contain Help topics, and topics are referenced by Help "context identifiers." The Help context ID is an integer that can be used programmatically to retrieve and display specific help topics in the help file. The author of the Help file defines both the topical content for the Help file and the associated context IDs for each individual help topic in the Help file.

Including the Help File

You can specify the Help file that your app will be using by clicking on the Project/Properties menu command and navigating to the General tab. There is a text box provided for you to specify the full path to the Help file. This is not a recommended approach (even though the facility is provided) because the result is a hard-coded pathname to the Help file. This could be a problem if your user installs to a nonstandard directory when installing your application. A better approach is to assign a string specifying the location of the Help file to the Helpfile property of the App object. The code that follows uses App.Path as the pathname when specifying the Help file that will be used in the application. This code is typically executed in the Load event of the first form, or from within Sub Main:

```
App.Helpfile = App.Path & "\MYAPP.HLP"
```

By specifying the Help file location at run time with an assist from the Path property of the App object, your application will always be able to find the Help file. The only caveat is that the Help file must reside in the same directory as the executable.

Applying Context Sensitivity

The Help file specified in App.Helpfile is the default Help file for the entire application until you change it. Once App.Helpfile points to a valid Help file name, your app can begin using it. If the user presses the F1 key from anywhere in the application, he or she will receive the first page of the Help file, which has a Help context ID of zero.

In addition, many controls also have a HelpContextId property. This property can contain the index of the Help topic in the Help file that is most pertinent to the control. For example, if App.Helpfile points to a valid Help file name and Command1.HelpContextId points to a valid Help context ID within that file, the topical text and graphics associated with the HelpContextId will appear when the user places the focus on Command1 and then presses the F1 key to obtain Help on that control. As the developer, you need a mapping of topic names and Help context IDs to apply the context sensitivity to your application. This is typically provided by the author of the Help file.

"Whats This" Help

Under Windows 95, a new kind of Help access is available: WhatsThis Help. This kind of Help floats above the item clicked, rather than appearing in the standard Help display window format. You can search Help for WhatsThis for more information on how to implement this variant of the standard context-sensitive Help. You can also search WhatThisMode method of the Form object to learn more about how WhatThis-type Help works.

Distribution Issues with the Help File

When you implement a Help file into your application, you must remember to alert the setup wizard to this fact when you build your installation. The setup wizard provides a step in the process for you to add files, and you must add your Help file to the list at that point. Also, be sure to install the Help file in your application's installation directory. By doing this, your application can use the App.Path property to find the Help file without any problems.

Incorporating the Windows API

Every user of your VB application is by definition a Windows user. As such, every one of your users has on his or her desktop the Windows API function libraries. These libraries are Dynamic Link Libaries (DLLs) containing over 800 individual function calls that are available to your VB application. Since VB is Windows programming language, almost all of the functions in the API are available to your program. Many of these functions have the potential to greatly enhance your application and reduce the number of files that must be installed with your EXE.

Theory of Operation

The Windows API is a vast store of functionality. Within the API you will find support for initialization files, Windows messaging, drawing, and memory management. The core functions of the API are contained in three libraries: KERNEL32.EXE, USER32.EXE, and GDI32.EXE. These files contain substantially all of the functions in the Windows API. Unfortunately, the documentation on how to use these functions does not ship with VB. This means you must learn the theory of operation on the API functions from some other source. You can find some help in the form of books and periodicals that focus on VB. You can also get some documentation from the Microsoft Developer Network CDROM, which includes all of the tools and documentation Microsoft publishes to create Windows applications. In this chapter, we will use the reading and writing of initialization files (INI files) as an object example of how to exploit the Windows API. We will also explore Windows messaging, a core technology in the Windows architecture that makes event-driven programming possible.

The Windows API is a huge storehouse of amazing functionality. Using it requires a knowledge of the overall Windows architecture, upon which the API is based. Within each domain of functionality in the Windows architecture, there are many functions that address the domain. Within the domain of reading and writing initialization files, for example, the Windows API provides a whole set of functions, not just one. You can see the scope of the Windows API by loading the API Text Viewer and scrolling the contents.

Check It Out

1. From your Windows or Windows NT desktop menu, click on the Programs/Microsoft Visual Basic 5.0/API Text Viewer menu command. This will load the API Text Viewer, which ships with VB. The API Text Viewer application will start up.

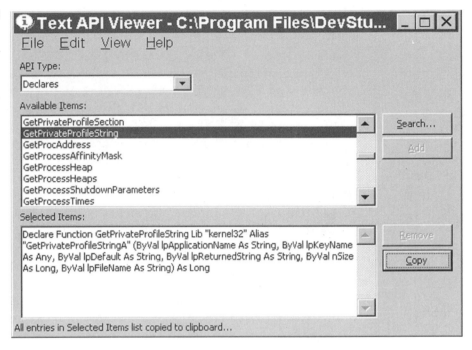

Figure 7-1
The API Text Viewer.

2. Click on the File/Load Text... menu command. In the dialog that appears, click on the file named WIN32API.TXT. If you encounter a dialog asking you if you want to convert text to a database, click on No and continue.

3. The upper list box in the API Text Viewer dialog will become populated with the names of functions you can call from Visual Basic. Scroll the list to get an idea of how big and huge the Windows API really is.

4. End the text viewer and return to VB.

Figure 7-1 shows the API Text Viewer. The API Text Viewer has options to show you the declare statements, messages, and constants that apply to a given API call. At this point, you need to understand the theory of operation of the Windows API as it applies to all Windows programs that may use it.

KERNEL, USER, and GDI

As stated previously, the Windows API resides in a set of Dynamic Link Libraries. These libraries usually have the .DLL extension as part of the filename. KERNEL32, USER32, and GDI32 all end with .EXE. There is a reason for this. In the early days of Windows programming, the .EXE extension was used for DLLs. Later, Microsoft adopted the .DLL extension for all Dynamic Link Libraries. But independent software vendors had created applications with calls to the KERNEL, USER, and GDI, all of which used the .EXE extension. So to this day, the core DLLs that hold the Windows API end with .EXE, not .DLL. As you gain more experience, almost every DLL you connect to with your VB programs will end with .DLL (except for KERNEL32, USER32, and GDI32, of course). Most DLL files are written in the C or C++ languages, but they can be written in Pascal, Delphi, or any programming system that supports DLL output.

You may think of Windows as the Windows 95 desktop, but that is not Windows at all. That is just a Windows application. For example, you could terminate the desktop app and other apps would still remain running. The desktop is just an instance of an application that calls into the functions found in the Windows API.

When Windows starts, it loads KERNEL32, USER32, and GDI32 into memory. These are Dynamic Link Libraries, or DLLs. A DLL can be shared by many applications. Thus, many applications make calls to the functions in these libraries that are loaded into memory. One copy of the DLL serves many applications; each application makes calls as needed to the functions in the library. This is accomplished through a process called dynamic linking. Unlike static linking, in which the functions you call are linked with your program to form an executable program, dynamic linking takes a different approach. With dynamic linking, each application that wants to call a function in a library can link to that library dynamically at run time. Each application contains a statement that is declarative in nature. This statement specifies the library name, function name, parameters passed in, and return value. Armed with this information, each application can link dynamically to the function library and call the functions that reside within the DLL. The first place an application looks for a DLL it wants to link to is memory. Thus, when your VB app attempts to link to USER32.EXE, it looks to memory, finds the library, links to it, and obtains services in the form of callable functions. (If the specified DLL is not in memory, your VB app will use the Windows search path to find the DLL and load it, looking first in the /SYSTEM directory of the Windows installation.)

Support for Windows API Calls in Visual Basic: The Declare Function Statement

Since VB is a complete language for Windows programming, it supports calls to functions in DLLs. This is accomplished via the Declare Function statement. The Declare Function statement specifies all the pertinent details needed to link to the DLL and call a specific function. The Declare Function statement typically is coded in the [general][declarations] of a standard module. Here is an example:

```
Declare Function GetPrivateProfileString Lib "kernel32" _
Alias "GetPrivateProfileStringA" _
ByVal lpApplicationName As String, _
ByVal lpKeyName As Any, _
ByVal lpDefault As String, _
ByVal lpReturnedString As String, _
ByVal nSize As Long, _
ByVal lpFileName As String) As Long
```

This statement specifies everything needed to successfully call the function named GetPrivateProfileString(), which is defined in the DLL named KERNEL32. This function retrieves string data from an initialization (INI) file, typically read at startup and written upon closedown of an application. These files hold user preferences and system settings in a text file that has a set format. The file name always ends in with .INI. Inside the file, sections are defined in square brackets, with keyword=value text listed under the section. For example, here is part of the INI file for Microsoft FrontPage:

```
[FrontPage 3.0]
FrontPageRoot=C:\Program Files\Microsoft FrontPage
FrontPageLangID=0x0409
UILangAbbrev=enu
```

FrontPage uses these values for various purposes. Guess how FrontPage retrieves the values? It probably uses the same API function you are studying. INI files have a predictable but flexible format that allows the developer to add new sections and keywords as the needs of the application change. The Windows API function GetPrivateProfileString() can be used to fetch values from INI files.

Here is the GetPrivateProfileString() declare statement explained:

```
Declare Function GetPrivateProfileStringA Lib "kernel32" _
Alias "GetPrivateProfileStringA" _
```

Declare Function specifies the function name as it will be used by the application and also specifies the module name where the function is defined. The Alias clause is optional and used only when the name used to refer to the function differs from the actual name. The actual name always follows the Alias clause, which may seem counterintuitive. The

Alias clause is useful in many situations when you cannot use the actual name in your program because of conflicts with reserved words, other procedure names, and so on.

Listed next are the parameter names and associated data types for each item passed to the function, as well as the return value:

```
ByVal lpApplicationName As String, _
ByVal lpKeyName As Any, _
ByVal lpDefault As String, _
ByVal lpReturnedString As String, _
ByVal nSize As Long, _
ByVal lpFileName As String) As Long
```

The prefix "lp" means "long pointer"; this is Hungarian prefixing, used by C programmers. The Windows API is for the most part written by and for C and C++ developers. As a result most of the documentation follows C coding conventions, right down to the prefixing.

All the parameters are passed by value, which is the C default for parameters passed in. Here are the parameters passed:

lpApplicationName specifies the section of the INI file where you want to retrieve a value. This would be [FrontPage 3.0] in the INI sample for the FrontPage INI file sample just shown.

lpKeyName is the keyword to get the value of, in the INI file within the section specified by lpApplicationName.

lpDefault specifies the value to return if there is no listing for the keyword specified in lpKeyName.

lpReturnedString is where the value will be placed when it is retrieved. This is a buffer. The default value specified by lpDefault will be written to this buffer if there was no value defined for lpKeyName. If the value is defined in the INI, it will be retrieved and copied to this buffer.

lpSize is the length of lpReturnedString in bytes. lpReturnedString is assumed to be a fixed length string, and watch out if len(lpReturnedString) and lpSize don't match! The function may try to write past the end of the string, resulting in a nasty error condition. This is always a danger when calling API functions.

lpFileName is the name of the INI file, including pathname if any.

Understanding and Using Window Handles

The INI file APIs such as GetPrivateProfileString() are a simple example of functions that perform a useful task, in this case INI file input/output.

Unlike the INI API functions, much of the Windows API relies upon a programming device known as a window handle. In Windows programming, all windows have a window handle, and the handle is used to refer to the window in question. Forms, for example, possess a read-only at run time property called hWnd, or handle to Window. The value in the hWnd can be used as a passed parameter to many API calls that require a window handle. You might not know this, but most controls on a form are considered windows, and they, too, have a hWnd property. The Listbox control, for example, will list the hWnd property in the Help file where the properties of the Listbox are detailed. This handle can be passed to many APIs. One detail to note is that when you use the hWnd, you should sample it directly from the control or form at the time you use it, rather than cache it in a variable. This is because the hWnd may be moved around in memory by Windows, making its validity less than durable. You should always get the current value of the hWnd property just before actual use for this reason.

Obtaining Window Handles from Outside Your Application

Although you do not know how to apply the Window handle to programming problems, you do know that forms and most controls have a window handle you can gain access to. This window handle can then be passed along as a parameter to many Windows API calls. But what if you want to perform an API operation on a window that is part of another application such as Excel or Windows NotePad? In this case, you'll need to call an API function to obtain the window handle. A function that performs this operation is called FindWindow().

As an example, let's say you have NotePad running and want to change the window caption of NotePad to the name of your application. It's actually quite simple to do; here are the steps:

1. Obtain the handle of NotePad's main window by calling the API function FindWindow(). This function returns the handle to the window you specify.

2. Use the NotePad window handle to call the SetWindowText() API function. This function places new text in the window caption of the specified window.

That's it. To get a flavor of what it is to work with the Windows API, follow the steps in the next section.

■   **Check It Out: Changing the Text of
an Application's Window Caption**

1. Start a new VB Standard EXE project. Minimize VB and navigate to
the desktop menu. Start the API Text Viewer.

2. Use the Text Viewer to find the Declare statements for the functions
FindWindow and SetWindowText. Copy these Declare statements
into the clipboard using the Copy button of the API Text Viewer.
This code will be pasted into your VB program from the Clipboard in
the next step.

3. Return to your VB application and add a standard module. Navigate
to the general declarations and click on Edit/Paste to add the Declare statements from the Clipboard to your application.

4. Now drop a CommandButton on Form1 and add this code under the
Command1_Click event. This will call the FindWindow() API func-
tion to retrieve the window handle. The code displays it with a Msg-
Box:

```
Dim hWndNOTEPAD As Long
hWndNOTEPAD = FindWindow(vbNullString, "Untitled - Notepad")
MsgBox hWndNOTEPAD
```

5. Run the application and click on Command1. You should see a
nonzero number displayed by the MsgBox. You never use this number directly; it is passed along to API functions that require the value. This step is just to verify that you are actually obtaining the
window handle. Also, note the use of the vbNullString constant as
the first parameter passed to FindWindow(). This parameter specifies the window class, something we do not cover here. Specifying
vbNullString tells FindWindow() that we are not specifying the window class name. The second parameter specifies the window title we
are looking for. When you run this test code, you should make sure
that you have NotePad started and see the window title "Untitled-
Notepad" as the window caption for this application's main window.
If you do not, be sure to use the precise text displayed by NotePad
when you run this sample code. If the text does not exactly match,
FindWindow() will return zero. Make sure you get a nonzero value
for the window handle before proceeding to the next step.

6. Now you will use the window handle to manipulate NotePad. Specifically, you will now change the window title of the NotePad window
to "VB BOOTCAMPVersion 5.0". Here is now you do it:

7. SetWindowText() is the function used to alter the caption of a window. This is analogous to saying Form1.Caption = "My title" in VB code, but you are using an API call to deal with a window that is not in your application. The first parameter to SetWindowText() is the handle of the window to edit. The second parameter is the new window text. Add this code to the code already found under Form1's Command1_Click:

```
Dim r As Integer
r = SetWindowText(hWndNOTEPAD, "VB BOOTCAMP")
```

8. Bring NotePad into view as you run the application. When Form1 appears, click Command1 and watch what happens. The window title of the NotePad window should change to VB BOOTCAMP.

9. Please note that you could "nest" the function calls to make the code more compact. This completed code would work under Command1:

```
Dim r As Integer
r = _
SetWindowText(FindWindow(vbNullString, "Untitled - Notepad"), "VB
BOOTCAMP")
```

As you can see, using the API can be very simple if you know what you want to accomplish, understand the mechanics, know the API calls to execute, and also know the details of the parameters passed to each function. What a short list of requirements! Teaching every aspect of the Windows API is beyond the scope of this book. However, you did just learn the details of the FindWindow() and SetWindowText API calls while learning more about window handles. API programming is an advanced topic. You need to understand the concepts and facilities of calling API functions, and especially the Declare Function statement, if you are to create or modify production VB applications.

Windows Messaging with the SendMessage() Function

The most powerful API call in the entire Windows API is probably the SendMessage() function. This function can make a window alter its appearance or perform some processing. A basic understanding of the messaging architecture of Windows is required to use the SendMessage() function. This section provides a primer on Windows messaging.

In the Windows environment, every window has a window handle. The window handle is like the location of a mailbox. If you know the window handle, you can communicate messages into the mailbox of the window. The "mailbox" of the window is known as the message queue. The "message" is nothing more than a code that is understood by the window. So messages get placed in the window's message queue. The message is sent by using the SendMessage() function. (There is also a function called PostMessage that is not discussed here.) The message code is specified as the second parameter in the SendMessage() API. The first parameter passed with SendMessage() is the handle of the window you are targeting with the message.

Every window has what is a called a message loop. This is a loop that is always checking message queue for messages. When a window receives a message, it either recognizes the message or it does not. The message loop takes care of determining if the message is recognized. If the message is recognized by the window, then the window will react to the message by performing some processing. List boxes, for example, get messages when the mouse skates over the list box, or the user clicks the list box. There are messages you can send to a ListBox window that will load it with items. You do not need to send this particular message using SendMessage(), since the Listbox provides the AddItem method, which performs the adding of text to the Listbox. In truth, when you execute the Additem method, VB sends the control a message (VB itself makes calls to SendMessage) indicating that a new item must be added. Additional information such as the text to add is also sent along with the third and fourth parameters to the SendMessage() function.

The SendMessage() function has these parts:

```
r = SendMessage(hWnd, Message, wParam, lParam)
```

Where

r is the return value indicating if the message was processed by the window. (Long)

hWnd is the window handle of the window to target with the message. (Long)

Message is a Long value indicating a message known to be recognized by the window.

wParam, the "word parameter," is an integer value. It varies according to the message sent.

lParam is the "long parameter," an integer value that varies in accordance with the message sent.

The message is typically expressed as a VB constant when calling the SendMessage() function. This aids in readability and makes maintenance easier. Message constants typically have meaningful names. For example, consider the following code:

```
r = SendMessage(Form1.List1.hWnd, LB_SELECTSTRING, -1,
Form1.Text1.Text)
```

This code uses the LB_SELECTSTRING constant to pass the "select string" message to a Listbox window. (The constant declaration is obtainable from the API Text Viewer.) The text to search for is indicated by the fourth parameter, which is the "long parameter." In this example, it is the text contained within a text control on the form.

When the Listbox receives the LB_SELECTSTRING message, it expects a certain format. The wParam (in the third position) indicates where to start the search; a value of _1 means "search the entire listbox." This is the value in the example for the wParam. The Listbox also expects the lParam to point to the search string. The Listbox searches itself for at least a partial match of the search string indicated by the lParam. If it finds the string, it selects that string as if the user clicked on the item. This is all documented as part of the default behavior of the Listbox window. (You can find this documentation in the Microsoft Developer CD.) This functionality has many applications, such as the ability to perform incremental search on a list box in much the same way as the VB Help system.

You will note that there is no SelectString method of the VB Listbox control. The author of the Listbox control chose not to implement this native ListBox functionality as a method exposed to the VB developer. However, it is available to you via the Windows API SendMessage() function. API programming is the next level of programming Windows with VB, after you have mastered the basic VB programming system. Visual Basic makes it easy to write Windows applications, but there is much, much more functionality available via the Windows API, should you require the additional power.

Knowing the basics of the window messaging architecture will allow you to extend your VB apps with powerful new features unavailable from "plain vanilla" VB. The bad news is, you have plenty to study to use the API effectively. The good news is, the API function libraries are already installed on every desktop, and you have over 800 functions to choose from. Every good VB programmer knows how to use the Windows API.

Distribution Issues with Dynamic Link Libraries

The files KERNEL32.EXE, USER32.EXE, and GDI32.EXE are not the only files that can contain callable functions. There are many additional DLL files that contain additional API functions that are not core to the operation of Windows. ODBC32.DLL, for example, contains the functions that make up the ODBC API.

When you are not absolutely sure that the DLL your application requires is already installed on the user's computer, you must ship the DLL and install it properly. This means you must include the DLL on the distribution disk and instruct the setup wizard to properly install it on the host computer. In this manner your application will have the support it needs to run. You should always install DLLs in the /SYSTEM directory of the Windows installation. When your app makes a call to a function the DLL, Windows will search memory, then the app's installation directory, then the /SYSTEM directory, and then the Windows search path, looking for the DLL. Since DLLs are shared by all applications, you should only have one copy of the DLL on the user's machine, and it belongs in the /SYSTEM directory.

Creating an Executable

Creation of an executable in VB is actually something of a misnomer. What comes out of VB requires the VB runtime file MSVBVM50.DLL. When you create an EXE, all of the forms, modules, and class modules in your project are used to create a single file with an EXE extension. The default name of your EXE is the name of your project. You can override the default and provide your own name if you wish. To build an executable, click on the File/Make [yourprojectname] EXE menu command. Visual Basic will insert the name of your project in the menu command. For example if your project name is Project1 you will see Make Project1 EXE... as a menu command from the File menu. Clicking on this option will show you the name of the EXE and where it will be written by VB when completed. To alter the properties of the EXE that will be created,

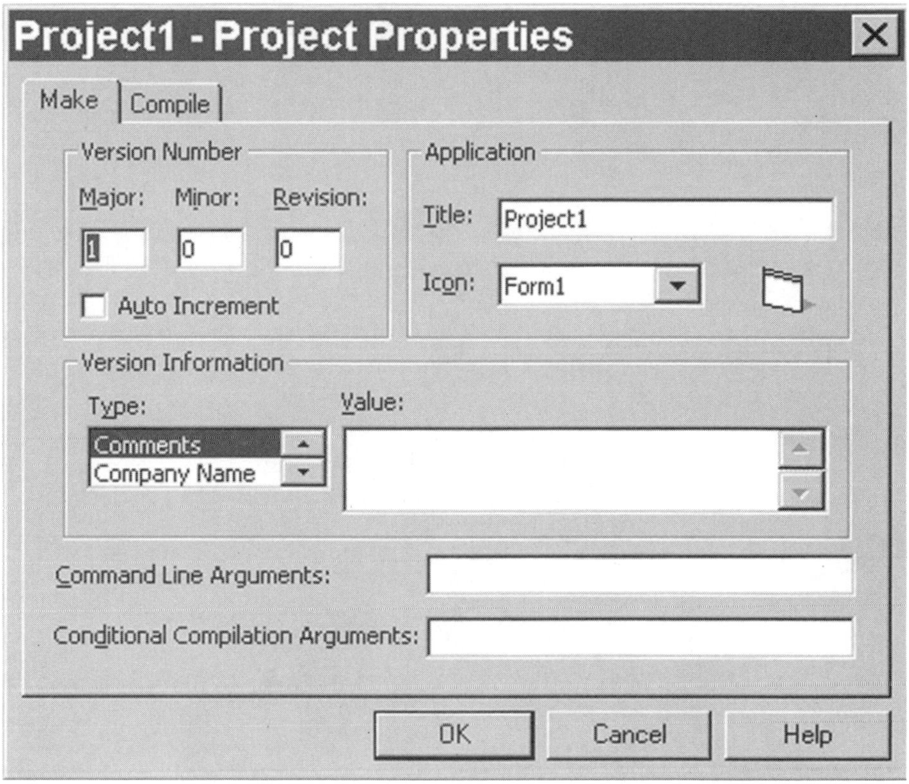

Figure 7-2
Setting the Make
options of the Pro-
ject1.EXE.

click on the Options... button that appears on the dialog. Figure 7-2 shows the Make options that are accessible from the Options... button of the Make EXE Dialog.

You will note that the dialog that appears as a result of clicking the Options... button displays two tabs, Make and Compile, under the window title [projectname] Properties. These are the same tabs that you see when you click on the Project/ [projectname] Properties menu command. The Make and Compile tabs allow you to alter the various properties of the executable before it is generated. You can alter these properties from either location within VB's menu structure.

The Make tab is where you set version information for the EXE as well as specify the project name and desktop icon for the application. The Compile tab is where you specify whether you want "p-code" or "native code" compilation of your forms, modules, and class modules. If you select native code, you can drill deeper into the options for the executable by clicking on the Advanced Optimizations... button to further fine-tune the EXE output. The Help... button on this dialog provides a quick way to

Figure 7-3

Setting the Compile options of the Project1.EXE.

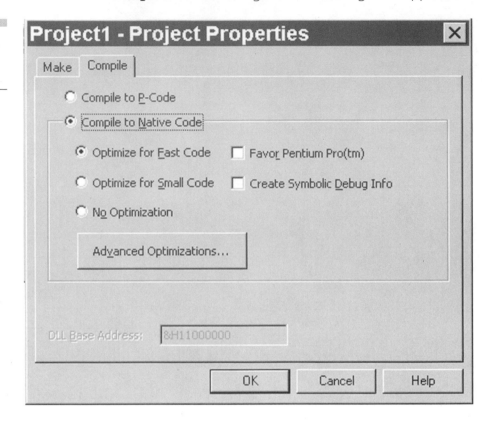

gain an understanding of all the options available when setting the properties for a VB EXE.

Anatomy of a VB Executable

You have two options when you compile: You can compile to p-code or you can compile to native code. P-code is actually a conversion from VB language to a set of tokens. Each token represents a particular keyword or coding construct of the language. This is pseudo-code, hence the slang term p-code.

Native code is actual machine language statements that are understood by the target CPU, meaning the Intel architecture in the case of Visual Basic. The default setting is to compile to native code, which is faster than p-code but is not platform independent.

Regardless of the compile option you choose, your VB EXE will require the distribution of MSVMVB50.DLL, the file containing the VB runtime module. Even the natively compiled EXE will require this file, since it will be making calls into the functions found in this DLL. This requirement means that your minimum VB application (as distributed to end users) must include both your EXE and the VB runtime module. VB does not create standalone EXEs.

Advanced Settable Features of the VB EXE

The Make Tab

Prior to building your EXE, you have several options. The first set of options pertains to versioning information about your executable. From the Project/Properties menu command you can access the Make tab of the Project Properties dialog as depicted in Figure 7-2. Here you can set the major version, minor version, and revision version numbers, as well as define the copyright information that will be compiled into your executable. Many of these values will become the default values for properties of your App object. For example, LegalCopyright is a property of the App object. The value of this property is set using the Make tab of the Project Properties dialog prior to compilation.

The Compile Tab

The Compile tab of the Project Properties dialog allows you to specify p-code or native code compilation. If you specify compilation to native code, you have further options. You can optimize the compile for speed or size, for example. You can also favor the Pentium Pro processor, but doing so will hurt performance on previous Intel processors. Likewise, favoring speed over size will result in a larger generated EXE, while favoring size will most likely result in a slower but smaller application executable.

Distributing Your Application and Creating Installations

When you are finished with generating your executable, it is time to consider the installation process. Simple applications simply require your executable plus the VB runtime module. Even so, creating a convenient and simple setup and installation disk is something users have come to expect. For this reason, you need to be familiar with the VB SetUp Wizard.

For most applications, you will have a large number of additional files beyond your EXE and the VB runtime module. If you are calling functions in a DLL that probably does not reside on the target user's machine, you will need to distribute and install it. There are INI files, data files, bitmaps, and any runtime DLLs such as the DLLs associated with Data Access Objects and the Data Control. The setup wizard helps you to manage the gathering, setup, and distribution of the many files that make up your VB application.

The Many Files of a VB Application

Here is a rundown of the many files you can expect to distribute with a typical VB application.

The VB Runtime Module

The VB runtime module needs to shipped with each VB application you create.

Your Executable

Your executable consists of all of your forms, modules, and class modules in your project.

OCX controls

OCX controls are added to your toolbox by clicking on the Project/Components menu command and selecting the Controls tab of the resulting Components dialog. These controls are not built in with the VB runtime module. They are also not compiled into your EXE. You simply check each item in the dialog off to include the tool in your toolbox. Each item in the dialog represents a file with an .OCX extension that must be shipped with your app if it is to run properly on the host computer.

Dynamic Link Libraries

If you are making calls to functions within KERNEL32, USER32, or GDI32, there is no need to ship any DLLs, since the target computer will already have them resident on the machine. If, however, you make calls to other DLLs that cannot be guaranteed to be resident on the user's machine, you must be sure to distribute them. Shipping DLLs can be tricky, because in many instances the fact that you need to ship a DLL can be hidden from you. One notable example is the use of the Crystal Reports custom control, which totally relies upon the Crystal Reports runtime engine, which is implemented as a set of DLLs. (The files you must include on your installation disk are enumerated in the VB documentation.) Please note that the OCX controls you are using in your project are listed as entries in the .VBP file, but the DLL upon which the OCX may depend is not. You need to carefully examine the documentation for each control your application utilizes, to be sure you ship all the DLLs that may be needed.

Reusable Components

Reusable components include ActiveX DLLs and ActiveX EXEs. These components can be thought of as containers that hold objects, with each object possessing properties and methods you can exploit in your programs. When ActiveX components are used by your application, you need to be aware that they will need to become part of your distribution of files to the end user.

Icons and Bitmaps

Most applications have some images and graphical content. In VB applications, you will typically be shipping some of icon (.ICO) and/or bitmap (.BMP) files. These must also be included on your installation disks or CDROM.

Resource Files

Resource files contain static data such as text that will be used in menu commands. You can factor static text and other data to resource files, which reduces the size of your EXE and makes it easy to support additional languages besides English. Resource files can be used to hold language-specific text data, and your VB app can retrieve the language-specific strings during run time. This is an advantage, because if you set up your application to utilize resource files, your app will not require a recompile when targeting a language other than English. You can search Help on ".RES" for more information on resource files. If your app is using any resource files, you'll need to ship and install them for your app to run.

Data Files and Initialization Files

Most applications use a database in some way, so you will be shipping one or more database files as part of your installation. You may also be providing INI files that support your application. These will also need to be distributed. Please note that even in the 32-bit world, INI files still play a role in how applications work. VB itself, for example, uses VBADDINS.INI to contain information about installable addins.

Setup Wizard Concepts and Facilities

The setup wizard is an application that ships with Visual Basic. The purpose of this wizard is to make it easy for you to create installations for your application that are simple for users to understand and use. The setup wizard is accessible from the desktop menu in the Visual Basic program group as depicted by Figure 7-4.

Overview

After a brief explanation as depicted by Figure 7-5, the wizard walks you through a series of steps. The first step is to select your project. The dialog for this step is depicted in Figure 7-6.

When selecting the project, you can specify the project you wish to distribute by clicking on the Browse... button and navigating to the directory where your project resides.

When the setup wizard works, it incorporates VB into the scheme of things. Also note the checkbox labeled Rebuild Project in the dialog shown in Figure 7-6. If you check off this option, you are instructing the setup wizard to create an executable automatically as the first step of the setup process. To do this, the setup wizard executes Visual Basic and uses VB to build the EXE. All of the Project Properties options will apply at the time this executable is created by the setup wizard. If you do not check off this option, you are indicating to the setup wizard that the EXE can be found in the directory you have specified.

There are several options for wizard output, including Create Internet Download Setup and Generate Dependency File Only. Since this tutorial

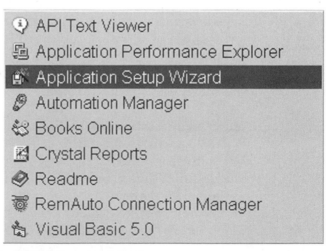

Figure 7-4   Starting the setup wizard.

Figure 7-5
Setup wizard intro-
duction.

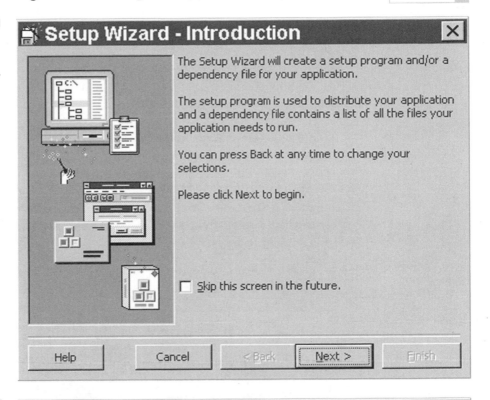

Figure 7-6
Setup wizard select
project step.

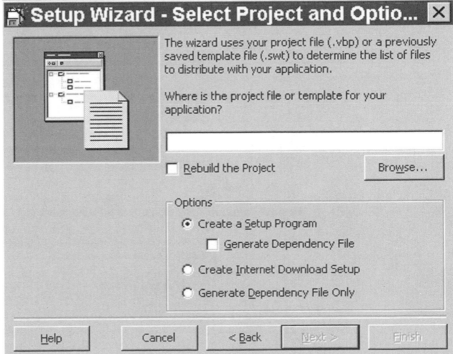

focuses only on the task of creating a setup program, the default Create a Setup Program option will be selected for this discussion. To learn about all of the many options of the setup wizard, you can access the Help file by clicking the Help... button, which is available at every step along the way when you use the setup wizard. The Help file is shown in Figure 7-7.

Using the Setup Wizard

For this example, an application that resides in the VB /SAMPLES directory will be used. The application that resides in /SAMPLES/MISC/CALLDLLS is the one we will use in this example of how the setup wizard works. CALLDLLS is pretty interesting if you are enthusiastic about the possibilities of using the Windows API: This appli-

Figure 7-7

Using the setup wizard Help file.

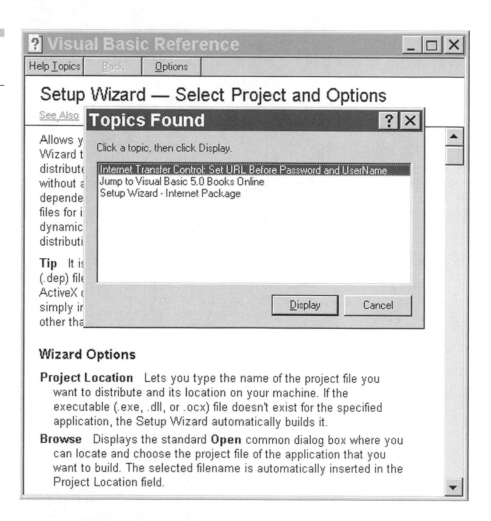

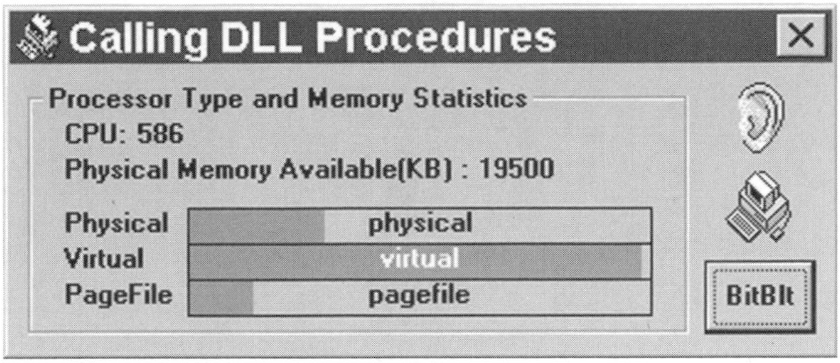

Figure 7-8 *Please reinsert caption text*

cation makes many calls for WinAPI functions, reporting system information such as CPU type, available memory, and other system metrics. It even makes calls to some of the functions that make up the Graphics Device Interface (GDI) subsystem. The user interface of the CALLDLL program is depicted in Figure 7-8. This application works well for demonstrating basic setup and installation concepts, since it contains no dependencies on reusable components.

Specifying How to Write the Setup Files
Figure 7-9 shows the option regarding how the generated setup will be stored. Your options are floppy disks, a single directory, or disk directories. Disk directories is a good option, since the wizard will create /DISK1, /DISK2, /DISKn directories such that each directory will never contain more than a floppy disk can handle. This allows you to copy to disk later if you wish.

Specifying Where to Write the Setup Files
Figure 7-10 shows how the setup wizard will ask you to specify where you want the setup files to be written. The default is the C:\TEMP\SWSETUP directory, but you can override this destination if desired.

Including Reusable Components
Figure 7-11 shows the next step: identifying the components upon which your application depends. In this example, no additional custom controls or reusable components are required. The setup wizard examines the .VBP file of the project to determine what additional custom controls or other components must be distributed. The .VBP file has all the information needed by the setup wizard to locate and copy the files to the destination directory for the setup wizard output.

Figure 7-9
Specifying how to
write the setup files.

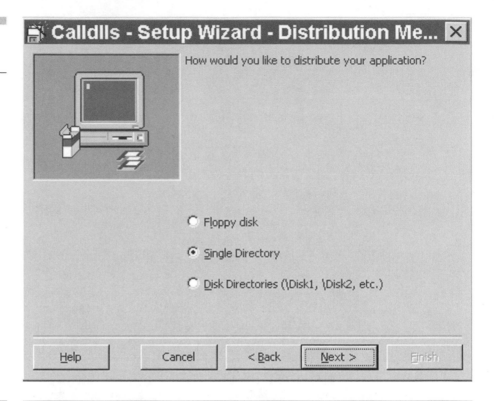

How would you like to distribute your application?

○ Floppy disk

● Single Directory

○ Disk Directories (\Disk1, \Disk2, etc.)

| Help | Cancel | < Back | Next > | Finish |

Figure 7-10
Specifying the desti-
nation directory for
the setup output.

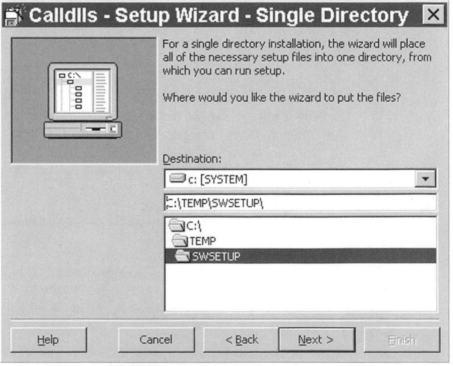

For a single directory installation, the wizard will place
all of the necessary setup files into one directory, from
which you can run setup.

Where would you like the wizard to put the files?

Destination:

c: [SYSTEM]

C:\TEMP\SWSETUP\

C:\
TEMP
SWSETUP

| Help | Cancel | < Back | Next > | Finish |

Figure 7-11
Adding ActiveX
Components.

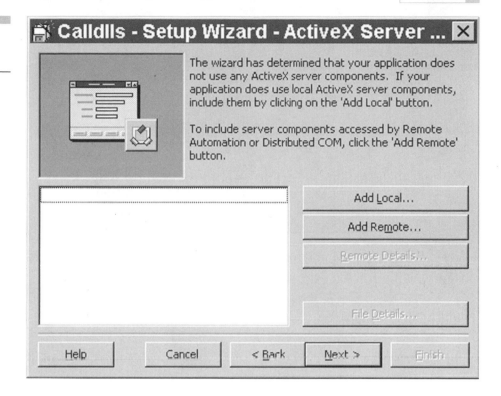

Viewing the Distributed Files and Finishing the Setup Process

Figure 7-12 illustrates the last step: reviewing the files that will become the content of the distribution package. This dialog provides you with the Add... button. Clicking on this button allows you to add more files to the installation process. You may have .ICO, .BMP, and perhaps .MDB files that need to be distributed. In this step the setup wizard provides you with the opportunity to include additional files in the installation.

Figure 7-13 shows that interactive step in the processing of the wizard: finishing the job. Like any other wizard, the setup wizard provides your with the ability to move forward and back through the wizard steps in a serial fashion.

Compressing Files for Distribution

The last thing the setup wizard does is compress the files you have selected for distribution and write them to your indicated destination: floppy disk, single directory, or multiple /DISK directories. Upon completion of the compression process, the wizard's processing is complete. Figure 7-14 shows the processing dialog that appears when the wizard is busy compressing all the distributable files.

Figure 7-12
Reviewing the set of
distribution files.

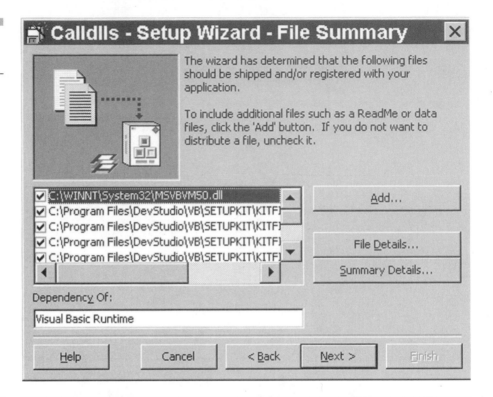

Figure 7-13
The setup wizard
finish dialog.

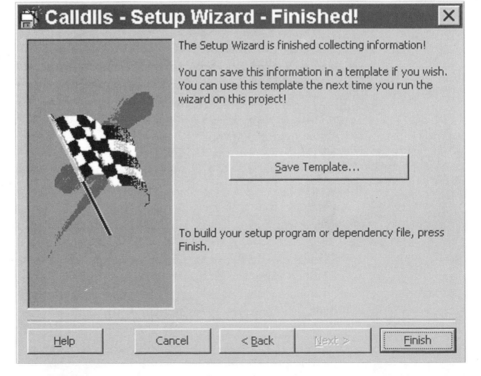

Figure 7-14
Compressing the
distribution files.

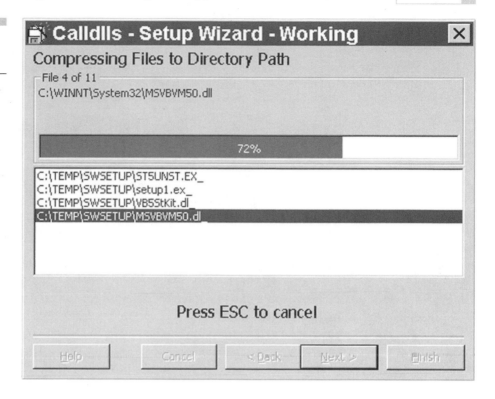

Using the Output from the Setup Wizard

Figure 7-15 shows how to test your installation by running the SETUP program from the destination directory. It is not really practical to use the development machine for testing the installation, since all of the components are already resident on that machine. You should write the files to floppies (or CDROM) and test the validity of the generated installation by performing the installation on a "clean" machine. If you are missing even a single file, your app will not function properly. The only way to test this completely is to use a machine that has not yet installed your application.

Running the Generated Installation Program

Figure 7-16 shows how your generated installation will look to the end user. Keep in mind that this is the default UI for the installations created with the setup wizard, and you are not required to do anything special to get this consistent look common to most installation programs.

Setup Wizard "Gotchas"

Although the example just explained had no OCX or ActiveX server components, it is important for you to understand that the setup wizard can install them automatically. This is accomplished by examining your pro-

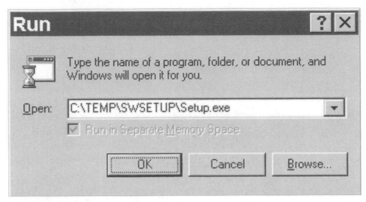

Figure 7-15   Using the setup wizard output.

ject's .VBP file. The setup wizard makes note of every OCX file (Object=) in your .VBP file and fetches it automatically during the setup procedure, placing it in the list of files to distribute. Likewise, the setup wizard also makes note of every ActiveX server (Reference=) in your .VBP file, and copies the server file(s) to the installation list also. This means your application should be actually using every OCX control in your toolbox and every ActiveX server selected in your references. Since the setup wizard simply examines the .VBP for Objects and References, all of these get copied to your installation as part of the setup wizard processing, whether you actually use them or not in the project. (The wizard is not smart enough to actually check your code for valid usage of each component in your References and Toolbox.) For this reason, be sure to unselect

Figure 7-16
Running the generated installation.

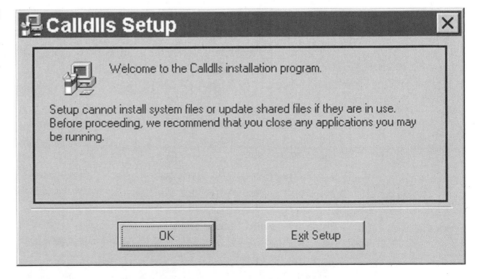

unused objects in the Project/References dialog. Also, remove unused custom controls from the VB Toolbox and save the project *before* you build your setup and installation

SUMMARY

In this part of the book we covered adding online Help and API calls to your VB apps. We also explained who to build your EXE, and how to create an installation and setup program for your completed VB app. At this point, you have the core knowledge needed to be effective using VB to create standalone applications.

Increasing Your Knowledge

- Help and Books Online Search Strings
- Declare
- Declaring
- API
- Setup
- COMPRESS
- Helpfile
- Helpcontextid
- WhatThisMode
- WhatsThisHelpId
- EXE
- Compiling
- hWnd

Mastery Questions

1. The primary purpose of compiling to native code as opposed to p-code is:
   a. security, since native code cannot be readily decompiled, making your code more secure.
   b. runtime speed of execution.
   c. the ability to optimize the code for speed or size.
   d. Both a and c

2. When calling a function in an API, you must:
   a. use a Declare Function or Declare Sub statement in the [general][declarations] of the form that contains the call to the API procedure.
   b. always pass any parameters "by value."
   c. use a Declare Function or Declare Sub statement to establish the dynamic link between your app and the API procedure.
   d. None of the above

3. The setup wizard relies upon what EXE that ships with Visual Basic to perform installations that are customized by the user?
   a. SETUP.EXE
   b. COMPRESS.EXE
   c. SETUP1.EXE
   d. All of the above

4. When you create a VB executable, to which directory by default does the EXE get written?
   a. The root directory
   b. The VB installation directory
   c. The project directory
   d. The /SWSETUP directory

5. The name of the runtime module for VB version 5 is:
   a. VMVBMS50.DLL
   b. VBVMMS50.DLL
   c. MSVMVB50.DLL
   d. MSVBVM50.DLL

6. When your application is looking for a DLL, the first place it looks is:
   a. the root directory.
   b. the /WINDOWS/SYSTEM directory.
   c. the working directory for the application.
   d. memory.

7. The VB project that can be customized by the user before beginning the setup creation process is located in which directory?
   a. The VB installation directory
   b. The /SWSETUP directory
   c. The /SETUPKIT/SETUP1 directory found under the VB installation directory
   d. The /SETUPKIT/SETUP directory found under the VB installation directory

8. Every VB application has an application title and is set from the Project/Properties menu command of VB. What is the title used for within the context of creating the setup for the application?
   a. It is used as the name of the setup's EXE.
   b. It is used in the setup screen to label the setup with the name of the application that is being installed.
   c. Both a and c
   d. None of the above

9. When calling a function found in KERNEL32.EXE, which files must be distributed with the application that makes the call?
   a. No files are required to be distributed.
   b. KERNEL32.EXE
   c. KERNEL32.EXE, USER32.EXE, GDI32.EXE
   d. None of the above

10. Which files that must be distributed with your VB application do not appear anywhere in your applications VBP file?
    a. Custom controls
    b. Reuseable ActiveX components
    c. Dynamic Link Libraries
    d. None of the above

Answers

1. d
2. c
3. d
4. c
5. d
6. d
7. c
8. b
9. a
10. c

EXERCISE: CREATING INSTALLATIONS

Hardware and Software Requirements
Visual Basic 5.0 Professional or Enterprise Edition

Required Files
None

Assumed Knowledge
Basic understanding of the Application Setup Wizard

What You Will Learn
- Creating an Installation with the Setup Wizard
- Using the Setup Wizard
- Shelling to DOS
- File Compression
- Understanding the System Registry
- Customizing Your Generated Installation

Summary Overview of Exercise

Creating installation disks is a critical part of distributing any Visual Basic application. Fortunately, VB provides an excellent utility for creating setup disks, called the Application Setup Wizard. This utility will help you create a professional setup quickly. However, there are a few things to keep in mind. In this exercise, you will use the setup utility to distribute a VB app and then customize the setup for your company.

Using the Application Setup Wizard

Step 1
Locate the Application Setup Wizard in your VB 5.0 program group and start the utility. The Application Setup Wizard is like any other wizard you may have used. It asks a series of questions and uses your answers to determine the nature of the final product. Distributing any Visual Basic application requires many files. At a minimum, you will have to distribute MSVBVM50.DLL. Many other files may also be required.

Step 2
Here is a list of some file types to keep in mind:

FILE TYPE	FILE NAMES
ActiveX Controls	*.OCX
Microsoft Access JET Engine:	DAO3032.DLL

	MSJT3032.DLL
	MSJTER32.DLL
	MSJINT32.DLL
	VBAJET32.DLL
	VBDB32.DLL
	MSRD2X32.DLL
	MSWNG300.DLL
	ODBCJT32.DLL
	ODBCJI32.DLL
	ODBCTL32.DLL
	VBAR2232.DLL
Help Files	*.HLP
Initialization Files	*.INI
Wave Files, Bitmaps, etc.	*.WAV, *.BMP, etc.

Step 3

As you can see from the table, a Visual Basic setup requires many files. In fact, these files will generally require two floppy disks before *any* of your code is distributed to the target machine! The good news, however, is that most of these files are "shared" files, which means that they only have to be on a computer one time and all Visual Basic applications can use them simultaneously. Shared files are generally stored in the \WINDOWS\SYSTEM directory.

Step 4

From the first step of the wizard, we can select the project for the installation. In the "Where Is the Project File or Template for your Application" field you will see an Browse... button. Push the Browse button and a dialog will appear allowing you to select the project directory. Navigate your way to the Calculator demo located in your VB directory under \SAMPLES\MISC\CALLDLLS.VBP. Select this project and push the OK button.

Step 5

Push the Next button. You will be asked for the target type: floppy disk, single directory or /DISK directories. Click on the Single Directory option.

Step 6

Now the wizard wants to know where to build the installation files. The default is \TEMP\SWSETUP. Take the default.

Step 7

Push the Next button. The wizard now shows you any ActiveX Servers required for this project. ActiveX Servers are reusable software components that can be created in Visual Basic. This project does not use ActiveX Servers, so simply press the Next button again.

Step 8

The next step shows you all the files that will become part of your installation. Here you will find the Add... button. This button is used to bring up a dialog allowing you to include more files such as bitmaps, icons, or data files. Since the wizard can't know everything about your project, you have to supply the names of additional files, if any. In this example, we have no additional files, so just check out the Add... button by clicking it. Don't press the Next button yet!

Step 9

For any file in the list, you may ask to see the File Details. For example, select MSVBVM50.DLL from the list and push the File Details. A dialog pops up showing you some information about this file. Notice that the destination directory is $(WinSysPathSysFile). This tells the wizard to install this component in the \WINDOWS\SYSTEM directory, which is the correct location for shared components. Push the Cancel button.

Step 10

Push the OK button. Now select the file CALLDLLS.EXE and push the File Details button. Notice that the application will be installed in $(App-Path). This is the variable that represents the installation directory selected when the user actually uses the setup. This is the correct location for your application and all supporting files like Help files, bitmaps, and wave files. Push the Cancel button.

To finish the installation setup, push the Finish button. At this point, the wizard will compress the files for your installation and write them out to the \TEMP\SWSETUP directory you took as the default destination path for the setup wizard output.

Step 11

When the installation setup is done, press the Exit button and leave the wizard. You are now ready to test the installation software!

Installing the Application

Step 12

Before running the setup, go to the Windows Explorer and navigate to the \TEMP\SWSETUP directory to view the files created by the wizard. There are several key files here. First, notice that most of the file have an underscore at the end. This character indicates a compressed file. These files were all compressed with a DOS utility called COMPRESS.EXE, which is run by the wizard for you when these files are compressed.

Step 13

Next, notice the file SETUP.EXE. This is the actual file that users will run to install the software. Just like any commercial installation.

Step 14

Notice the file SETUP.LST. This file contains all of the information necessary to properly install the files. Using the setup wizard is what generates this file! If you want to change the application name that appears in the setup, realize that the setup gets the name from SETUP.LST. During the processing of the setup wizard, it picks up the Application Title you specify in your project (via the Project Properties/Make tab) and writes the app title to SETUP.LST.

Later, SETUP1.EXE reads this file. This file is a simple text file that can be viewed in any text editor. To view this file, run NotePad from the Accessories group and open the file SETUP.LST. Be careful not to change the contents.

Step 15

Once you have examined the contents of the \TEMP\SWSETUP directory, run the file SETUP.EXE from the command line. The setup utility will walk you through all of the steps necessary to install the software. Follow the directions and when you are done, run the CALLDLLS app to verify it works.

Uninstalling Software

Step 16

Windows 95 and Windows NT both support the concept of uninstalling software. Uninstall is supported by Visual Basic applications with no special work from you. Windows 95 or Windows NT 4.0 only: Go to the Settings\Control Panel and select Add\Remove Software. A dialog appears with several programs listed. To uninstall the calculator, double click on

CALLDLLS. The uninstall runs and removes the program. Notice that all of the files are removed for you.

Customizing Setup

Step 17

Although the setup wizard gives you a pretty good setup right out of the box, quite often we want to customize the look of an installation program. In this section, we will add a custom banner to the setup.

You will perform the following steps:

a. Alter the main form of SETUP1.VBP to add a new banner

b. Recompile to update SETUP1.EXE

c. Recompress SETUP1.EXE to the new name SETUP1.EX_.

d. Move the recompiled, recompressed file to the \TEMP\SWSETUP directory, where all of the output files from the setup wizard were written.

Step 18

When the wizard builds a setup for you, it actually creates a Visual Basic application. In VB 5.0 the file is SETUP1.EXE. Start up Visual Basic. Locate the Visual Basic project built by the wizard. The project is called SETUP1.VBP and is located in the SETUPKIT\SETUP1 directory. Open this project.

Step 19

The designers of this SETUP1 project have created it so that all customization can be accomplished using the form named frmSetup1. Select this form in the project window and push the View Form button. You will now see a form as a rectangle with a blue background, with some label controls on it. Size this to make it bigger.

Step 20

With the form frmSetup1 visible, drop a new label control on the blue portion of the form. Select the label control and change its properties as follows:

Left	100
Top	100
Height	45
Width	600

BackStyle	0-Transparent
ForeColor	White
Font-Size	18
Caption	Your Company Name

Close the form and save the SETUP1 project.

Step 21

Now we have to build the distribution file by hand, since we are overriding the files created by the wizard. Build an executable of your custom setup by selecting the File/Make EXE menu command. When prompted, select to replace the existing EXE file with your new one.

Step 22

Close Visual Basic. What you just did was alter the file SETUP1.EXE, which is gathered during the processing of the setup wizard. The startup form of SETUP1.EXE now has your company name.

Step 23

Now we have to compress the file by hand. This is done from DOS using the COMPRESS.EXE utility that the wizard uses to create our files. We just have to do it manually now. Shell out to DOS from Windows by selecting the DOS icon from the Programs/Command Prompt menu command.

Step 24

Change your current directory to the \SETUPKIT\KITFIL32 directory under your Visual Basic directory. This is where COMPRESS.EXE is located. Manually compress the new EXE file by typing the following on the command line:

```
COMPRESS.EXE -R drivename\VB\SETUPKIT\SETUP1\SETUP1.EXE
```

In this line, the argument "-R" tells the compression utility to replace the last letter in the file with an underscore. The output of this utility is a file that will be named SETUP1.EX_ .

Step 25

The last step in customizing the setup is to replace the setup file created by the wizard with our new one. Leave DOS and return to Windows. In the Windows Explorer, copy the new compressed setup file to your \TEMP\SWSETUP directory chosen by default earlier as the destination for all setup files. Overwrite the existing file with your new one.

Step 26

Now you have customized your setup. Rerun the installation by double clicking \TEMP\SWSETUP\SETUP.EXE to view the customized banner!

Exercise Summary

The setup wizard can also generate setups appropriate for downloading from the Internet and much more. You need to examine the Help file and Books Online to learn more about the advanced functionality of the setup toolkit provided by VB.

Exercise Extra Effort

Use an image control to add a custom bitmap to the setup.

CHAPTER 8

Understanding Classes and Instances

What You Need to Know

You should be familiar with the fundamentals of forms and standard modules including loading and unloading forms.

What You Will Learn

- Fundamental Class Concepts
- Collections
- Object Variables

Visual Basic has been around since 1991 and in that time has obviously gained wide popularity as a Windows development platform. In the early days, VB could be used to create applications through forms and functions. That, of course, is still true today. However, VB has continued to grow in power by adding new features. Most of the new features have been directed toward improving VB's object orientation, but in spite of the improvements, many developers are mystified by object concepts-particularly classes and instances. In this chapter, we will introduce you to the concepts of classes and instances using familiar VB components such as forms. You may be surprised how much power is available from a simple VB form.

A VB History Lesson

Consider the simple act of loading a Visual Basic form. This, of course, can be accomplished using the Load command. If we want the form to both load and be displayed, we might use the Show command. Examine the following code carefully.

```
Form1.Show
```

This incredibly simple code contains in it a tremendous amount of complexity and power. This may seem ridiculous at first, but it is true. When most VB developers look at the code and try to imagine what must be happening inside VB, they think about in more or less the following manner.

Form1.Show causes VB to go to the hard drive, find the form named Form1, load it into memory, and display it.

Wrong! This is not at all what happens. But this is what VB allows you to think has happened. Why? The reason actually goes back to the beginnings of Visual Basic. When VB was first created, its purpose was to make Windows programming more accessible to developers. You see, in 1991 Windows development was hard. If you wanted to write a Windows application, you used the C language and something called the Windows SDK to get the job done. Very hard. You know that little X in the upper-right hand corner of a VB form-the ControlBox? We know that when it is clicked, a form is summarily dismissed without regard for state. Well, back in 1991, you had to code that functionality yourself. Hard! It felt like months just to create a "Hello, World!" program in Windows.

Visual Basic made life a lot easier. VB takes care of little items like the ControlBox for us. This is good, but in creating VB, the people at Microsoft made some strategic decisions about VB's functionality. They decided to hide many of the more gruesome details of Windows programming from the VB developer. This made life easier but less powerful. Now, as VB gains more and more object-oriented features, understanding some of the hidden programming behavior becomes critical.

So what really happens when you show a form? The truth is that when you execute the code Form1.Show, Visual Basic does not load and display Form1, but rather uses Form1 like a blueprint to create a new form.

Specifically, Visual Basic uses the Form1 you created at design time as a template to produce an exact duplicate of the form in memory. Then, just to hide all of this from you, VB promptly names the form it built in RAM with the same name as the template you built at design time-Form1. Confusing, huh?

Classes and Instances

All of the issues surrounding VB form creation speak to a very important topic in VB-classes and instances. The classic metaphor for classes and instances is that of a cookie cutter and cookies. If you have a bunch of cookie dough spread out on your counter, you can use your favorite Santa Claus cookie cutter to make as many cookies as you want. Each cookie you create looks exactly the same and has the characteristics of the cutter you are using. Change the cookie cutter to a Christmas tree and the look of the cookie changes, but all the new cookies look alike and are based on the new cutter. Because we are developers and have to have our own lingo, we call the cookie cutter a "class" and the cookie an "instance."

This metaphor applies exactly to how VB creates a new form when you execute the code Form1.Show. Visual Basic uses the Form1 that you created at design time like a cookie cutter or class. This class is taken to the dough, your memory, and a new cookie is made, the form instance. The beauty of this process is that once we understand it, we can make as many cookies as we want.

The Life Cycle of a Form

When you use a form, it goes through some definite steps. Normally, we are not bothered by the details, but we can actually take manual control over the life of a form if we want. This allows a great flexibility in both design and function. As an example, we will examine the life of a form step by step through custom code.

Declaring Object Variables

Before we can manually create a form, we must have a variable declared that can represent the form in our code. This is called an object variable. The purpose of the object variable is to act as a surrogate for the form instance we will create. Object variables have the same scoping rules as any variable in VB. Therefore, we could declare them as Public, Private, or Dim. The following code declares a variable capable of representing an instance of Form1.

```
Public MyForm As Form1
```

When you look at this code, it may seem strange at first. After all, last time we checked, VB did not have an intrinsic data type known as Form1. But here we are declaring a variable As Form1. The reason this is legal is because Form1 is in the same project as my code. We can use any object we want in this way, causing a variable to represent all kinds of things in VB. Here are some other legal object variables. Study each carefully.

```
Public MyForm As Form
Public MyTextBox As TextBox
Public MyControl As Control
Public MyScreen As Screen
Public MyFont As StdFont
Public MyPicture As StdPicture
```

Object variables can be declared As Form, which allows the variable to represent any form in the application. You can also declare variables against specific types of controls like As TextBox or As ListBox. These declarations can then be used to pass controls to custom functions and subroutines. Additionally, you can declare variables as any system object, such as Screen, Printer, the standard font object, and the standard printer object. Object variables are very useful ways to code. For example, the following code accepts any ListBox control as an argument and searches it for a given text entry from a TextBox.

```
Public Function SearchList(MyList As ListBox, MyText As TextBox) As
Boolean
  SearchList = False
Dim I As Integer
  For I = 0 To MyList.ListCount-1
    If MyList.List(I) = MyText Then SearchList = True
Next
End Sub
```

Creating Instances

Once an object variable is declared (say as Form1), we can use it to represent an existing object as in the preceding example or create a brand new object from the class. Creating a brand new object from a class is called instancing. Instancing is accomplished through the use of the keyword New. The New keyword is perhaps the most important keyword in VB. The reason why is that the New keyword clearly shows the relationship between classes and instances. The following code is used to create a new instance from an existing class called Form1.

```
Public MyForm as Form1
Set MyForm = New Form1
```

Notice the use of the Set keyword with the New keyword. Set is always required when setting the value of an object variable. This line of code sets the object variable MyForm equal to a new instance of class Form1. Form1 is the cookie cutter and MyForm is the cookie. The important part here is that you can use this line of code to create as many instances of a form as you need-just declare a variable for each instance and you can have multiple copies of Form1.

Whenever a form is instanced using the New keyword, the form receives the Initialize event. The Initialize event is used to notify a class that a new instance has been created. In this way the class can take any steps necessary to prepare the new instance for use. After the event, you may choose to load and show the form with this code:

```
Load MyForm
MyForm.Show
```

Destroying Instances

When you have finished using a form, you might think of unloading it. But unloading a form is not enough to recover the memory used by the instance. VB provides a special keyword for destroying instances called Nothing. When an instance is set to Nothing, all of the memory used by the instance is returned to the heap. VB also fires the Terminate event to notify the class that the instance memory is being reclaimed.

```
Unload MyForm
Set MyForm = Nothing
```

Understanding Collections

Whenever you create multiple instances of a class, it makes sense that you want to have a way to manage the instances. Visual Basic provides just such a mechanism, known as a collection. Collections are similar to arrays except that they hold objects and provide methods and properties that manipulate all of the objects at once.

Visual Basic supports two built-in collections-Forms and Controls. The Forms collection is the collection of all forms currently loaded in the project. You do not have to do anything special to get a form into the Forms collection. Visual Basic puts a form in the collection as soon as it is loaded. Once loaded, you can use the Forms collection to loop through every form in the project. The following code adds the name of every form in the project to a ListBox called List1.

```
Dim MyForm As Form
For Each Form In Forms
  List1.AddItem MyForm.Name
Next
```

This same technique applies to controls. For any given form you select, VB supports a Controls collection that has references to every control on the form. Just as you can see all the forms in a project, you can use Controls to see all the controls on a form. VB also supports building your own custom collections with the Collection Object. You'll learn more about this later.

EXERCISE: INSTANCING FORMS

Forms in Visual Basic are just like all other VB objects, they are actually templates or classes that can be instantiated at run time. This means that a form created at design time can be used at run time to stamp out copies of the form. In this exercise, you will use form instancing to create a simple Visual Basic Word processor.

Multiple Document Interface

Step 1

Create a new directory using the File Explorer called \INSTANCES. Start a new Standard EXE project in Visual Basic.

The Word Processor you will construct will be modeled after Microsoft Word, which is a Multiple Document Interface (MDI) application. MDI applications are apps that have one large parent window with several smaller child windows contained inside of them. Visual Basic allows you to create MDI applications quite easily. To place a parent window in your application, select Project/Add MDI Form from the menu. Only one MDI

Form is allowed per application. After placing the MDI Form in your app, look at the Project menu and you'll notice the menu selection is now gray.

With most MDI applications, the parent window covers the entire screen. You can cause this to happen by changing the WindowState property of the MDI Form to 2-Maximized. Change the WindowState property now.

Step 2

Making a regular form part of the MDI application is as simple as setting one property. Select Form1 and press F4 for the properties. Change the MDIChild property to True, which makes the form a child of the MDI Form. Now save and run your application to see the fundamental MDI behavior.

Sub Main

Step 3

In this application, we want to control when a child form is displayed. We will do this by starting our application from Sub Main. From the menu select Project/Add Module to add a standard module. With the code window visible, select Tools/Add Procedure and add a Public Sub named Main. This is the routine we will use to start the application. To start from Sub Main, select Project-Properties from the menu. In the General tab, set your startup object as Sub Main and close the options dialog.

Step 4

When the application loads, all we want to do is display the large parent window. To accomplish this, add the following code to the Sub Main routine:

```
MDIForm1.Show
```

Save and run your application. You should now only see the parent window covering the entire screen.

Instancing the Form

Step 5

In order to create a word processor, we will have to make a File menu. The menu will be attached to the parent window so that it can be used with any active document. Select the MDI Form in the design environment and open the menu editor by selecting Tools-Menu Editor from the menu.

In the menu editor, build a file menu according to the following data:

```
Menu  File
Caption  &File
Name  mnuFile
MenuItem  New
Caption  &New
Name  mnuFileNew
MenuItem  Open
Caption  &Open…
Name  mnuFileOpen
MenuItem  Save As
Caption  Save &As
Name  mnuFileSaveAs
MenuItem  Menu Bar
Caption  -
Name  mnuBar
MenuItem  Exit
Caption  E&xit
Name  mnuFileExit
```

Step 6

The Exit menu choice will destroy all open instances of forms and end the application. Add the following code to the Click event of the Exit menu to end the application:

```
Dim f As Form
'Unload all the forms except the MDI Form
For Each f In Forms
  If TypeOf f Is Form1 Then
    Unload f
    Set f = Nothing
  End If
Next

'Unload MDI Form
Unload Me
End
```

Step 7

To simulate a blank document for our word processor, we will use a textbox to cover the entire front of Form1. The user can type into the textbox, which will automatically wrap the text. Add a textbox to Form1. In the Resize event of Form1, add the following code to resize the text box when the form is resized:

```
'Resize the Text Box
Text1.Height = Me.ScaleHeight
Text1.Width = Me.ScaleWidth
Text1.Left = 0
Text1.Top = 0
```

Step 8

In order to make the text wrap, change the following properties of Text1:

```
MultiLine  True
ScrollBars  2-Vertical
```

Step 9

Every time the user selects the New menu item, we will create an instance of Form1 as the new document. Add the following code to the Click event of mnuFileNew to generate the new blank document:

```
Dim f As Form1
Static n As Integer

'Create a new instance
Set f = New Form1
f.Text1.Text = ""
n = n + 1
f.Caption = "Document " & Format$(n)
f.Show
```

Save and Run the project. Test the instancing of forms using the menu.

File Operations

Step 10

The menu supports saving and opening files in the word processor. In order to access the file system, we will use the Common Dialog control. Check your toolbox by using the tool tips to determine if the Common Dialog control is loaded. If not, place the Common Dialog control in your project by selecting Project-Components from the menu and checking the Microsoft Common Dialog Control entry.

When the Common Dialog Control is in your toolbox, add it to the MDI-Form. Select the control and press F4 to reveal the properties. The Filter property is used to filter out all files that do not match the selected filter. In the case of our word processor, we filter out all files that do not have a .TXT extension. Set the properties as follows:

```
FileName  *.txt
Filter  VB Word Processor(*.txt)|*.txt||
```

Step 11

When the SaveAs menu is selected, we want to let the user save the document. We will use the Common Dialog Control to save the file. The Common Dialog Control is used to present built-in Windows dialogs to the user. The Common Dialog Control can show dialogs that allow file saving and opening. Add the following code under the Click event of the mnu-FileSaveAs menu:

```
Dim strFile As String

'Open common dialog
CommonDialog1.ShowSave
strFile = CommonDialog1.filename

'Save the File as text
Open strFile For Output As #1
  Write #1, MDIForm1.ActiveForm.Text1.Text
Close #1
```

Step 12

When the Open menu is selected, we want to allow the user to open the text file. Add the following code to the mnuFileOpen Click event to open the file.

```
Dim strFile As String
Dim strText

'Open common dialog
CommonDialog1.ShowOpen
strFile = CommonDialog1.filename

If UCase$(Right$(strFile,3))<>"TXT" Then Exit Sub

'Create a New Instance for the Document
'This line of code makes a call to the
'same code that is used when you select
'new from the menu. It's perfectly legal
'to call an event-handling subroutine
'directly from code this way!
mnuFileNew_Click

'Open the File as text
Open strFile For Input As #1
  Input #1, strText
Close #1
MDIForm1.ActiveForm.Text1.Text=strText
```

Step 13

Save and Run the project. Create, save, and open several documents.

Improving Your Knowledge

Help File Search String

■ Set

■ New

- Nothing
- Initialize
- Terminate
- Forms
- Controls

Book Online References

- "Creating a Reference to an Object"

Mastery Quest142ions

1. What code creates a new instance of Form1 into the variable My-Form?
   a. Set Form1 = New MyForm
   b. Set MyForm = New Form1
   c. Set New MyForm = Form1
   d. Set New Form1 = MyForm

2. What code destroys the object MyForm?
   a. Unload MyForm
   b. Set MyForm = New Form1
   c. Set MyForm = Nothing
   d. Destroy MyForm

3. What variable decalaration can represent any form?
   a. Public MyForm As Form
   b. Private MyForm As Form
   c. Public MyForm As Forms
   d. Private MyForm As Forms

4. Given the code Form1.Show, what is the firing order of events?
   a. Load, Initialize, Paint, Resize
   b. Load, Initialize, Resize, Paint
   c. Initialize, Load, Paint, Resize
   d. Initialize, Load, Resize, Paint

5. Given the following code, what is the firing order of events?

```
Unload Form1
Set Form1 = Nothing
```

   a. Unload, Terminate
   b. Terminate, Unload
   c. Terminate
   d. Unload

Answers to Mastery Questions

1. b
2. c
3. a, b
4. d
5. a

CHAPTER 9

New Features of Visual Basic 5.0

What You Need to Know

You need to have some familiarity with the fundamentals of Visual Basic. This is an ideal chapter to start with if you are an experienced VB4 programmer looking to transition to VB5.

What You Will Learn

- The New Visual Basic Integrated Debugging Environment
- New Data Types
- New Programming Features
- New Class Features
- New Compilation Features

Visual Basic 5.0 incorporates a completely new Integrated Debugging Environment (IDE). Although many of the features of previous versions still exist, the IDE reflects a new philosophy regarding the construction of distributed object applications. This chapter will investigate the new features of the VBIDE as well as the Visual Basic for Applications (VBA) language set used in the product.

The VBIDE

Figure 9-1 shows the new VBIDE and its set of dockable windows. Unlike the previous versions of VB, the IDE no longer has the floating windows surrounding a form. Instead, the windows are dockable alongside the border of a multiple document interface (MDI) design. The MDI design is a major change in VB and represents more than a simple shift from float-

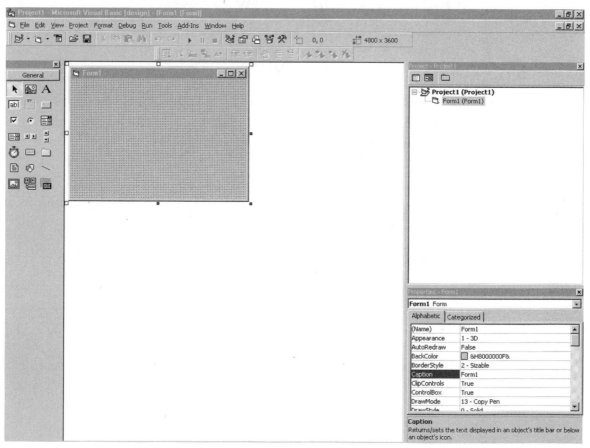

Figure 9-1   The VBIDE.

ing windows to dockable windows. The MDI design, as its name implies, means that more than one Visual Basic project may be open at a time.

At first, the advantages of multiple projects may be unclear. After all, why would I want to work on more than one project when I am having plenty of trouble with just one? The answer lies in changing your definition of an application. Normally, most developers think of an application as a monolithic pile of code that is compiled as a single entity and distributed. This is no longer true. Actually, an application is now defined as a set of distributed software components that work together to perform a function. In fact, many of these components may participate in more than one "application" at a time by performing common function like database services.

Visual Basic 5.0 reflects the new reality of distributed object applications introducing a new concept known as the program group. Program groups are made up of many Visual Basic projects. Each project is a separate software component that performs a function in a distributed application. The complete application is the collection of all these projects, the program group. In Visual Basic 5.0, you create individual components as VB projects and run them together as a group when complete.

Check It Out: Using Program Groups

1. Start a new Visual Basic Standard EXE by selecting File/New Project... from the menu and choosing Standard EXE from the project dialog.

2. This project, with its form support, might be used as a front end for an application.

3. Now add a new project, by selecting File/Add Project... from the menu. This time add an ActiveX DLL from the project dialog.

4. Notice that both projects are visible in the Project window. Use your mouse to toggle between the projects.

5. Right-click on one of the projects and note the menu that appears. Visual Basic lets you set a project as the startup project. This means that the selected project is the one to run first when you select Run/Start from the menu. All of the other projects can be accessed by the startup project.

6. Save the group by selecting File/Save Project Group... from the menu. The project group allows you to keep the projects together that work together.

The Components Dialog and Toolbox

Managing ActiveX Controls in Visual Basic 5.0 is accomplished through the toolbox and components dialog. Viewing the toolbox is done by selecting View/Toolbox from the menu whereas selecting Project/Components reveals the Components dialog. VB provides several new features and ActiveX controls for projects.

The toolbox is used to manage the components that are currently part of a program group. Components appear in the toolbox because one of the projects in the group uses the control. The contents of the toolbox remain unchanged as you switch from project to project as long as you remain within the same program group. Controls may also be grouped in Visual Basic 5.0 using the new tabbed toolbox feature.

The toolbox supports grouping by adding up to four additional tabs along with the General tab that exists by default. You may use the new tabs to sort existing or new controls within the program group. The tabs remain as part of the toolbox permanently, but the controls you place in the tabs are only present as long as their respective projects are open.

The components dialog lets you add new controls to a Visual Basic project. Visual Basic 5.0 supports more than just ActiveX controls, however.

Figure 9-2   Toolbox tabs.

The components dialog also lists the Insertable objects available at design time. Insertable objects are not new, they existed in VB 4.0, but many developers are still not familiar with them. Insertable objects are ActiveX components that represent objects available from other applications on the operating system. This includes Excel spreadsheets, Word documents, Paintbrush images, and so on. All of these components can be embedded into a Visual Basic form.

▌ Check It Out: Insertable Objects

1. Start a new Standard EXE project in Visual Basic by selecting File/New Project… from the menu.

2. Open the Components dialog by selecting Project/Components… from the menu.

3. In the Components dialog, select the Insertable Objects tab.

4. Locate the Insertable object titled Bitmap Image, which is part of the Paint program. Select the checkbox and push the OK button. An icon should appear in your toolbox for the Insertable object.

5. Select the new icon from the toolbox and draw it onto Form1 so that it covers the entire form.

6. Run the project by selecting Run/Start from the menu.

7. Double-click on the Insertable object to start the Paint program. Now you can draw inside the work area.

Visual Basic 5.0 also supports a completely new type of object known as a Designer. Designers are additional components that allow you to create software graphically. Visual Basic ships with two designers: the User-Connection and Forms 2.0 designer. These may be added to the VBIDE by simply selecting them in the Components dialog. When added to the VBIDE, the designers do not appear in the toolbox. Instead, they appear on the menu under Project/Add ActiveX Designer. Once added, the designers are a permanent part of your environment.

The Forms 2.0 designer is a form that can be used inside the Office VBA environment. This designer allows you to create and use forms from Excel, Word, and PowerPoint in Office 97. The UserConnection designer is an object that allows you to connect to an ODBC data source. The User-Connection designer is discussed in detail in Chapter 14.

In addition to the new component features, VB also ships with a number of new ActiveX controls for creating applications. These controls provide data access features, Internet capabilities, and new GUI features. The following are all of the controls that ship with the Enterprise edition of Visual Basic 5.0.

Figure 9-3
The Components
dialog.

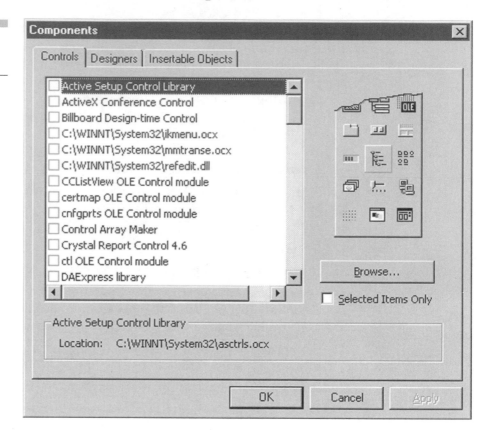

Remote Data Control

The Remote Data Control provides a graphical interface for connecting to ODBC-compliant data sources. This control is a version 2.0 upgrade from the version that shipped with VB 4.0.

Data Bound Controls

Visual Basic provides a number of controls capable of receiving database information from a data control like the Remote Data Control. Visual Basic includes data bound grids, list boxes, and combo boxes.

Windows Common Controls

Visual Basic provides ActiveX controls you can use to create interfaces that look and feel just like Windows. These controls include Tabstrips, Toolbars, StatusBars, ProgressBars, TreeViews, ListViews, ImageLists, and Sliders. The controls are identical to the controls used in Windows interfaces like the File Explorer and Control Panel.

MAPI Controls

VB provides two controls for use with the Microsoft Mail system. These controls, MAPISessions and MAPIMessages, allow you to create and receive mail.

Internet Controls

Visual Basic has two controls for use with networks and the Internet. The WinSock control gives access to Windows sockets, which can be used to access any network including the Internet. The Internet control implements FTP and HTTP protocols to give your VB applications direct access to file transfer and web server facilities.

Miscellaneous Controls

Several other controls are provided for various uses. The SSTab control is used for tabbed dialogs. The CommonDialog control is used for accessing files. The Animation control is used to create animations with AVI files for effects during operations. The UpDown button is a spin control. The RichTextBox is a TextBox that accepts Rich Text Format (RTF). The MSChart control is used to create graphs. The MMControl is used to play multimedia files. The PicClip control is used to manage many bitmaps as one contiguous whole. The MSComm control is used for modem communications. And the MaskedEdit control is used to restrict user input to a certain format.

Project Templates and the New Project Dialog

Every time you start a new project in Visual Basic, you notice that VB prompts you to select from a collection of predefined projects. The project templates are presented in the New Project dialog. The New Project dialog contains templates for all the different projects that VB can create, but it can also contain templates for projects that you create. All of the templates are kept in the templates directory, which is C:\PROGRAM FILES\DEVSTUDIO\VB\TEMPLATE by default. You can change the templates directory in the options dialog found under the Tools/Options menu selection.

▇ Check It Out: Creating a Template

1. Start a new Standard EXE Project in Visual Basic.
2. Place a command button on Form1.

Figure 9-4
The New Project
dialog.

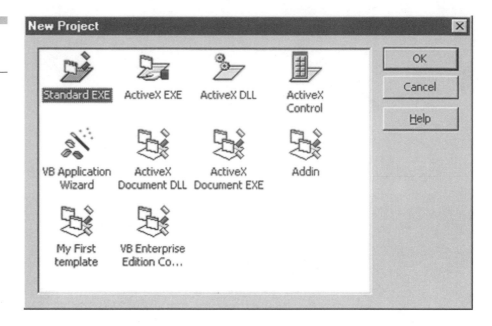

3. In the Click event of Command1, add the following code:

```
MsgBox "This is My First Template!"
```

4. Change the name of Form1 to frmFirst.
5. Save the project by selecting File/Save Project into the template directory as specified in your options dialog. Use the following file names:

```
First.frm
My First Template.VBP
```

6. After saving, select File/New Project from the menu. Examine the New Project dialog. You should see your new template. Select this project.
7. Run the new project and verify that your template works correctly.

The Code Window

The code window has many new features in Visual Basic designed to speed coding. These features begin with the most noticeable of all the

Figure 9-5 Auto Quick Tips.

code window improvements: Auto List Members and Auto Quick Tips. Auto List Members and Auto Quick Tips provide hints to programmers as they write code. If an object or function is selected, then these features give the programmer a list of choices to complete the line of code.

The VB code window now supports drag-and-drop coding. With this feature, you can highlight a line of code in the code window and drag it to another location in the window. You can make a copy of the code by holding the Ctrl key while you drag the code. The code can even be dragged to the Debug window during break mode for evaluation.

The code window also has a gray margin area where breakpoints and bookmarks are set. Setting a breakpoint is now a simple matter of clicking in the margin next to the line of code where you want the breakpoint. Bookmarks are set using the Edit menu. Bookmarks can be turned on and off from the Edit menu as well as allowing you to navigate between the bookmarks that you have set. Neither bookmarks nor breakpoints are saved when a project is saved. They are destroyed when the project is closed.

During Break mode, you can now gain access to the values of variables by simply putting your mouse over the variable in the code window. You can also highlight expressions and see their value in the same way. A Locals window is also available for viewing variable values in addition to the Watch window.

Form Design Features

Visual Basic 5.0 provides many new formatting features for use with forms. Most of these new features are available directly off the Format menu. Visual Basic provides alignment tools that allow you to align groups of controls on left, top, right, center, middle, or bottom edges. You can use these options to quickly line up buttons and text boxes on forms. You can also choose to make groups of controls the same size either horizontally, vertically, or both.

Project Properties

Many key aspects of a Visual Basic project can be controlled through the Project Properties dialog. The Project Properties dialog is access through the Project menu. Select Project/Properties... to display the tabbed dialog. Each tab of the dialog controls a specific set of project features.

The General Tab

The General tab configures many fundamental aspects of your project. The Project Type drop-down box is used to set the type of project-Standard EXE, ActiveX EXE, ActiveX DLL, or ActiveX Control. This value is set when you choose a project template from the New Project dialog, but you can change it any time.

The Startup Object box allows you to select the component within your project that will be the first to execute. In a Standard EXE this can be any form in the project or Sub Main. With ActiveX components, you can choose to have no startup object and simply activate the component when it is called by a client.

The Project Name, Description, and Help file can also be set in this tab. This information is important, particularly in ActiveX components, since the name and description will appear in tools such as the object browser.

Selecting Upgrade ActiveX Controls will convert any controls in your project that conform to the VBX standard with the equivalent OCX. This will only affect a project if the project was previously created in VB 4.0, 16-bit version, or VB 3.0. In these cases, you must upgrade the controls or they will not be recognized in VB 5.0. Selecting Require License Key allows you to require a license for the ActiveX controls you create. Selecting Unattended Execution allows components to support multithreading.

The Make Tab

The Make tab contains information about the current build of the project. In this tab you can set the version number and ask Visual Basic to automatically increment the build version each time you compile the project. Automatic incrementing is limited to the smallest revision number only. Changing the major and minor values must be done by hand.

Along with the version number, you can embed version information into the compiled project. All of the embedded information is available at run time through properties of the App object. This allows you to easily build splash screens and About boxes that can be used in every project.

Any VB application you create can be executed not only from an icon, but also from the command line. When an application is executed from a

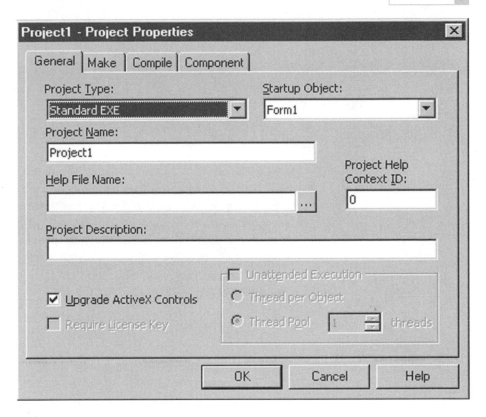

command line, Visual Basic accepts command line parameters. These command line parameters are passed in to a special variable in VB called Command$. Command$ contains all of the parameters that you can read and then parse to modify your application's behavior. The Make tab gives you a way to pass these parameters at design time by simply typing them into the Command Line Arguments textbox.

Check It Out: Using Command Line Arguments

1. Start a new Visual Basic Standard EXE.
2. Open the Project Properties dialog and click on the Make tab.
3. In the Command Line Arguments dialog, enter the following test arguments:

```
/s /u
```

4. Close the Project Properties dialog.

5. In the Load event of Form1, add the following code to read the arguments:

```
MsgBox Command$
```

6. Run the project and verify that Command$ receives the command line arguments you specified.

In addition to command line arguments, Visual Basic also supports Conditional Compilation. Conditional Compilation allows you to designate certain parts of code to compile based on an argument you specify. This allows you to create a number of different editions for a component from just a single code base. Imagine, for example, that you wanted to create a Standard, Pro, and Enterprise edition of your software. To accomplish this, you might specify Conditional Compilation constants and set their values in the Make tab as follows:

```
Standard = -1:Pro = 0;Enterprise = 0
```

The values for the constants are set as either True or False. Visual Basic recognizes _1 as True and 0 as False. In this case, only the Standard constant is True. You can then use the constants in code to block out functions reserved only for the Pro or Enterprise editions.

```
#If Standard Then
  'Code for Standard Edition
#Else
  'Code for Pro and Enterprise
#End If
```

The number/pound sign (#) indicates to the VB compiler that the If...Then statement is a preprocessor command. In other words, VB looks at the Conditional Compilation blocks before the compiler runs. Any code that does not meet the text is not compiled. The code left out is simply treated as a comment.

The Compile Tab

The Compile tab allows you to control all of the native compile features of Visual Basic. Projects can be compiled into p-code or native code using the appropriate options. Several options and optimizations are available to enhance to project speed or size.

Optimize for Fast Code is perhaps the most frequently used optimization. This optimization will produce the fastest possible executable, but

Figure 9-7
The Make tab.

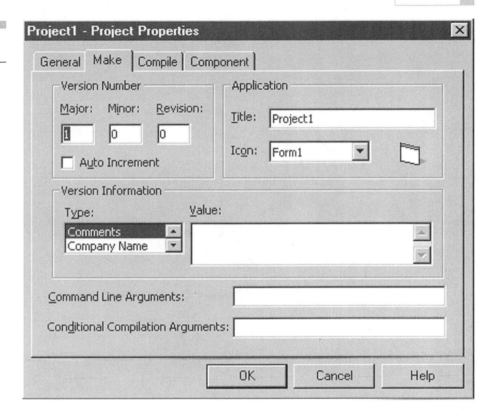

the size of the program in memory will be larger than with other optimizations. The memory footprint may not be a concern in most applications, but it may be an issue for clients that are constrained by available resources.

Optimize for Small Code causes the memory footprint of the application to be as small as possible. The small memory footprint will cause the program to run slower than code optimized for speed, but will use fewer resources. This may help components load and initialize faster.

No Optimization removes all optimizing features, while Favor Pentium Pro produces software specifically targeted at the Pentium Pro processor. Creating Symbolic debug info will allow the application to be debugged in Visual C++ and creates a special file containing debug information to go along with the component.

The Component Tab

The Component tab is used to specify features of the software component that affect its use and compatibility. The Start Mode option is used by VB to identify projects in the design environment that are dependent on other projects such as DLLs. Selecting the ActiveX component allows the

Figure 9-8
The Compile tab.

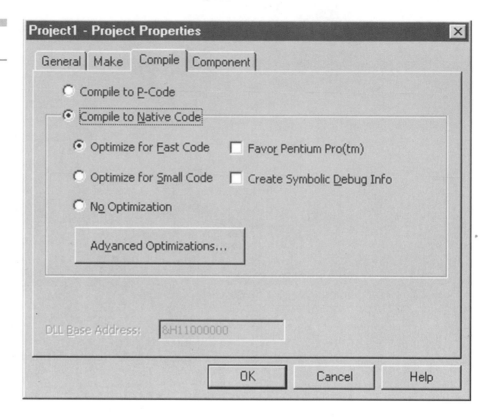

projects to stay running in the design environment so they can be debugged. This option has no effect after the project is compiled.

The Remote Server files option creates a special file for use with servers that will be distributed on a network. The created file, with extension .VBR, contains information required by clients connecting to distributed ActiveX components. Distributed ActiveX components are covered in detail in Chapter 15.

Version compatibility determines how ActiveX components behave as they are modified, compiled, and distributed. Components that set the Project Compatibility option allow clients to maintain a reference to them in the References dialog even after the component code is modified. This is primarily for debugging purposes. The Binary Compatibility option ensures that when the component is modified and compiled, it is completely backwardly compatible with clients that have used a previous version of the component. Binary compatibility can only be maintained if the function signatures of all properties and methods are left unchanged across versions.

Figure 9-9
The Component tab

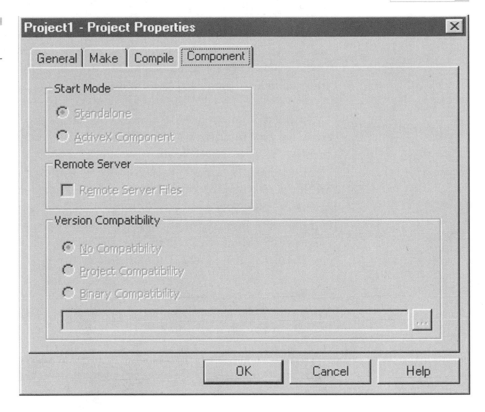

VBIDE Options

Project properties are responsible managing features of the project. The VBIDE options are used to manage options associated with the Visual Basic environment itself. The options are accessed by selecting Tools/Options... from the menu.

The Editor Tab

The Editor tab sets options that affect the code window behavior. The Auto Syntax Check option causes Visual Basic to check the syntax on a line of code when you exit the line. If the line is incorrect, VB pops an error dialog. At first, the Auto Syntax Check seems like a good idea, but actually it's not. If the Auto Syntax Check is off, then VB still checks your line of code, but when it's wrong, no message box is displayed-the line of code simply turns red. This is a much more useful way to identify problematic code. Our recommendation is to turn Auto Syntax Check off.

The Require Variable Declaration option causes Visual Basic to place the keywords Option Explicit at the top of every form and module in your project when first added. Option Explicit tells the Visual Basic compiler that every variable in the application must be explicitly declared before it can be used. This is an excellent feature that eliminates errors due to typographical errors. Our recommendation is to have Require Variable declaration on.

Auto List Members, Auto Quick Tips, and Auto Data Tips enable the popup lists for the code window discussed earlier. Our recommendation is for all of the features to be on.

Auto Indent tells the VB text editor to automatically indent the next line of code to the level of the previous line. This feature allows you to create "block indented" code that enhances readability and nesting. Our recommendation is Auto Indent on.

The Window settings features enable display features of the code window discussed previously. These useful features should all be on.

The Editor Format Tab

The Editor Format tab is where the text editor display features are configured. In this dialog, you can set the font style, size, and color for the

Figure 9-10
The Editor tab.

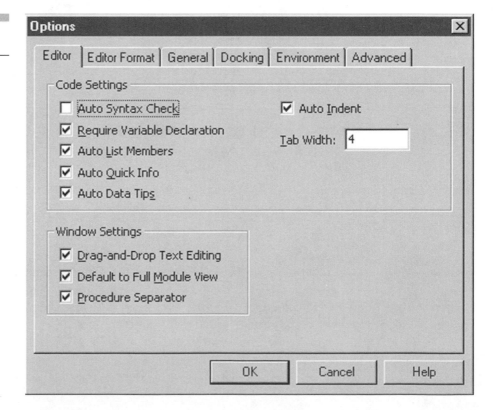

Figure 9-11
The Editor Format
tab.

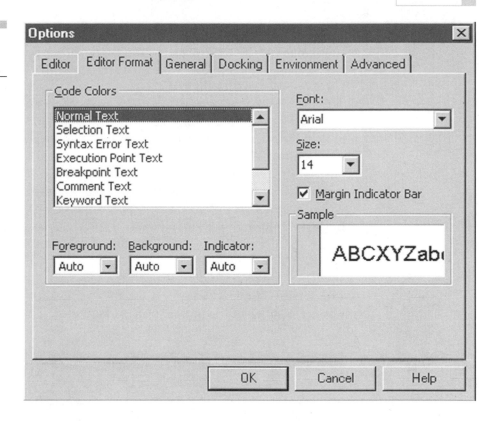

Text Editor. You can also enable the code window margin. The margin area supports placing breakpoints and bookmarks in your code.

The General Tab

The General tab contains a number of options that are not easily grouped under another tab. The Form Grid Settings allow you to configure the grid pattern that appears on a form. The grid pattern is used to help align controls when building a GUI. You can alter the pattern density with the Height and Width numbers. If you disable the grid, then you can move controls one pixel at a time.

The form grid also supports two useful keyboard shortcuts for resizing and repositioning controls. After a control is selected, you can move it by holding down the Ctrl key and pressing the arrow keys. Resizing occurs when you hold the Shift key and press the arrows.

The error trapping options are also available from this tab. Break on All Errors disables all of the On Error GoTo statements in your code. This can be very useful when debugging since it forces VB to enter Break mode rather than handle the error in a trap. Break in Class Module causes errors raised by ActiveX components to force VB into Break mode inside the component instead of the calling client. Break on Unhandled Errors is the normal error handling mechanism that will only break if no error handler is coded into the routine.

The Compile On Demand option is used only while running code in the VB environment and tells Visual Basic to compile just the next few lines of code necessary to continue running an application. With this option unchecked, VB will fully compile the entire project before running it. If you select Background Compile as well, VB will not only compile on demand, but also do it in the background while you are not interacting with the app. Our experience is that Compile On Demand can be confusing since it often raises compile time errors, such as undeclared variables, while the program is executing. Our recommendation is to turn this option off.

Show Tool Tips simply turns on the tool tips for the toolbox. Collapse Proj. Hides Windows causes the code modules and forms to disappear when you collapse the project tree in the project window.

Figure 9-12
The General tab.

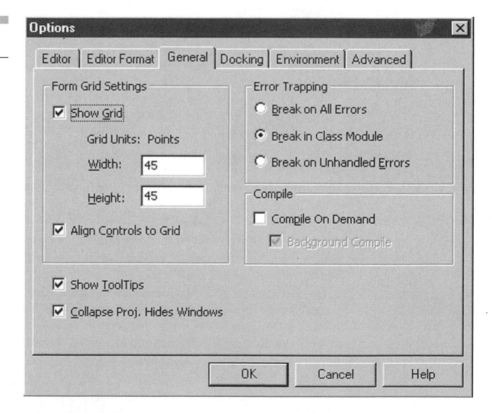

Figure 9-13
The Docking tab.

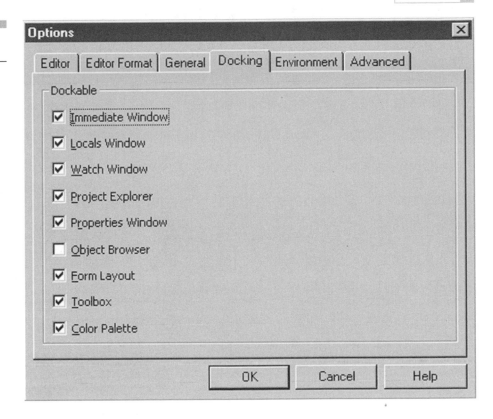

The Docking Tab
The docking tab is fairly simple. In this tab, you can specify which windows are dockable in the VB environment. All dockable windows can also be easily undocked by grabbing them and pulling with the mouse.

The Environment Tab
The Environment tab specifies options that affect the entire VB environment. When you start VB, you can choose to have it create a default project (Standard EXE) or prompt you with the New Project dialog. When you subsequently run a project, you can instruct VB how to handle unsaved changes. You may either ignore them, have VB prompt you to save the project, or automatically save the changes without prompting. We highly recommend at least setting this option to receive a prompt from VB when your project is dirty.

Templates were discussed previously and they can be managed from this tab. Select the available templates and template directory in this tab.

Figure 9-14
The Environment tab.

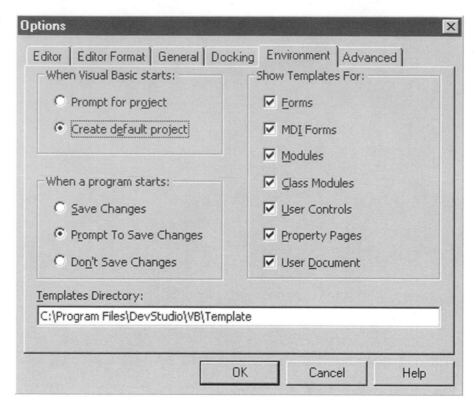

The Advanced Tab

The Advanced tab contains only three options. Background Project Load instructs VB to load your project files in the background. This feature allows you to get to work faster. "Notify when changing shared project items" allows you to receive notification when a shared item such as a form or class is changed in one project. The same form or module can be used in many projects, but if one project changes the component, all other projects must be synchronized with the change.

New Language Features

This section takes a look at new features of the Visual Basic language. Many of the features covered in this section are explained in detail elsewhere in the book. Where appropriate, we have provided references to the explanations in other chapters. For features not examined anywhere else, this section provides detailed coverage.

Figure 9-15
The Advanced tab.

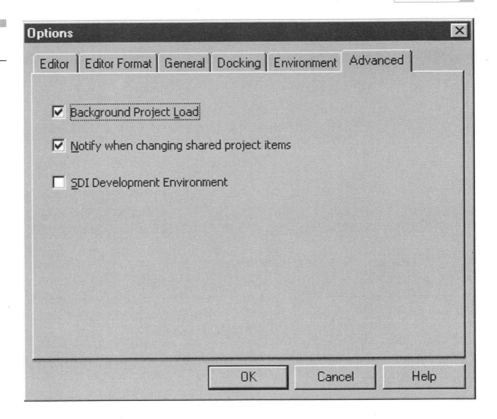

Data Features

Visual Basic contains several new features for variables and members. Visual Basic supports completely new kinds of data type known as the Decimal data type. The Decimal type is a data type is actually a subtype of the Variant. Decimal data types provide a 12-byte unsigned data type that can provide exponents up to _28. Numbers as large as 7.9E28 and as small as 1E_28 are possible.

Enumerations are now supported in Visual Basic. Enumerated data types allow you to specify a set of values for a data type instead of a range. Properties like MousePointer are a good example of enumerations. This property provides several discreet possible values that translate to different mouse pointers. You can create your own enumeration with the Enum keyword.

```
Enum GolfClub
   Iron
   Wood
```

```
   Putter
   Chipper =100
End Enum
```

In this enumeration, the entries Iron, Wood, and Putter correspond to values 0, 1, and 2 respectively. Chipper is given a specific value of 100. Variables may now be declared as type GolfClub and will support values of 0, 1, 2, and 100. If you declare the enumeration as Public these constants become available to outside of the project just like intrinsic constants you find in Visual Basic.

Another new data feature is the addition of the scoping keyword Friend. The Friend keyword is used instead of Public or Private to denote a special scope for a Function, Sub, or Property procedure. Friend provides project-level scope. This means that a procedure declared as friendly can be accessed by all components in a particular project, but cannot be accessed by any component outside of the project.

Class Module Features

Visual Basic provides a number of new features to enhance the construction of Class Modules. As VB continues to mature, more and more object-oriented features are finding their way into the product. The latest is polymorphism. Polymorphism is explained in detail when we discuss Class Modules, but essentially polymorphism provides a mechanism that allows software objects to communicate without absolutely identifying themselves.

Polymorphism seems distant and strange when you first encounter it, but it is extremely powerful. Visual Basic provides polymorphism through "interfaces." These interfaces are predefined sets of properties and methods that guarantee the existence of certain features in a Class Module. When these features are guaranteed to exist, other software can make a call to a Class without worrying whether the key features are present. Interfaces are used in Visual Basic through the Implements keyword. A complete discussion of this advanced topic occurs in Chapter 10. For now, just recognize that this feature is valuable in establishing communication between software components.

Visual Basic also supports the addition of user-defined events to Class Modules. VB 4.0 introduced the concept of the Class Module, but only provided support for properties and methods. VB 5.0 extends that to events. Again, Classes are discussed in more detail later in the book.

Visual Basic also provides a new way to use classes. This new way allows you to use Classes as function libraries similar to an actual DLL.

This is accomplished by setting the Instancing property of a class module to Global. The Global selection makes the functionality of a Class Module available without specifically instancing the class.

Windows API Support

Visual Basic supports a new keyword for accessing the Windows API called AddressOf. The AddressOf keyword allows you to pass a pointer to a Visual Basic function as an argument to the Windows API. This feature enables callbacks and subclassing features never before available to VB. AddressOf is discussed in detail in Chapter 13.

Data Access Features

Visual Basic supports an upgraded version of Remote Data Objects (RDO), is version 2.0. Version 2.0 of RDO supports standalone objects that ease coding and a complete event model that allows you to easily create asynchronous data access activities that provide event-based notification of completion. Visual Basic also provides a new design time data access feature called the UserConnection object. RDO 2.0 is discussed in detail in Chapter 14.

System Object Features

The system objects have been enhanced with some new features. The App object now supports logging to the Windows NT event log through the LogEvent method. In theory, this method is designed to write events to the event log or to a text file under Windows 95. However, there is a "gotcha" here.The logging support in the App object is well documented, but the features won't actually work for you until you compile the project to an exe. beware!!

The Debug object also has a new method. The Assert method forces a break on a line if the expression evaluates to False. This allows you to build conditional breaks into your code based on the Assert. This is an excellent enhancement to variable watches.

Increasing Your Knowledge

Help File Search String

- Project Properties Dialog Box
- Options Dialog Box

Books OnLine References

▪ "What's New in Visual Basic 5.0"

Mastery Questions

1. Which of the following are new designers that ship with VB 5.0?
   a. UserBox
   b. UserConnection
   c. UserList
   d. UserForm

2. Visual Basic supports templates for:
   a. forms.
   b. projects.
   c. classes.
   d. functions.

3. Which of the following are features of the code window?
   a. Auto List Members
   b. Auto Quick Tips
   c. Conditional Compilation
   d. Project Compatibility

4. Depending on the type, a Visual Basic project may begin from:
   a. any class module.
   b. any form.
   c. Sub Main.
   d. None

5. A Conditional Compilation constant named Debug may be defined by:
   a. setting Public Const Debug = True.
   b. setting #Const Debug = True.
   c. setting #Private Const Debug = True.
   d. Placing "Debug = _1" in the Project Properties dialog.

6. The setting used in Project Properties to ensure the complete backward compatibility of a DLL is:
   a. No Compatibility.
   b. Project Compatibility.
   c. Class Compatibility.
   d. Binary Compatibility.

Answers to Mastery Questions

1. a, c
2. a, b, c
3. a, b
4. b, c, d
5. b, d
6. d

CHAPTER 10

Fundamental and Advanced Class Modules

What You Need to Know

- Fundamentals of Properties, Events, and Methods for Forms and Controls
- Form Instancing and Instancing Concepts
- Forms and Controls Collections

What You Will Learn

- Class Module Construction and Usage
- Key Object-Oriented Principles

According to Microsoft, there are millions of Visual Basic programmers worldwide, and in many cases, these programmers have several years of experience with the product. However, the majority of these same programmers have little or no experience with classes and object-oriented principles. The gap between VB programmers who use forms with functions and those who use classes with methods is startlingly wide. As a Visual Basic programmer, you should understand that classes represent the gateway to every advanced feature in the product. You cannot get very far without them.

As with every technology a programmer tries to learn, the first road-block to understanding classes is the terminology. In fact, the terminology associated with classes is particularly daunting. Words like "polymorphism" and "encapsulation" strike fear into the hearts of students, and instructors do not always have the patience to demystify the concepts. The result is that object-oriented (OO) principles are often guarded like a secret scroll of knowledge that can only be read by the worthy.

We say nonsense! Object-oriented principles are not difficult. They are sensible implementations in software of attributes possessed by entities in the real world. That is what makes OO so powerful-the ability to model the real world. That's why you care!

Consider the state of programming without the benefit of OO. In Visual Basic, this limits you to Forms, Standard Modules, Functions, and Subroutines. In this world, you create applications (say a database app) that consists of forms with textboxes as front ends mapped to fields in a database. Good enough. The primary tools you have for building your application logic are the procedures-functions and subs. So ask yourself this question, "Can you identify something in the real world that is a function or subroutine?" Maybe, but it's hard to do. The best answer we have heard is a "play" in football. The play is reusable, and it is a procedure. It defines precisely how to process a sequence of events. But this is an extremely limited example that does not seem to generalize well to all kinds of business applications. Therefore, we conclude that reusable procedures do not model real-world problems well.

Now consider the same problem using software objects. Can you identify anything in the real world that is an object? You bet! Everything is an object. Look around you-cups, pens, this book, computers, etc., etc., etc. Objects model the real world extremely well. They provide the kind of one-to-one mapping that allows you to identify and manipulate the key entities of a business problem such as customers and invoices. Classes are also highly reusable entities that should ease development and maintenance of even complex systems.

With all of these advantages, you should wonder why many business programmers are slow to adopt classes in their development efforts. In fact, some developers have very strong negative reactions to the concept. We have seen COBOL programmers stand up in the middle of lecture and heatedly ask, "This is all wonderful, but can't I just do this with a function?" Our experience is that many developers have been using poor tools for so long, they are nearly incapable of recognizing the magnitude of OO as a tool. This chapter will get you over the gap.

Encapsulation

Many times when we teach programmers OO principles for the first time, they respond as if OO is a lightening bolt that radically changed software instantaneously. We prefer, however, to think of OO as a natural evolutionary step that grew logically from existing data structures. Imagine if you will, the dawn of programming (well, maybe not that far back) when data could only be managed through single variables. Somewhere along the lines, a programmer must have said "Hey, these variables are cool, but I'd really like to be able to manage lots of these variables together in a nice package." The response to this request was the variable array.

Variable arrays are great because you can take similar data and manage it in one place. So, if you have a series of customer names, you can create an array called Names and index it. You know the drill-Names(1), Names(2), etc. This, of course, satisfied the ancient programmer for only a short period before he or she said, "Hey, these arrays are cool, but I have arrays that are related like Names() and Addresses() that I want to manage in a nice package." The response to this request was the user-defined type.

User-defined types are great because you can define several different types of data and keep them together. Additionally, you can create an array of user-defined types to manage the many occurrences of data. So our customer array might look like this:

```
Type Customer
  Name As String
  Address As String
End Type
Public MyCustomers As Customer(10)
```

This, of course, satisfied the ancient programmer for only a short period before he or she said, "Hey, these user-defined types are cool, but I still have to pass them to the functions I want to call. I want to keep the data and the related functions together in one nice package." The response to this request was the class.

Classes are great for managing data and related functions because they are kept together in one neat package. This concept of bundling together all the related functions and data is known as encapsulation. Encapsulation is a natural outgrowth of the desire to manage related items that has its roots in the declaration of the very first variable. In Visual Basic, encapsulation appears through the implementation of properties, events, and methods. Each property, event, and method is directly associated with a class whether that class is a Form, an ActiveX Control, or a Class Module.

Check It Out: Classes Are Everywhere

1. Visual Basic supports classes in many ways-not just Class Modules. In fact, every Form and ActiveX Control is an instance of a class.

2. Start a new Standard EXE project in Visual Basic.

3. Place six TextBoxes on Form1.

4. Open the Object box in the Properties window and look at the entries for the six text boxes. Each entry has a name (e.g., Text1, Text2, etc.) and each entry shows the control class (i.e., TextBox). Every control you use in VB is an instance of a class that is defined in the toolbox.

Building a Simple Class

When utilizing classes in Visual Basic, it is important to recognize the difference between two very different skills-object building and object using. Object building is the design and construction of classes; object using is simply exercising the functionality of a class. In this section, we'll examine the fundamentals of class construction.

Classes are built around the Visual Basic Class Module. Class Modules can be added to any project by selecting Project/Add Class Module from the menu. Initially, Class Modules appear remarkably similar to Standard Modules in that they are code windows with no Graphic User Interface (GUI). The difference between Class Modules and Standard Modules lies in the fact that Class Modules can be instantiated at run time whereas Standard Modules cannot. In fact, the only way you can use Class Modules is to instantiate them. You simply cannot call them directly like a function or variable in a Standard Module. Class Modules also do not support Form concepts like loading or unloading.

In Visual Basic the Form, Standard Module, and Class Module make up the three major components from which projects are built. Each one of these components has a special role to play in an application. Much of their respective roles are defined by two characteristics: instantiation and user interface. Forms obviously have a user interface. You also know that they can be instantiated at run time. Standard Modules, on the other hand, do not have a GUI nor can they be instantiated at run time. Final-

TABLE 10-1:

Component attributes

Component	GUI	Supports Instancing
Form	Yes	Yes
Standard Module	No	No
Class Module	No	Yes

ly Class Modules do not have a GUI, but they can be instantiated at run time. These fundamental differences allow each component to perform different tasks in a VB application.

Creating Properties

In a Class Module, you may create your own properties. Properties are data members of the class much like the declared members of a user-defined type. Consider the following user-defined type that represents a customer:

```
Type Customer
   FirstName As String
   LastName As String
   Address1 As String
   Address2 As String
   City As String
   State As String
   Zip As String
End Type
```

This user-defined type can easily be converted into a class by taking the definitions for all the variables in the type and placing them in the [general][declarations] section of a Class Module named Customer. The only difference is that the variables must be scoped with the Public keyword. The result is the following code:

```
Public FirstName As String
Public LastName As String
Public Address1 As String
Public Address2 As String
Public City As String
Public State As String
Public Zip As String
```

Because the variables are declared with the Public keyword, they are available to the entire project. However, since classes must always be instanced before they can be used, the variables are always accessed through full qualification. This means that if an instance of the Customer class is named MyCustomer, we can access the variables using the following code:

```
MyCustomer.FirstName = "John"
MyCustomer.LastName = "Smith"
MyCustomer.Address1 = "444B Washington Ave"
MyCustomer.Address2 = "Suite 100"
MyCustomer.City = "North Haven"
MyCustomer.State = "CT"
MyCustomer.Zip = "06473"
```

Just as for user-defined types, we can create an array of class instances to hold the many different customers we may need to access. The big difference here is that classes are typically managed in a collection rather than an array. These collections can function exactly like the Forms and Controls collections provided by VB.

Creating Methods

If Class Modules function similarly to user-defined types, why use them? Wouldn't it be simpler just to create user-defined types that do not require instancing? Well, if we were just managing data, this argument might have some validity, but classes manage more than data, they also manage the procedures that work with the data. In Visual Basic, we know these functions as methods.

Methods are really just functions and subroutines defined inside a Class Module. The routines are defined using the Public keyword as we did for the variables we used to store data. Because of the Public declaration, they also have project scope and require full qualification when addressed. Thus a function called SendInvoice is called for an instance of the Customer class with the following code:

```
MyCustomer.SendInvoice
```

The Life Cycle of a Class Module

Once created, a Class Module can be accessed by the project. Now is when we put on our "object using" hat. This skill is fundamentally different from design because Class Modules are used in a manner similar to any object in Visual Basic. Therefore, more senior developers are typically object builders while junior personnel can be object users. This allows senior people to develop strong functional blocks that can be used by anyone on the team.

Using a class consists of three simple steps: Declare a variable, instantiate the object, and call properties or methods. These steps are always the same regardless of how complicated the object designs become. You should bear this in mind as you learn to use classes-*object using never changes*. The only things you will learn that are new pertain largely to object construction. This is another significant advantage because once people learn to use objects, they can use any object.

Declaring Object Variables

The first step in using any class is to declare an object variable against the class that you want to use. Visual Basic provides many different ways to declare these variables, and each syntax has implications for the performance and behavior of your application. The simplest syntax is to declare the variable as the name of the class you want. So for a class named Employee, we might use the following code:

```
Public MyEmployee As Employee
```

Declaring a variable as the class you want to use allows Visual Basic to closely examine your code and verify that the properties and methods you address in code are actually members of the class you declared. Visual Basic can do this for you only when you declare the variable specifically as the class of interest. This is known as early binding. Early binding has a positive impact on not only code maintenance, but also performance. Early bound variables allow the Visual basic compiler to implement optimizations for the particular class because VB knows at compile time what properties and methods are available. Early binding is good.

Many times when we review code written by newcomers to OO principles, we find mistakes in variable declaration. For example, many developers declare object variables with the Object data type. So for a Customer class, we might find this code:

```
Public MyCustomer As Object
```

This is extremely bad coding! When variables are declared As Object, they may represent any object-not just the intended class. Because VB doesn't know what object you want, it cannot implement any compile time error checking for property and method calls or implement any optimizations. In fact, VB can't check the property and method calls until run time when the object is actually instantiated. This is called late binding. Late binding is evil.

In addition to declaring variables As Object, VB also supports many other built-in object types that you can declare. You may, for example, declare any variable as the class name of any control. So declarations As TextBox or As ListBox are perfectly legal. If you want more generic declarations, you can declare variables As Control or even As Form to represent any form or control in your project. Visual Basic also supports font and picture objects that can be declared As StdFont or As StdPicture. All of these examples are early bound declarations. Only As Object or As Variant declarations result in late binding.

Instantiating the Object

The second step in the life of a class is instantiating the object. Instantiation is done in several different ways. Our recommendation is to instantiate all of your object variables using the New keyword on a separate line. This syntax uses the object variable and class name to create an instance. Thus, the Customer class might be instantiated as follows:

```
Public MyCustomer As Customer

Private Sub Form_Load()
  Set MyCustomer = New Customer
End Sub
```

This technique is known as manual instantiation because you must explicitly create the object instance on a separate line. The value of manual instantiation is that if you call any property or method of an object before it is properly instantiated, you receive a trappable runtime error. Specifically, you get Error 91, "Object Variable or With Block Variable not Set."

Many VB developers do not use manual instantiation. Instead, we often see code in which the New keyword is used on the same line as the variable declaration. At first, this syntax seems compact and beneficial. After all, it saves a line of code required to instantiate the object. This syntax, however, has significant impact on the behavior of the application. When object variables are declared with the New keyword, they are instantiated as soon as any code in the application addresses a property or method of the object. This is known as automatic instantiation. Therefore, the following code is sufficient to declare and instantiate an object:

```
Public MyCustomer As New Customer
Private Sub Form_Load()
  MyCustomer.Name="John Smith"
End Sub
```

The problem with automatic instantiation is that you can easily cause instances to be created when you don't want them. Remember, under manual instantiation, you receives a trappable runtime error if you try to address an object that is not yet instantiated. Under automatic instantiation, you would not receive an error, instead, VB would create a new instance. This may not be what you want.

Finally, we also see the function CreateObject() used to make object instances. CreateObject is very similar to the New keyword in that it returns a new object instance. The companion function GetObject() can be used to retrieve a reference to an object that has already been instantiated by another process.

Regardless of how an object is instantiated, Visual Basic notifies your class through the Initialize event. The Initialize event always fires when a class instance is created. In the Initialize event, you can set the default values for properties of the class or perform other functions necessary to initialize the instance.

Calling Properties or Methods

Once the object is instantiated, you may call any of the available properties and methods. Of course, when you are finished with the object, you should set it equal to Nothing to release the memory used by the object. The following code would destroy an instance of the Customer class:

```
Set MyCustomer = Nothing
```

When an instance is destroyed using the Nothing keyword, your class is notified through the Terminate event. The Terminate event allows your class to clean up before it is destroyed. You can use this event to destroy other objects and forms that were used by your class.

Understanding Object Variable Scope

Object variables, like all variables you declare, are affected by the scoping rules. Public object variables are available to an entire project and Private variables are reserved for the component in which they are defined. Understanding how scoping impacts the functionality of an application is critical to building robust classes.

As we discussed earlier, an object variable can be destroyed by setting it equal to the keyword Nothing. Not only can they be destroyed this way, but all declared object variables that have not yet been instantiated are also equal to nothing. You can test for the existence of an object instance by using the Is operator. For example, the following code tests to see if an object variable has been instantiated:

```
If MyCustomer Is Nothing Then
  MsgBox "No Customer!"
End If
```

The Is operator is an extremely powerful feature that you will use often when programming objects. This operator can also be used in conjunction with other keywords that return information about the class of an object. Visual Basic supports two operators for determining an object's class: TypeOf and TypeName. TypeOf returns a Boolean base on a test class; TypeName returns a string.

```
If TypeOf MyCustomer Is Customer Then
  MsgBox TypeName(MyCustomer)
End If
```

The object operators are useful, but beg some questions regarding the exact nature of an object variable. Although we use variables such as MyCustomer here to reference an instance of the Customer class, the object variable itself does not contain the actual class information. Instead, it contains only a "reference" to the class-that is, a long integer pointer to an address in memory.

Consider what happens when you create an instance of an object using the New keyword. When the line of code is executed, Visual Basic reads the class information from the Class Module you constructed at design time and looks for a place in memory large enough to hold an instance. When it finds a suitable place in memory, an instance is created and the address of the instance is stored in your object variable. This reference then gives you access to all the features of the class.

This is not, however, the end of the story. Whenever an instance is created from Visual Basic, the object itself keeps track of how many variables in your program are referencing the instance. This is known as reference counting. So suppose we create an instance and then reference it twice-the reference count for the object would go up to two.

```
Public MyReference1 As Customer
Public MyReference2 As Customer

Private Sub Form_Load()
'Instance Created and First Reference set
  Set MyReference1 = New Customer

'Set a second reference
Set MyReference2 = MyReference1
End Sub
```

Setting multiple references in a project is quite common as objects access one another, but the real impact of this code occurs when the objects are destroyed. The reason for concern is that the Terminate event of any Class Module only fires after all of the reference variables in code are set to Nothing. This means that setting one of the preceding references to Nothing is not enough to fire the Terminate event.

```
Private Sub Form_Unload()
  'The Terminate Event will not yet fire
  Set MyReference1 = Nothing

  'Now the Terminate Event will Fire!
  Set MyReference2 = Nothing
End Sub
```

The impact of reference counting is that the memory used by an object instance is not released until all of the object variables are set to Nothing. Fortunately, Visual Basic helps out considerably in the management of instance memory through the scoping rules. When an object variable loses scope, it has the same effect as setting the variable to Nothing. By definition, when a Visual Basic application terminates, all variables lose scope. Therefore, exiting a VB application will always destroy all objects created by the application even if you never set the instances to Nothing. This feature, known as "garbage collection," is one of the best features of Visual Basic and completely relieves the VB programmer from the burden of strict memory management.

EXERCISE: CLASS MODULE FUNDAMENTALS

The class module allows you to create your own objects in VB. These objects may then be reused in every project you build. In this exercise, you will examine the fundamental concepts associated with object construction and usage.

Building a Class

Step 1

Start a new Visual Basic Standard EXE project. Insert a Class Module into your project by selecting Project/Add Class Module from the menu. Change the name of your class module from the default Class1 to Employee. You will use this object to track employees in a mythical company. Close the Properties window.

Step 2

Every class that you create will have properties and methods. Properties describe the object you are creating; methods allow you to manipulate the object. For the Employee class, you will define several properties to describe the employee. These properties are added by creating Public variables in the [general][declarations] section of the class module. Add the following code to the [general][declarations] section to describe the employee:

```
Public Name As String
Public Salary As Currency
Public Title As String
```

Step 3

Once the properties are defined, you will need a way to interact with the employees that you create. Interacting is done through methods of the class. Methods are defined by creating Public subroutines or functions in the class. Add a new subroutine to your class by selecting Tools/Add Procedure from the menu. In the Insert Procedure dialog select the following options:

Name:	Promote
Type:	Sub
Scope:	Public

Step 4

In the Promote subroutine, we will pass an argument that specifies the new position for the employee. At our mythical company, we use numeric codes to identify the position and salary of an employee. Therefore, we need to add an argument to the Promote method to indicate the new position for the employee. Add the argument to the routine so that the sub-routine definition looks like this:

```
Public Sub Promote(intPosition As Integer)
End Sub
```

Step 5

Once the argument is defined, you have to take action to establish the new position for the employee. You will change the salary and title of the employee based on the new position code. Add the following code to the Promote method:

```
Select Case intPosition
  Case 1
    Title = "Programmer"
    Salary = 35000
  Case 2
    Title = "Team Leader"
    Salary = 40000
  Case 3
    Title = "Manager"
    Salary = 60000
  Case 4
    Title = "Department Head"
    Salary = 100000
  Case 5
    Title = "CIO"
    Salary = 300000
End Select
```

Step 6

Whenever an instance of a class is created in Visual Basic, the class receives the Initialize event. The Initialize event allows you to initialize the values of any property in the class. In this way, you can establish default values for the class. Simply set the variables in the Initialize event. You can find the Initialize event by selecting Class from the Object box and Initialize from the Procedure box. Add the following code to the Initialize event to set default values for the class.

```
Title = "Programmer"
Salary = 35000
```

Using the Class Module in a Project

Step 7

Now that the class is built, you will create a GUI to utilize the class. The GUI will be built on Form1. The GUI for Form1 will consist of one list box, three labels, and two buttons. Place these controls on the form and set the properties as follows:

```
Form1
Caption                          "Classes"
Height                           4485
Left                             1665
Top                              1830
Width                            6840
Button
Caption                          "Promote Employee"
Height                           510
Left                             3285
Name                             cmdPromote
Top                              2610
Width                            3120
Button
Caption                          "New Employee"
Height                           510
Left                             3285
Name                             cmdNew
Top                              1890
Width                            3120
ListBox
Height                           3630
Left                             180
Name                             lstEmployees
Top                              180
Width                            2490
Label
BorderStyle                      1'Fixed Single
Height                           420
Left                             3015
Name                             lblName
Top                              225
Width                            3570
Label
BorderStyle                      1'Fixed Single
Height                           420
Left                             3015
Name                             lblTitle
Top                              765
Width                            3570
Label
BorderStyle                      1'Fixed Single
Height                           420
Left                             3015
Name                             lblSalary
Top                              1305
Width                            3570
```

Step 8

When you work with objects, you will have to keep track of many instances. Visual Basic provides a special built-in Collection object that you can use to store objects. Add the following code to the [general][declarations] section of Form1 to define the collection:

```
Private Employees As New Collection
```

Step 9

When the application is first loaded, no employee objects are created. The application will start with a blank list. Add the following code to the Form_Load event of Form1:

```
'Clear the List box
lstEmployees.Clear

'Clear the Labels
lblName.Caption = ""
lblTitle.Caption = ""
lblSalary.Caption = ""

'Disable the Promote button
cmdPromote.Enabled=False
```

Step 10

After initializing the controls, the user can create a new employee. Creating the employee will generate an instance of the Employee class, add the new employee to the collection, and place the employee in the list box. Add the following code to the Click event of cmdNew to generate a new employee:

```
'Instantiate the Employee
Set m_Employee = New Employee
Employees.Add m_Employee

'Get a Name for the employee
Employees.Item(Employees.Count).Name = InputBox("Enter a Name.")

'Place the Employee in the List
lstEmployees.AddItem Employees.Item(Employees.Count).Name
```

Step 11

When an employee is selected from the list box, the information for that employee is displayed in the label controls. Additionally, we want to be able to promote the currently selected employee. Add the following code to the Click event of lstEmployees to display the employee information and enable the promote button:

```
'Display Info
lblName.Caption = Employees.Item(lstEmployees.ListIndex + 1).Name
lblTitle.Caption = Employees.Item(lstEmployees.ListIndex + 1).Title
lblSalary.Caption = Format$(Employees.Item(lstEmployees.ListIndex _
+ 1).Salary, "Currency")

'Enabled Promote Button
cmdPromote.Enabled = True
```

Step 12

Promoting an employee is a simple matter of calling the Promote method. In the Click event of cmdPromote, add the following code to call the Promote method:

```
Dim intPosition As Integer

'Get new position code
intPosition = Val(InputBox("Enter New Position Code"))

'Promote employee
Set m_Employee = Employees.Item(lstEmployees.ListIndex + 1)
m_Employee.Promote intPosition

'Refresh Display
lblName.Caption = Employees.Item(lstEmployees.ListIndex + 1).Name
lblTitle.Caption = Employees.Item(lstEmployees.ListIndex + 1).Title
lblSalary.Caption = Format$(Employees.Item(lstEmployees.ListIndex _
+ 1).Salary, "Currency")
```

Step 13

Save and run the application.

Increasing Your Knowledge

Help File Search String

- Classes
- Set
- New
- Nothing

Books OnLine References

- "Creating Your Own Classes"

Mastery Questions

1. Class Modules support which of the following events?
   a. Load
   b. Unload
   c. Initialize
   d. Terminate

2. Name the components that can be instantiated.
   a. Standard Module
   b. Class Module
   c. StdFont
   d. Form

3. Which of the following could be a method of a class?
   a. Public Sub Walk()
   b. Private Sub Form_Load
   c. Private Sub Talk()
   d. Public Function Talk() As String

4. Which Declarations are early bound?
   a. Public MyObject As StdPicture
   b. Public MyObject As Control
   c. Public MyObject
   d. Private MyObject As New Customer

5. In the following code, on what line does the Initialize event fire?
   a. Public MyCustomer As New Customer
   b. Private Sub Form_Load()
   c. MyCustomer.Name = "John"
   d. End Sub

6. In the following code, on what line does the Initialize event fire?
   a. Public MyCustomer As Customer
   b. Private Sub Form_Load()
   c. Set MyCustomer = New Customer
   d. End Sub

7. Given the following code, how many instances are created?

```
Public MyForm1 As Form
Public MyForm2 As Form

Set MyForm1 = New Form1
Set MyForm2 = New Form1
```

   a. 0
   b. 1
   c. 2
   d. 3

8. Given the following code, how many instances are created?

```
Public MyForm1 As Form
Public MyForm2 As Form

Set MyForm1 = New Form1
Set MyForm2 = MyForm1
```

   a. 0
   b. 1
   c. 2
   d. 3

9. Given the following code, how many instances are created?

```
Public MyForm1 As Class1
Public MyForm2 As Class1

Set MyForm1 = CreateObject("Project1.Class1")
Set MyForm2 = GetObject("Project1.Class1")
```

    a. 0

    b. 1

    c. 2

    d. 3

10. Given the following code, how many instances are created?

```
Public MyForm As Class1
Public MyCollection As New Collection
Set MyForm1 = New Class1
MyCollection.Add MyForm
```

    a. 0

    b. 1

    c. 2

    d. 3

Answers to Mastery Questions

   1. c, d

   2. b, c, d

   3. a, d

   4. a, b, d

   5. c

   6. c

   7. c

   8. b

   9. b

  10. b

Advanced Class Modules

What You Need to Know

▪ Class Module Fundamentals

What You Will Learn

- Property Procedures
- Default Members
- Custom Events
- Object Model Fundamentals
- Office Automation
- ActiveX Components

Like so many features of Visual Basic, classes can be used at many different levels. In the previous chapter, we examined the fundamentals of classes, including properties and methods. Classes have many more features, however, that make them more robust and functional. In this chapter, we will cover the advanced features of classes and how to implement them.

Using Property Procedures

You have already learned that classes are similar to user-defined types with the addition of functions known as methods. You also know that in the simplest construction of a class, you can use Public variables to represent properties of your class. Although Public variables can be used, they present some difficulties when constructing more complex classes.

Performing Data Validation

Imagine you want to create an Employee class for tracking information about employees in a database. This class might have several properties such as Name, Age, Height, Weight, and HairColor. If you wanted, you could simply implement these in a Class Module as Public variables.

```
Public Name As String
Public Age As Integer
Public Height As Integer
Public Weight As Integer
Public HairColor As String
```

When using the object, all you would have to do is declare a variable, instantiate the object, and set the properties. As long as things go right, you will have no problem. But suppose that a user of your application accidentally enters the data for the Name property into the field reserved for the HairColor property. The results can be problematic as you try to use the data. In fact, the HairColor property is declared As String, so it can take any legal string value-not just valid colors for hair. The problem, then, with Public variables as properties is that we cannot adequately validate the entered data.

In order to properly validate property values, we need to introduce a new type of procedure known as a Property procedure. Property procedures are special procedures in Visual Basic designed to read and write property values to a class. Adding property procedures to a class is done from the Tools/Add Procedure dialog.

When Property procedures are added to a class, you should name the procedure with the name of the property you are trying to create. When you close the Add Procedure dialog, Visual Basic actually generates two different procedures. This procedure pair contains one procedure for reading the variable called Property Get and one procedure for writing the variable called Property Let. The following code shows the HairColor property declared with Property procedures.

```
Public Property Get HairColor() As Variant
End Property
Public Property Let HairColor(ByVal vNewValue As Variant)
End Property
```

Notice that Visual Basic declares both procedures as Public, which means they are accessible from outside the class. This is exactly the behavior we want, since the procedures will allow reading and writing to a class property. Since these procedures provide all of the public access required, we can change the scope of the Public variable we declared earlier to Private. Additionally, we might want to give it an appropriate prefix to indicate that it is now a private variable. Private variables are often called "members" of a class, so we typically use an "m_" prefix.

```
Private m_HairColor As String
```

The Property Get routine can be thought of as a function-it returns a value. In Visual Basic, the Property Get defaults to returning a type Variant, but we want to return the proper type for HairColor, which is String. Similarly, the Property Let behaves like a subroutine that takes an argument. Visual Basic also defaults the argument to Variant, but we'll

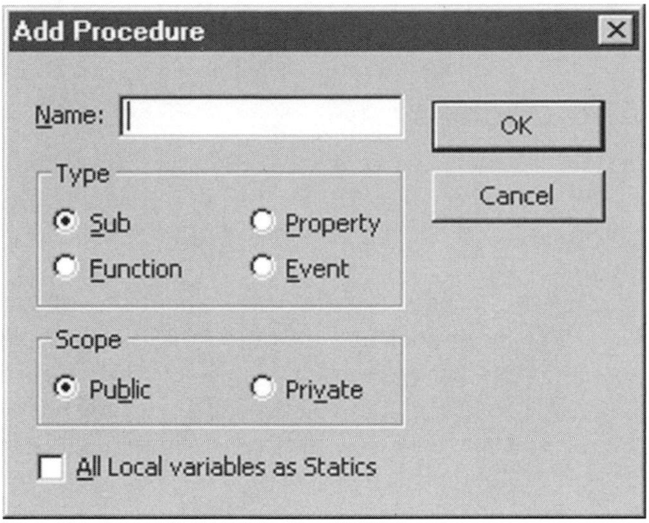

Figure 10-1   The Add Procedure dialog.

change that as well. The final code should appear as follows to properly declare the HairColor property with Property procedures.

```
Public Property Get HairColor() As String
End Property
Public Property Let HairColor(ByVal vNewValue As String)
End Property
```

When Property Get is called, we intend to return the property value. This is done by setting the function equal to the value of the HairColor variable in our code. In this way, Property Get returns a value just as a function does. When Property Let is called, we want to store the entered data in the Private data member. Thus, the final code for a basic property declaration would appear as follows.

```
Public Property Get HairColor() As String
  HairColor = m_HairColor
End Property
Public Property Let HairColor(ByVal vNewValue As String)
  M_HairColor=vNewValue
End Property
```

The interesting thing about Property procedures is that the calling client does not need to know that Property procedures have been used in the class. The client's calling syntax remains the same. Simply declare a variable, instantiate the object, and call the properties. The following code would create an object and set HairColor using the Property procedures.

```
Dim MyEmployee As Employee
Set MyEmployee = New Employee
MyEmployee.HairColor = "Brown"
```

If you stop and think about it, although we have made the class code more complex, we really have not changed the fundamental problem. Nothing in our efforts with Property procedures will prevent a wayward client from trying to send the Name data to the HairColor property. However, now we can place validation code in the Property Let procedure to deal with the problem.

```
Public Property Let HairColor(ByVal vNewValue As Variant)

  If vNewValue = "Black" Or vNewValue = "Brown" Then
    M_HairColor = vNewValue
  End If

End Property
```

This strategy allows us to perform validation on the submitted data and only save it to the private data member when it is valid. This technique is known as data hiding. You can use any feature of VB you want to validate this data. It can be simple as shown here or more complex, such as reading a database table. The point is that you have all of the power and flexibility of VB to perform your data validation.

Incorporating Business Rules

Data validation is not the only reason to use Property procedures. Property procedures also support the implementation of business rules. These rules can trigger actions based on the changes to properties. For example, suppose we had created an Order class for a computer repair shop to track work orders. This class might have a property called Model that tracks the type of computer we are repairing. In our business, if the model is an x386 or earlier, we charge a premium for the work of 15%. Creating the property could be done with the following code.

```
Private Enum enmModel
  x286
```

```
    x386
    x486
    Pentium
End Enum

Private m_Model As enmModel

Public Property Get Model() As enmModel
  Model = m_Model
End Property

Public Property Let Model(ByVal NewModel As enmModel)
  m_Model = vNewValue
  If m_Model < x486 Then Price = Price * 0.15
End Property
```

The business rule changes the Price property if Model is less than x486. This is typical of a business rule. All businesses have these rules, and they generally take the form of If...Then statements. These statements are easily created in a Property procedure.

Controlling Reading and Writing

Property procedures provide one last feature to your classes-the ability to create read-only and write-only properties. By default, Visual Basic always creates Property procedures as Public. However, nothing stops you from changing the scope to Private for either Property Let or Property Get. Changing the scope of Property Let from Public to Private creates a read-only property. Changing Property Get creates a write-only property. In either case, if a client attempts to read or write data in an inappropriate manner, Visual Basic will raise a trappable runtime error in the client. This means your classes work exactly the same as other controls and classes in VB.

Setting Procedure Attributes

Properties and methods of objects inside Visual Basic often have special behavior. Perhaps the best-known example of special behavior is the default property. The default property is a property of a class that receives an argument when no property is explicitly called. The Text property of a TextBox or the Caption property of a Label are common examples. These properties can be accessed without explicit code because they are designated as the default.

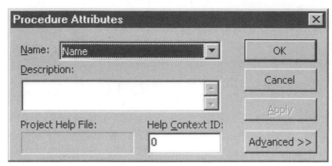

Figure 10-2   The Procedure Attributes dialog.

```
Text1 ="Hello"
Label1 = "World!"
```

Designating special behavior for properties and methods of your classes is supported in Visual Basic through the Procedure Attributes dialog. The Procedure Attributes dialog provides a way to tag a procedure-either a property or a method-as the default member of a class. In the dialog, simply select the procedure to designate as the default. Click on the Advanced button for the selected procedure and set the ProcedureID dropdown to (default). Now you can access this member implicitly through code.

Generating Events

So far, we have spent a lot of time discussing properties and methods, but very little has been said about creating your own custom events. Visual Basic supports a simple and powerful set of keywords for creating and using custom events in classes. These keywords allow you to build events into your classes and receive notification through event-handling subroutines in your clients.

Custom events begin inside a Class Module. In the [general][declarations] section of a Class Module, you may define events that you want your class to generate. Defining the classes is done with the Event keyword. Thus, if you wanted to create an event inside the Employee class that fires whenever a property is set to an invalid value, you could use the following code.

```
Event PropertyError()
```

Once defined, this event can be fired inside any client using the Employee class. Firing the event is done with the RaiseEvent keyword. RaiseEvent is used in code wherever you need the event to fire. Imagine that we wanted to improve the data validation for the HairColor property to include raising an event when HairColor is incorrectly set. The following code uses the RaiseEvent keyword to notify clients when entered data is not valid.

```
Public Property Let HairColor(ByVal vNewValue As Variant)
  If vNewValue = "Black" Or vNewValue = "Brown" Then
    M_HairColor = vNewValue
  Else
    RaiseEvent PropertyError
  End If
End Property
```

On the client side, we would like to see an event-handling subroutine appear in the code window so that we can code some action when the error occurs. Normally, we expect to see event procedures in the Procedure box associated with an object selected from the Object box. However, this does not happen automatically for classes. To get Class Module custom events to show in a client's code window, you must declare the object variable WithEvents. WithEvents is a special keyword that causes all of the events in a class to be made available to a client. The following code is sufficient to access all of the Employee class events.

```
Private WithEvents MyEmployee As Employee
```

Once declared WithEvents, the variable MyEmployee appears in the Object box of the code window. All of the events associated with the object appear in the Procedure box. You can then code to these events just like any other event handler.

Polymorphism

Paramount in any discussion of classes is the concept of communication between the classes. Classes can often be thought of as "black boxes" of functionality. No one is quite sure what's inside, but we know how to make them work. This is exactly analogous to objects in the real world. Consider a television set. How many people are qualified to explain in detail how the internals of a television work? And yet, even though we don't know how they work, we can still use them. This is because we understand the controls that are on the front of the set. We understand the effect of adjusting the volume or changing the channels. In object-oriented terms we say that we understand the "interface" of the television.

Interfaces are a critical concept in dealing with software objects just as they are in dealing with television sets. If we understand the interface of an object, then we understand how to make it work even if we don't understand how it works. On televisions, functions are performed through knobs and switches. In software objects, functions are performed through properties and methods. If we had a software class called Television, it might have a method called On.

Interestingly, objects often have methods with the same names. This is not surprising and also reflects real-world objects. How many appliances do you have with an On/Off switch? And yet each switch does something different. If you turn the television on, you get a picture and sound. If you turn a light on, you get, well, light. The point is, you do not have to learn a new skill to turn an appliance on. You already know it. Even if you were presented with an object you had never seen before, say a Zorkometer that just fell off an alien vessel, you would be confident in your ability to turn it on and off if you saw a power switch on the device. Furthermore the label on the switch implies a guarantee to the user that the expected functionality is built into the object.

The concept of software objects that have the same methods but perform different functions is called polymorphism. This word, which is routinely misdefined, simply states that objects can have identical interfaces, but perform different function-exactly like a television and a Zorkometer. As a developer, why do you care? The primary reason is that polymorphic behavior allows one object to manipulate the interface of another without exactly knowing the nature of the function to perform. Just as you know how to operate switches on devices you haven't seen before, polymorphism allows manipulation of software objects that have not been seen before.

Polymorphism is primarily implemented through inheritance in languages like Java and C++. Inheritance is the ability to define a new class based on an old class. This allows you to define an Employee class from a Human class. Unfortunately, Visual Basic does not support inheritance. Instead, VB implements polymorphism through interfaces. Interfaces are nothing more than the set of all properties and methods of a class. Using this definition, every class has at least one interface because every class has at least one set of properties and methods. In Visual Basic, the set of properties and methods defined in a class is called the default interface. While every class has at least one interface, VB now allows a class to have more than one using the keyword Implements.

The Implements keyword allows you to specify additional interfaces that you want your class to use. That means that you can grab predefined sets of properties and methods and immediately make them part of your own class. Where do these predefined interfaces come from? Well, from other classes, of course. Implements allows you to take the default interface from any class and use it in your own. Why would you want to do this? Simple, to factor out common interface elements that define classes much like switches on appliances.

Suppose we wanted to create software objects that represented a television and a light bulb. We already know both have a switch used to turn them on, so it seems reasonable that both should have a method called SwitchOn. Without interfaces, we would have to provide a Public method to each class to implement the SwitchOn method. Furthermore, if we wanted to turn on one of the appliances, we would have to specifically know which one we wanted to turn on. The following code could be used to execute the SwitchOn method for both classes.

```
Public MyTelevision As Television
Public MyLightBulb As LightBulb

Set MyTelevision = New Television
Set MyLightBulb = New LightBulb

MyTelevision. SwitchOn
MyLightBulb. SwitchOn
```

If we wanted, we could simplify this code somewhat by placing all of the appliances into a Collection object and using the For Each...Next loop to execute the On method for each class. But, because each class is different, we are forced to declare a variable As Object so that it can represent any class. This means that our code will have a late-bound object variable that prevents the compiler from verifying our method call as well as implementing optimizations at compile time.

```
Dim MyAppliance As Object
For Each MyAppliance in MyCollection
  MyAppliance.SwitchOn
Next
```

What we really want is a way to factor out the common interface elements and use early binding when dealing with object variables. Interfaces provide that and more. Sticking with the same example, we will add a class to our project called the IAppliance class. The IAppliance class contains the definition of the common interface elements for all appliances. This class will be our interface class, and by convention interfaces always begin with a capital I. In the class, we define the properties and methods for the interface, but we do not put *any* code in the routines.

```
Public Sub SwitchOn ()
End Sub
```

When defining an interface, you rarely put code in the property and method declarations. The interface serves only as the definition for which properties and methods are in the interface-not the functionality. To define the functionality along with the interface, Visual Basic would have to support inheritance. This concept of properties and methods with no code may seem strange, but it is absolutely driven by the lack of inheritance. If VB had inheritance, we would be explaining polymorphism quite differently. When you define an interface with no code, this is also called an "abstract" class.

Once the interface is defined as a separate class, you can use the collection of properties and methods in another class. This is done with the Implements keyword. To use the interface, simply declare it in the [general][declarations] section with the Implements keyword.

```
Implements IAppliance
```

Once you implement the IAppliance interface in the Television and LightBulb classes, a funny thing happens. The interface name appears in the Object box and all of the properties and methods of the class appear in the Procedure box. In this way, you have a complete list of all the properties and methods available to you. The following code shows the declaration for the SwitchOn method as it might appear in the Television or LightBulb class.

```
Implements IAppliance
Private Sub IAppliance_SwitchOn()
End Sub
```

Notice that when you implement an interface, the properties and methods appear as Private in the implementing class. This is because the IAppliance class contains the Public elements for any class implementing IAppliance. If the properties and methods appeared as Public in the implementing class, then they could be called directly without the interface at all.

Implementing an interface represents a contract between you and the compiler. When an interface is implemented, you are promising the compiler that *all* of the properties and methods that appear in the interface will appear in the Television or LightBulb class. Remember when we talked about the label next to the On/Off switch? We said that represented a guarantee that the switch would perform the intended function. Well, interfaces are the same guarantee. You must place some code (at least a comment mark) in every property and method of the interface or it will not compile. Once implemented, however, interfaces greatly improve code performance. First of all, interfaces allow you to use early binding easily. Remember the example of using a collection to hold all appliances? Well, with interfaces, the code that used late binding becomes early bound.

```
Dim MyAppliance As IAppliance
For Each MyAppliance in MyCollection
  MyAppliance.SwitchOn
Next
```

Now the object variable is declared As IAppliance. This variable is capable of representing any object that implements IAppliance. Furthermore, the compiler recognizes the interface and makes the variable early bound. The compiler error checks your method calls and improves your speed-a big plus for you.

Additionally, the code is now able to deal with objects it has never seen before. If we build a Zorkometer class, we can be assured that the preceding collection code will work without recoding provided our new Zorkometer class implements IAppliance. Our code knows how to turn the Zorkometer on even though it has never seen it before.

Check It Out: Using Interfaces

1. Start a new Visual Basic Standard EXE project.

2. Add a Class Module by selecting Project/Add Class Module from the menu.

3. Name the new Class Module IHuman. This class will be used as an interface to define the common properties and methods for humans. We will implement this class in two other classes-Adult and Baby. Add the following code to IHuman to define the interface:

```
Public Property Get Name() As String
End Property

Public Property Let Name(ByVal strName As String)
End Property

Public Sub Eat(strFood As String)
End Sub

Public Sub Move()
End Sub
```

4. Add a new class to the project and call it Adult. In the [general][declarations] section, implement the IHuman Interface with the following code:

```
Implements IHuman
```

5. Using the Object and Procedure boxes, add each of the members from IHuman to the Adult class. Add code to the Adult class members such that the result looks like this:

```
'Implements the Abstract Class Creature
Implements IHuman

'Private Data Member for Name Property
Private m_Name As String

'THE FOLLOWING ARE IMPLEMENTED
'BECAUSE OF THE IHuman INTERFACE

Private Property Get IHuman_Name() As String
  IHuman_Name = m_Name
End Property

Private Property Let IHuman_Name(ByVal strName As String)
  m_Name = strName
End Property

Private Sub IHuman_Eat(strFood As String)
  MsgBox "MMMMM " & strFood & " was good!"
End Sub

Private Sub IHuman_Move()
  MsgBox "I'm walking."
End Sub

Private Sub Class_Initialize()
  MsgBox "Adult Created!"
End Sub

Private Sub Class_Terminate()
  MsgBox "Adult Destroyed!"
End Sub
```

6. Add a new class to the project and name it Baby. Implement the IHuman interface and add code to create the following result:

```
'Implements the Abstract Class Creature
Implements IHuman

'Private Data Member for Name Property
Private m_Name As String

'THE FOLLOWING ARE IMPLEMENTED
'BECAUSE OF THE IHuman INTERFACE

Private Property Get IHuman_Name() As String
  IHuman_Name = m_Name
End Property

Private Property Let IHuman_Name(ByVal strName As String)
  m_Name = strName
End Property

Private Sub IHuman_Eat(strFood As String)
  If strFood <> "Milk" Then
    MsgBox "WAAAAAAA!!!"
  Else
    MsgBox "MMMMM — BURP!!"
  End If
End Sub

Private Sub IHuman_Move()
  MsgBox "I'm crawling."
End Sub

Private Sub Class_Initialize()
  MsgBox "Baby Created!"
End Sub

Private Sub Class_Terminate()
  MsgBox "Baby Destroyed!"
End Sub
```

7. Add the following code to Form1 to call the classes you have created.

```
Private Sub Form_Load()
  Dim MyHuman As IHuman

  Set MyHuman = New Adult
  MyHuman.Move
  MyHuman.Eat "Steak"

  Set MyHuman = New Baby
  MyHuman.Move
  MyHuman.Eat "Steak"

End Sub
```

8. Run the project and note how the variable MyHuman can be used to represent any class that implements IHuman and still retain early binding.

Component Object Model

During our discussion of Class Modules, we examined how to use classes inside your own projects. The classes we have constructed so far have been for the private use of a single project. This use of classes, however, is extremely limited and does not take advantage of all that classes have to offer. Not only can classes be used within an application, they can also be used across applications.

Although the idea of creating software out of classes may be new to many VB programmers, it is certainly not new to software in general. Most modern software is constructed through classes. In fact, most of the programs that you are already familiar with are based on classes. This includes not only Visual Basic itself, but also Word, Excel, PowerPoint, and Microsoft Access. All of these programs-and most other Windows applications-are constructed through classes.

The reason that you care about classes inside other applications is that Visual Basic actually has the ability to use the classes located in these other applications. That's right. You can use classes defined in other applications directly in your VB code as if they were part of the Visual Basic language. This is extremely powerful. If you need a spell checker, don't bother writing your own; simply borrow it from Excel. The true power of classes comes from using them across applications.

The reason that we can use objects across applications is that all of these objects support a common set of interfaces that allow them to communicate. Just like the interfaces you created in VB, interfaces to other classes support polymorphism through standard interfaces. This set of standard interfaces that allow objects to communicate in Windows is called the Component Object Model (COM). Nearly all of the classes in all of the applications that run under Windows support COM. It allows your Visual Basic application to instantiate and call any class from any application.

In order to use classes from a COM-enabled application, you need to be able to tell Visual Basic about the applications and its classes. In the Visual basic environment, you may use any COM-enabled application by setting a reference to the program. References are used to allow Visual Basic to examine all of the classes, properties, events, and methods that a COM-enabled application has to offer. You can see all of the COM-enabled applications and set references to them using the References dialog, which is accessed by selecting Project/References… from the menu.

Once you have selected an application in the References dialog, you can use all of the objects contained inside the application. You can declare variables against the objects. You can instantiate the objects. You can call

Figure 10-3
The References
dialog.

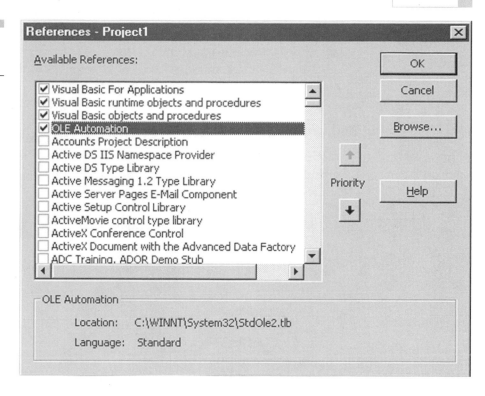

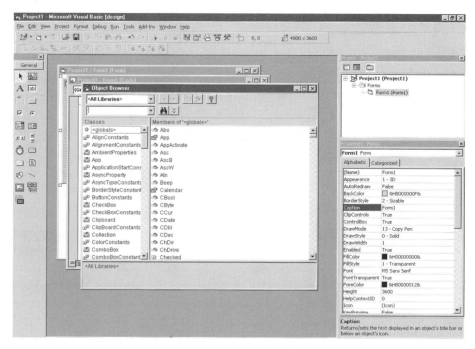

properties and methods as well as receive events. All of the functionality of the object becomes available to you as if it were built in to Visual Basic.

The problem, of course, is that the References dialog does not show you the objects inside an application, it merely lists all of the applications that are COM enabled. So how does a developer learn about the objects that are actually inside the application? The answer is to use another utility called the Object Browser. The Object Browser is a utility that shows all of the classes, properties, events, and methods for any given COM-enabled application. You can access the Object Browser by selecting View/Object Browser from the menu.

▊ Check It Out: Using the Object Browser

1. Start a new Visual Basic Standard EXE project.

2. Open the References dialog by selecting Project/References fro the menu.

3. In the References dialog, set a reference to the Microsoft Visual Basic 5.0 Extensibility model. This reference will allow you to see many of the COM-enabled objects that are inside Visual Basic itself. Close the References dialog.

4. Open the Object Browser by selecting View/Object Browser from the menu.

5. At the top of the Object Browser, you will see an entry marked <All Libraries>. Drop this box and select VBIDE. You are now examining the objects inside VB.

6. In the Classes list, locate the class VBE. This is the class that represents the entire Visual Basic product. Select this object and note the entries in the Members of VBE list. These members represent all of properties, methods, and events of the VBE object.

7. Locate the Name property of VBE. Notice the small blue globe that appears next to the property. This globe indicates that the Name property is the default member for class VBE.

The Object Browser is a good utility for examining objects, but it does not provide all of the necessary information to understand the relationships between objects in an application. The objects in the browser are typically listed in alphabetical order, which is the least helpful presentation. Far better than an alphabetical representation is a hierarchical, graphical representation. This style of presentation is called the object model of an application. Figure 10-5 shows part of the object model for Microsoft Word.

Figure 10-5
Need caption and art

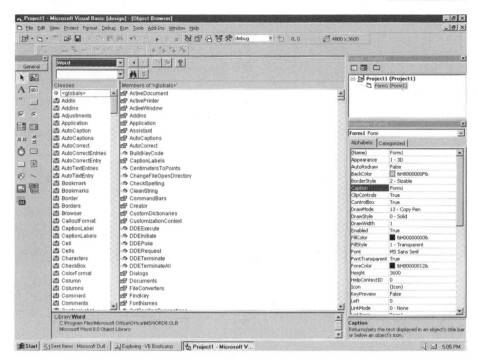

This presentation style is much more meaningful to a developer. You can easily see the logical relationship of classes. Notice that the Document class falls under the Application class in the model. This tells you that Application objects contain Document objects. This makes perfect sense since when you actually use Word, the application contains documents. Typically, you'll find this level of documentation in the Help file for the target product.

Check It Out: Using the Word Object Model

You must have Word 8.0 installed for this exercise.

1. Start a new Standard EXE project in Visual Basic.
2. Open the References dialog by selecting Project/References from the menu.
3. In the References dialog, set a reference to Microsoft Word 8.0 Object Library.
4. Declare variables for the Application and Document classes by placing the following code in the [general][declarations] section of Form1:

```
Private objApplication As Word.Application
Private objDocument As Word.Document
```

5. Place a CommandButton and a TextBox on Form1. You will use these controls to print the contents of the TextBox through Word when the CommandButton is pressed.

6. Create the new Word document by placing the following code in the Load event of Form1:

```
'Create the New Document
Set objApplication = New Word.Application
Set objDocument = objApplication.Documents.Add
```

7. Print the contents of the TextBox by placing the following code in the Click event of Command1

```
'Print the contents of the TextBox
objDocument.Content.Text = Text1.Text
objDocument.PrintOut
```

8. Run the project.

Creating ActiveX Components

You have seen that you can create classes as private entities inside an application or you can use the classes available from other applications. In Visual Basic, however, you can also create your own classes that can be used by other applications. These projects allow you to construct your own classes with their own object models and deploy them to other developers.

In Visual Basic, classes that can be used by other applications are constructed using the ActiveX DLL or ActiveX EXE project type. The ActiveX DLL and ActiveX EXE types differ in performance and behavior. ActiveX DLLs are also called in-process servers because they run in the same memory space as the calling client. ActiveX EXEs are out-of-process servers and run in a separate memory space from the client. This is important, because running in the same memory space as a client is much more efficient. In fact, ActiveX DLLs can run over twenty times faster than ActiveX EXEs! The single biggest performance impact in Visual

Basic development is the selection of ActiveX DLL versus ActiveX EXE. No other choice in any aspect of VB has such a profound impact on performance.

With the tremendous advantages of ActiveX DLLs, you may wonder why an ActiveX EXE is even an option. The answer is that only ActiveX EXEs can run on separate machines from the calling client. By definition an in-process server must be on the same machine as the client. These issues are discussed in detail in Chapter 15.

Creating and using ActiveX components is a simple matter of building your classes into a separate project and setting a reference to the project inside a Standard EXE project. This works the same for your ActiveX components as it does for any other COM-enabled application like Excel. To other developers, your objects work exactly the same way as any others.

EXERCISE: ADVANCED CLASSES

Class Modules are extremely powerful components in VB. They support reusable features like DLL and EXE creation as well as polymorphism and interfaces. This exercise examines some of these advanced features.

Building an ActiveX DLL

Step 1
Start a new Visual Basic ActiveX DLL project. Change the Name of your Class Module from the default Class1 to Employee. You will use this object to track employees in a mythical company.

Step 2
Every class that you create will have properties and methods. Properties describe the object you are creating; methods allow you to manipulate the object. For the Employee class, you will define several properties to describe the employee. These properties are added by creating Property procedures in the class. Add the following code to the [general][declarations] section to declare the private data members that will be used by the Property procedures:

```
Private m_Name As String
Private m_Salary As Currency
Private m_Title As String
```

Step 3

Each of the private variables has to be wrapped in a Public property procedure. Using the Tools/Add Procedure dialog, create the following code to access the private data members. Be careful to note the type declarations of the procedures.

```
Public Property Get Name() As String
  Name = m_Name
End Property

Public Property Let Name(strName As String)
  m_Name = strName
End Property

Public Property Get Salary() As Currency
  Salary = m_Salary
End Property

Public Property Let Salary(curSalary As Currency)
  m_Salary = curSalary
End Property

Public Property Get Title() As String
  Title = m_Title
End Property

Public Property Let Title(strTitle As String)
  m_Title = strTitle
End Property
```

Step 4

Many objects have a default member. For the Employee class, we will make the Name property the default. This is done through the Procedure Attributes dialog. From the menu select Tools/Procedure Attributes. In the dialog, select Name and push the Advanced button. In the ProcedureID box, select (Default). Close the dialog by pushing OK.

Step 5

This class will have an event called BadCode that will notify calling clients when they enter a bad promotion code. This event is declared in the [general][declarations] section of the Employee class. Add the following code to declare the event.

```
Event BadCode()
```

Step 6

Once the properties are defined, you will need a way to interact with the employees you create. Interacting is done through methods of the class. Methods are defined by creating Public subroutines or functions in the class. Add a new subroutine to your class by selecting Tools/Add Procedure from the menu. In the Insert Procedure dialog select the following options:

```
Name:                      Promote
Type:                      Sub
Scope:                     Public
```

Step 7

In the Promote subroutine, we will pass an argument that specifies the new position for the employee. At our mythical company, we use numeric codes to identify the position and salary of an employee. Therefore, we need to add an argument to the Promote method to indicate the new position for the employee. Add the argument to the routine so that the subroutine definition looks like this:

```
Public Sub Promote(intPosition As Integer)
End Sub
```

Step 8

Once the argument is defined, you have to take action to establish the new position for the employee. You will change the salary and title of the employee based on the new position code. If any promotion code is illegal, the BadCode event is fired. Add the following code to the Promote method:

```
Select Case intPosition
  Case 1
    Title = "Programmer"
    Salary = 35000
  Case 2
    Title = "Team Leader"
    Salary = 40000
  Case 3
    Title = "Manager"
    Salary = 60000
  Case 4
    Title = "Department Head"
    Salary = 100000
  Case 5
    Title = "CIO"
    Salary = 300000
  Case Else
    RaiseEvent BadCode
End Select
```

Step 9

Whenever an instance of a class is created in Visual Basic, the class receives the Initialize event. The Initialize event allows you to initialize the values of any property in the class. In this way, you can establish default values for the class. Simply set the variables in the Initialize event. You can find the Initialize event by selecting Class from the Object box and Initialize from the Procedure box. Add the following code to the Initialize event to set default values for the class.

```
Title = "Programmer"
Salary = 35000
```

Calling the DLL from a Client

Step 10

Now that the ActiveX DLL is built, you will create a Graphic User Interface (GUI) to utilize the class. The GUI will be built in a Standard EXE. Add a Standard EXE to your application by selecting File/Add Project from the menu.

We want the application to start from the Load event of Form1. You therefore have to set the Standard EXE as the startup project. Right-click on Project2 from the Project window and select Set As Startup from the menu. Next, select Project/Properties from the menu and set Startup Object to Form1.

We also need to reference the Employee class in Project1 from Project2. With Project2 active, set the reference by selecting Project/References... from the menu. In the References dialog, set a reference to Project1.

Step 11

The GUI for Form1 will consist of one list box, three labels, and two buttons. Place these controls on the form and set the properties as follows:

```
Form1
Caption                              "Classes"
Height                               4485
Left                                 1665
Top                                  1830
Width                                6840
Button
Caption                              "Promote Employee"
Height                               510
Left                                 3285
Name                                 cmdPromote
Top                                  2610
Width                                3120
Button
Caption                              "New Employee"
Height                               510
Left                                 3285
Name                                 cmdNew
```

Top	1890
Width	3120
ListBox	
Height	3630
Left	180
Name	lstEmployees
Top	180
Width	2490
Label	
BorderStyle	1'Fixed Single
Height	420
Left	3015
Name	lblName
Top	225
Width	3570
Label	
BorderStyle	1'Fixed Single
Height	420
Left	3015
Name	lblTitle
Top	765
Width	3570
Label	
BorderStyle	1'Fixed Single
Height	420
Left	3015
Name	lblSalary
Top	1305
Width	3570

Step 12

The Employee class supports an event called BadCode that notifies call-ing clients when they have entered a bad promotion code. We want to receive this event, so we declare the Employee object WithEvents. Add the following code to the [general][declarations] section of Form1.

```
Private WithEvents m_Employee As Employee
```

Step 13

When you work with objects, you will have to keep track of many instances. Visual Basic provides a special built-in Collection object that you can use to store objects. Add the following code to the [general][decla-rations] section of Form1 to define the collection:

```
Private Employees As New Collection
```

Step 14

When the application is first loaded, no Employee objects are created. The application will start with a blank list. Add the following code to the Form_Load event of Form1:

```
'Clear the List box
lstEmployees.Clear
'Clear the Labels
lblName.Caption = ""
lblTitle.Caption = ""
lblSalary.Caption = ""

'Disable the Promote button
cmdPromote.Enabled=False
```

Step 15

After initializing the controls, the user can create a new employee. Creating the employee will generate an instance of the Employee class, add the new employee to the collection, and place the employee in the list box. Add the following code to the Click event of cmdNew to generate a new employee:

```
'Instantiate the Employee
Set m_Employee = New Employee
Employees.Add m_Employee

'Get a Name for the employee
Employees(Employees.Count) = InputBox("Enter a Name.")

'Place the Employee in the List
lstEmployees.AddItem Employees(Employees.Count)
```

Step 16

When an employee is selected from the list box, the information for that employee is displayed in the label controls. Additionally, we want to be able to promote the currently selected employee. Add the following code to the Click event of lstEmployees to display the employee information and enable the promote button:

```
'Display Info
lblName.Caption = Employees(lstEmployees.ListIndex + 1)
lblTitle.Caption = Employees(lstEmployees.ListIndex + 1).Title
lblSalary.Caption = Format$(Employees(lstEmployees.ListIndex _
+ 1).Salary, "Currency")

'Enabled Promote Button
cmdPromote.Enabled = True
```

Step 17

Promoting an employee is a simple matter of calling the Promote method. In the Click event of cmdPromote, add the following code to call the Promote method:

```
Dim intPosition As Integer
'Get new position code
```

```
intPosition = Val(InputBox("Enter New Position Code"))

'Promote employee
Set m_Employee = Employees(lstEmployees.ListIndex + 1)
m_Employee.Promote intPosition

'Refresh Display
lblName.Caption = Employees(lstEmployees.ListIndex + 1)
lblTitle.Caption = Employees(lstEmployees.ListIndex + 1).Title
lblSalary.Caption = Format$(Employees(lstEmployees.ListIndex _
+ 1).Salary, "Currency")
```

Step 18

If the entered promotion code is bad, we expect a call to BadCode. We will place a simple message box in this event to notify us of the problem. Add the following code under the m_Employee_BadCode event.

```
MsgBox "Bad Promotion Code entered!"
```

Step 19

Save and run the application.

Increasing Your Knowledge

Help File Search Strings

- Property Procedures
- Procedure Attributes
- Event, RaiseEvent, WithEvents
- References
- DLL

Books Online References

- "Programming with Objects"

Mastery Questions

1. Which of the following are advantages of Property procedures?
   a. Allows data validation
   b. Allows events to be raised
   c. Allows business rule creation
   d. Allows read-only property creation

2. Class events can be declared:
   a. in a Procedure.
   b. at the project level.

    c. in [general][declarations].

    d. in the Object Browser.

3. What keyword causes a custom class event to fire?

    a. GenerateEvent

    b. RaiseEvent

    c. FireEvent

    d. LoadEvent

4. Polymorphism in VB is best defined as:

    a. deriving one class from another through inheritance.

    b. classes with similar properties but different types.

    c. the default interface of a class.

    d. classes with the same methods, but different functions.

5. Abstract classes in VB:

    a. cannot be created.

    b. have properties, but no methods.

    c. have members, but no code.

    d. have code, but no members.

6. When a class implements an interface, it:

    a. must put code in every member of the interface.

    b. can use all of the code in the interface parent.

    c. becomes an abstract class.

    d. no longer has a default interface.

7. Interfaces have which of the following advantages?

    a. They allow early binding.

    b. They allow reuse.

    c. They support guaranteed callback methods.

    d. They help enforce standard coding practices.

8. An object model:

    a. shows object names.

    b. shows object relationships.

    c. shows object members.

    d. shows object size.

9. The technology that supports communication between objects is called:

    a. Object Linking and Embedding.

    b. Component Object Model.

    c. Distributed Component Object Model.

    d. ActiveX.

10. Which project type runs in-process?
   a. Standard EXE
   b. ActiveX EXE
   c. ActiveX Document EXE
   d. ActiveX DLL

Answers to Mastery Questions

1. a, c, d

2. c

3. b

4. d

5. c

6. a

7. a, b, c, d

8. a, b, c

9. a, b, c, d

10. d

CHAPTER 11

Creating
ActiveX Controls

What You Need to Know

You need to have complete familiarity with Visual Basic Class Modules including constructing properties, events, and methods.

What You Will Learn

- UserControl Objects
- Extender Object
- Ambient Object
- Custom Properties, Events, and Methods
- Invisible, Container, and Data Bound Controls
- Property Pages

Ever since Visual Basic was first released, it has contained reusable elements. These elements have undergone many changes over the years, but the philosophy of reuse has been a cornerstone of VB development since day 1. Throughout this book, you have reused elements that exist inside VB's toolbox.

Over time these elements have gone by many names, including such generic descriptions as "objects" or "controls."

Stepping back a couple of versions, you may have been familiar with the toolbox elements in Visual Basic 3.0 that were known as Visual Basic Extensions or VBXs for short. Under Visual Basic 3.0, VBXs were proprietary reusable components that could only be used in a VB application. When Microsoft developed Visual Basic 4.0, VBXs were changed radically. The value of reusable components in Windows had been proven by the VBX, and Microsoft wanted to extend the reuse to all of its visual development tools. Hence the VBX standard was scrapped in favor of an open standard known as OLE Custom Controls or OCXs. OCXs are unique because although VBXs can only be used in Visual Basic, OCXs can be used by all visual tools and any third-party tools that support the standard, such as Borland's Delphi. OCXs ushered in a new era of reuse in Windows development, but more changes were close behind.

Not long after Visual Basic 4.0 was introduced, the drumbeat of the Internet was starting to be heard at Microsoft. Conventional wisdom had Microsoft dead at the hands of the Net like another archaic Big Blue. In response, Microsoft looked around at their current suite of tools and tried to determine which ones could easily map to Internet development. One of the first candidates for conversion was the OCX. Microsoft reasoned that what the OCX did for cross-language development, it could now do for Internet development. The OCX standard was modified slightly to allow controls to be hosted by a web browser, and the ActiveX control was born.

Visual Basic now has the ability to create these ActiveX controls, which is a significant advance for VB developers. When you create controls, not only can you subsequently use them in Visual Basic as you can any other visual tool such as FoxPro or C++, you can also use them on the Internet. This section teaches you the fundamentals of control development in VB.

What You Need to Know
You must be familiar with all of the object-oriented principles of Visual Basic as they are used in Class Modules. This includes properties, events, methods, encapsulation, and polymorphism.

What You Will Learn
You will learn to create ActiveX controls and package them for distribution.

Control Creation Fundamentals

Creating an ActiveX Control begins by selecting the ActiveX Control project template in Visual Basic. This template contains the structure required to create a control. When the project is first started, a single gray area appears. This area resembles a Visual Basic form, but it has no border, no title bar, and no control buttons. In fact, this area is not a form at all, but rather one of the new family of visual elements called a Designer.

Designers are used in Visual Basic as visual elements that help programmers create objects. The ActiveX Control Designer is embodied in a special object called the UserControl object. The UserControl object contains all of the plumbing necessary to create a true ActiveX Control. In fact, if you compiled your control right now, you would have a real ActiveX control that you could distribute-it would just be invisible and have no functionality!

When you create an ActiveX Control using VB, you may choose from three different approaches. Your first choice is to draw your own control. The UserControl object contains custom drawing methods that will allow you create circles, lines, and points. In this way you can create a custom look for your control.

Figure 11-1
The ActiveX Control Designer.

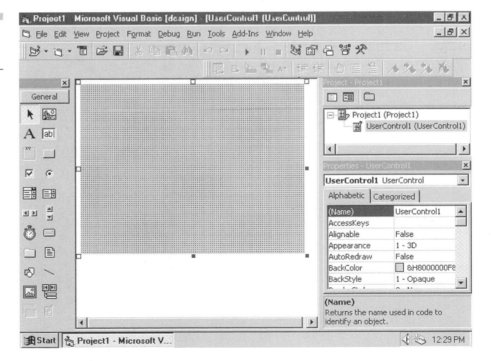

If you do not draw your own, you may choose to use existing Visual Basic controls to create a new control. For example, you may use a TextBox and a Vertical Slider to create a spin button or a TextBox and ListBox for a new ComboBox. Any existing control may be used as a component in a new control. Existing controls that are incorporated into new controls are called Constituent Controls.

Constituent Control: *any control that is placed on a UserControl object for the purpose of creating a new ActiveX Control in Visual Basic.*

Finally, you may simply choose to extend the functionality of a single control. For example, you may want to take a TextBox, retain all of its existing features, but add a new Mask property for restricting data input and create a Masked Edit Control. Extending one control in your ActiveX Control project is called subclassing a control.

Subclassing: *creating a new object that has all of the original properties and methods of an existing object while adding some custom features.*

Regardless of which way you choose to create your new control, the process of coding the project remains largely unchanged. The only issue becomes properly redrawing a control when it is resized by the user. Resizing is a major issue in control design that you should tackle as a first order of business for any new control.

Whenever a user of your control resizes it, the UserControl object receives a UserControl_Resize() event. Constituent controls do not automatically resize when the UserControl object resizes. You must code that functionality yourself in the Resize event. If you are subclassing a single control, this can be done by simply forcing the value of the constituent control's Height and Width properties to match the new Height and Width of the UserControl object. However, if you are using multiple constituent controls, you will have to carefully plan how to resize them for the correct effect.

Check It Out: Control Resizing

1. Start a new ActiveX Control project in Visual Basic 5.0.

2. Drop a TextBox on the UserControl object, and place it in the upper left corner.

3. Resize the UserControl object so that it is completely covered by the TextBox.

4. Close all open windows in the project.

5. Add a new project to the group by selecting File/Add Project... from the menu. In the New Project dialog, select to add a new Standard EXE project.

6. Look in your toolbox for the new project and identify the icon for your new control. Put your mouse over the control and the ToolTip should read UserControl11. This is your new control in action!

7. Add the control you created to Form1 in the Standard EXE project by double clicking on the control in the toolbox.

8. Try resizing the control on Form1. Does the TextBox resize or just the UserControl designer?

9. Now add the following code to the UserControl_Resize event:

```
Text1.Height=UserControl.Height
Text1.Width=UserControl.Width
```

10. Close all of the open windows and then reopen Form1. Now resize the control on Form1.

Adding Properties

ActiveX Control properties are created in your project much as properties are created in VB Class Modules. To create new properties, you must use property procedures. These properties can then show up in the Properties window of the Visual Basic environment. Properties for your controls come from several sources. Some properties are provided to a control by its container, others come from Constituent controls, and still others may be custom properties you create.

Extender Properties

Not all properties of an ActiveX Control come from the control itself. Some properties are actually given to the control by the container that hosts it. This is true of the ActiveX Controls you create in Visual Basic. Containers often provide properties that are meaningful to the container, but not necessarily the control. Consider the TabIndex property, which specifies the order in the tab sequence for a control. The control itself is not particularly concerned with its tab order, but the hosting Form is definitely interested. The Form must respond to events that are related to the tab order, such as GotFocus and LostFocus. Therefore, the Form extends the TabIndex property to a control to track these events.

In Visual Basic 5.0, you have access to all of the properties extended to your ActiveX control through a special object called the Extender object. The Extender object can be used in your Visual Basic ActiveX control project to affect any of the extended properties. In most cases, you will never address the Extender directly since the Extender properties are generally not a concern to you.

Check It Out: Extender Properties

1. Start a new ActiveX Control project in Visual Basic 5.0.

2. Drop a CommandButton on the UserControl object, and place it in the upper left corner.

3. Resize the UserControl object so that it is completely covered by the CommandButton.

4. In the Resize event of the UserControl object, add the following code:

```
Extender.Left=0
Extender.Top=0
```

5. Close all open windows in the project.

6. Add a new project to the group by selecting File/Add Project... from the menu. In the New Project dialog, select to add a new Standard EXE project.

7. Look in your toolbox for the new project and identify the icon for your new control. Put your mouse over the control and the ToolTip should read UserControl11.

8. Add the control you created to Form1 in the Standard EXE project by double clicking on the control in the toolbox.

9. Try resizing the control on Form1. Does the control resize?

Constituent Control Properties

When you first create a control with Visual Basic, you may think that your control will contain all of the properties, events, and methods of the Constituent controls that are contained on the UserControl object. This, however, is not true. All controls used as part of the UserControl are considered private to the control project. Therefore, none of them appear as properties of your control unless you explicitly write code to expose them. This is done by wrapping the property of the Constituent Control in a Property Procedure. For example, if you want to expose the Text property of a Constituent Control Text1, you would use the following code:

```
Public Property Get Text()As String
  Text=Text1.Text
End Property

Public Property Let Text(strText As String)
  Text1.Text=strText
End Property
```

The process of creating a set of Property Procedures for each Constituent Control property can be quite laborious. Fortunately, Visual Basic provides a wizard to ease the work. This wizard exists as an Add-In that you can find under the Add-In Manager dialog in Visual Basic. The Add-In is called the ActiveX Control Interface Wizard.

Figure 11-2

The ActiveX Control Interface Wizard.

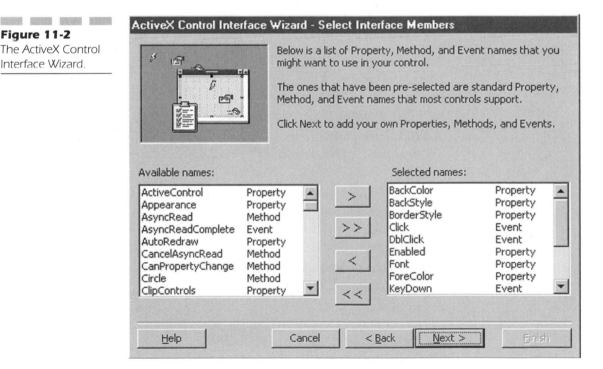

Custom Properties

Custom properties are as simple to create in your ActiveX control project as they are in any Visual Basic Class Module. Custom properties are created by simply adding a pair of Property Procedures to your control and providing a Private variable to retain the property state. Every time you add a Property Procedure to your control project, Visual Basic automatically adds the property to the Properties window.

▪ Check It Out: Adding Custom Properties

1. Start a new ActiveX Control project in Visual Basic 5.0.

2. Drop a TextBox on the UserControl object, and place it in the upper left corner.

3. Resize the UserControl object so that it is completely covered by the CommandButton.

4. Add a Property Procedure called MyProperty. Create the property procedures with the following code:

```
Private m_MyProperty As String
Public Property Let MyProperty(strMyProperty As String)
M_MyProperty = strMyProperty
End Property

Public Property Get MyProperty() As String
MyProperty = m_MyProperty
End Property
```

5. Close all of the project windows

6. Add a new project to the group by selecting File/Add Project...from the menu. In the New Project dialog, select to add a new Standard EXE project.

7. Look in your toolbox for the new project and identify the icon for your new control. Put your mouse over the control and the ToolTip should read UserControl11.

8. Add the control you created to Form1 in the Standard EXE project by double clicking on the control in the toolbox.

9. Examine the Properties window for your new property. Try changing its value.

Read-Only and Write-Only Properties

Often ActiveX controls support a number of special types of properties. For example, some properties are available only for reading or writing, but not both. You may easily create these types of properties in your ActiveX controls as well. As is the case for many features in ActiveX controls, you can create read-only or write-only properties exactly as you would for Class Modules.

As an example, suppose we wanted to create a read-only property for a custom TextBox called Length. The Length property will return the length of the text in our control as a read-only property. Users of our control can call this property at run time, but they cannot write to it. Creating the property requires a pair of Property procedures, but to make the property read-only, we must alter the scoping qualifier of Property Let from Public to Private. The following code shows how to create the Length property:

```
Private Property Let Length(intLength As Integer)
End Property

Public Property Get() As Integer
End Property
```

Runtime Only and Design Time Only Properties

Creating properties that are available only at design time or run time is accomplished through the use of a special object called the Ambient object. The Ambient object is responsible for providing information to your ActiveX control regarding the state of the form that is hosting your control. In the case of properties, you can use the UserMode property of the Ambient object to determine if your control is being hosted at design time or run time. If the application hosting you is in Visual Basic, then Ambient.UserMode is False. If the hosting application is currently executing, then Ambient.UserMode is True.

In order to enforce design time or runtime behavior, you must check the value of UserMode and then either raise an error or execute the requested functionality. Suppose you wanted to create a design time only property called Style that accepts an integer to change the look of your control. Creating the property would require two Property procedures, as we have discussed. However, creating Property procedure pairs will make the same property available at run time as well. Therefore, we must check Ambient.UserMode in Property Let as follows:

```
Private m_Style As Integer

Public Property Let Style (intStyle As Integer)
  If Ambient.UserMode = False Then
    m_Style = intStyle
  Else
    MsgBox "Not available at Run-Time"
  End If
End Property
```

You may use the same strategy to create runtime only properties as well. If you raise an error at run time, then the control container is responsible for trapping and handling the error just like any VB control. If you raise an error at design time, just use a message box as a display. No other special code is required.

Procedure Attributes

Along with special property behaviors such as read-only or write-only, many properties also exhibit special characteristics beyond those already described. These special behaviors are not easily categorized and can exist at both design time and run time. As examples, consider default property behavior and the Caption property behavior.

All VB programmers should be familiar with the behavior of default properties. Default properties are properties that receive values from code implicitly even when not called explicitly. The Text property of a TextBox is a good example. You can easily access the Text property of any TextBox implicitly with the following code:

```
Text1 = "New Technology Solutions"
```

If you want this type of behavior in your controls, you can implement it with Procedure Attributes. Procedure Attributes is a catch-all dialog that allows you to designate certain properties in your control as having special behaviors. To designate a property as the default for your control, select Tools/Procedure Attributes... from the menu. In the Procedure Attributes dialog, select the property to make as the default and press the Advanced button. Under the advanced setting, locate the ProcedureID. Then simply select default as the ProcedureID.

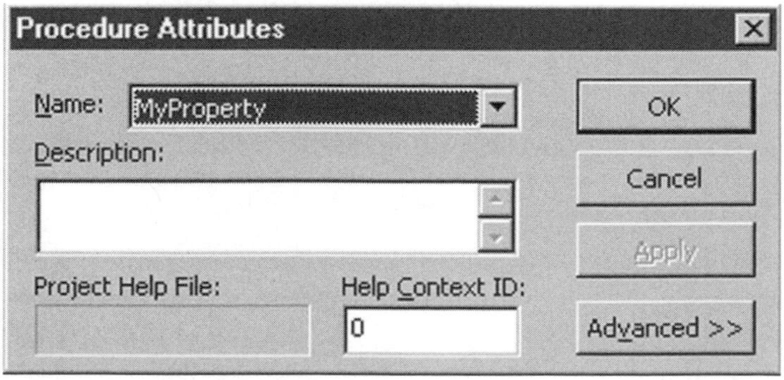

Figure 11-3 Procedure Attributes dialog.

The Caption property has special behavior that can also be implemented with the Procedure Attributes dialog. At design time, the Caption property automatically sends characters one at a time to the target control as you type. If you have not noticed this behavior before, try dropping a Label control and changing the Caption property. Notice how each letter enters the Caption independently. The controls we make in Visual Basic do not have this behavior by default. To simulate the Caption property, you must set the appropriate procedure attribute to Caption in the Procedure Attributes dialog.

Saving Property Values with the Property Bag

As your controls are used in the Visual Basic design environment, the values of their properties will be changed from the defaults that you specify. This, of course, is a natural part of the design process. Like other controls in Visual Basic, you will want to save these changes as part of the project so that they can be remembered when the project is saved and later opened. Saving properties for your controls is the function of a special object called the Property Bag. Using the Property bag correctly depends on a complete understanding of the life cycle of a control.

When your control is used in the Visual Basic environment, it receives several events throughout its life. The most important of these events are InitProperties, WriteProperties, and ReadProperties. These events are fired in the UserControl object, and they allow you to set default property values, save properties, and retrieve properties respectfully.

InitProperties

When an ActiveX Control is first dropped on a VB form from the ToolBox, your control is notified through the InitProperties event of the UserControl object. This event allows you to set the default property values for the control. Setting these values is a simple matter of coding them in the event:

```
Public Sub UserControl_InitProperties
  Border = 1
  Size = 2
End Sub
```

Any properties you set in InitProperties reflect the defaults that will always be seen when the control is first dropped. Some properties, however, have special behavior when dropped on a form. Certain properties such as those associated with Fonts and Colors actually accept the values currently in use by the form. The properties are called Ambient properties.

Check It Out: Ambient Properties

1. Start a new Standard EXE project in Visual Basic.

2. Change the Font property of Form1 to Times New Roman size 16.

3. Change the ForeColor of Form1 to Green.

4. Drop a Label Control on Form1.

5. What happened to the Font and ForeColor properties of Label1?

6. Change the Font of Form1 to Arial 10.

7. Change the ForeColor of Form1 to Red.

8. Drop another Label control on Form1.

9. What happened?

Ambient properties are controlled in your ActiveX Control project through the Ambient object. We used the UserMode property of the Ambient object earlier to help create runtime only and design time only properties. We can also use it to synchronize our controls with the hosting container. The Ambient object allows you to read the Font, ForeColor, Back-Color, and DisplayName of the control on the container. To synchronize your control with the container, you could use the following code:

```
UserControl_InitProperties
  On Error Resume Next
  Set Font = Ambient.Font
  ForeColor = Ambient.ForeColor
  BackColor = Ambient.BackColor
  Text = Ambient.DisplayName
End Sub
```

In this code note the use of error handling. You should always use error handling before addressing the Ambient object because the properties of the object may not be available for every container you use. For example, when using this code for a downloaded component in Internet Explorer, you can expect to receive runtime errors. For the most part, we can discard any errors that occur here, so the code just resumes on the next line.

Write Properties

Whenever a user saves the current project, your control is notified through the WriteProperties event. This event is your signal that it is time to save the properties of your control to persistent file storage. Visual Basic saves the properties of controls as plain ASCII text inside the FRM file that represents the hosting form. It will also save your properties to the FRM file, if you tell Visual Basic when your control properties are dirty and write the correct code to save them.

Telling Visual Basic that your control has dirty properties that need saving is done in the Property Let procedure for any property you need to save. Inside a Property Let for your control, you tell Visual Basic that the property has been changed by calling the PropertyChanged method of the UserControl and Providing the name of the changing property. The following code notifies the Visual Basic design environment that the Text property of a custom control has changed:

```
Public Property Let Text(strText As String)
  Text1.Text = strText
  PropertyChanged "Text"
End Property
```

When you call PropertyChanged, Visual Basic will make sure that the user of your control is prompted to save any changes before they leave the current project. This generates the familiar "Save Changes?" dialog box. When the user saves changes, you receive a call to WriteProperties, where you use the PropertyBag object to write to the FRM file of the hosting form. The WriteProperty method of the PropertyBag takes as arguments the name of the property to save, the value of the property, and the default value:

```
Public Sub UserControl_WriteProperties
  PropBag.Write Property "MyProperty",MyProperty,"MyDefault"
End Property
```

Visual Basic writes the property to the FRM file only if the current value is different from the default. In this way, VB can save some file space by not writing properties that will already be set correctly next time in the InitProperties event. When writing to the PropBag, you should always use error handling because FRM files are saved as ASCII text and can be edited. Therefore, you cannot guarantee the integrity of the FRM file when you read and write to it.

ReadProperties

The ReadProperties event is fired when the control is created because a project was opened in Visual Basic. In this case, your control is created in the opening project and you are prompted to read your properties out of the Property Bag. This is a simple matter of using the ReadProperty method of the Property Bag and selecting your properties accordingly. The ReadProperty method takes as arguments the name of the property to read and its default value. Once again, you should use error handling when reading from the Property Bag:

```
Public Sub UserControl_ReadProperties
  MyProperty = PropBag.ReadProperty("MyProperty","MyDefault")
End Sub
```

Adding Events

Just like other ActiveX controls you have used in Visual Basic, the controls you create also support event generation. Once defined inside your control, the events become available to any hosting application and will even appear automatically in the Visual Basic code window at design time. Furthermore, event definition in ActiveX controls follows the same rules as event definition in any VB Class Module.

Adding a new event definition is done in the [general][declarations] section of the UserControl object with the Event keyword. Simply name your event and any required arguments using the following syntax:

```
Event EventName(Arg1, Arg2, Arg3,…)
```

Once the event is defined, you can fire it from your control using the RaiseEvent keyword. This command can be used from any sub or function and immediately causes the event routine in the hosting form to execute. The following syntax fires the event:

```
RaiseEvent EventName
```

Check It Out: Defining Events

1. Start a new ActiveX Control project in Visual Basic 5.0.

2. Drop a TextBox on the UserControl object, and place it in the upper left corner.

3. Resize the UserControl object so that it is completely covered by the CommandButton.

4. Add an event declaration to the [general][declarations] section of the UserControl object with the following code:

```
Event TextChanged(strText As Text)
```

5. In the Changed event of Text1 raise your custom event with the following code:

```
RaiseEvent TextChanged
```

6. Add a new project to the group by selecting File/Add Project... from the menu. In the New Project dialog, select to add a new Standard EXE project.

7. Look in your toolbox for the new project and identify the icon for your new control. Put your mouse over the control and the ToolTip should read UserControl11.

8. Add the control you created to Form1 in the Standard EXE project by double clicking on the control in the toolbox.

9. Open the code window for Form1 and select UserControl11 from the Object box. Note that the events declaration for your custom event is visible in the code window.

Adding Methods

Methods in ActiveX control projects are generated by defining Public Subs or Functions in the UserControl object. The Public routines may be subsequently called from any container hosting your control. This technique is exactly the same as defining methods for Class Modules.

Building Special Feature Controls

In addition to the standard control creation features, Visual Basic supports the construction of special features into your controls such as data binding or invisible controls. These features are generally created by changing properties of the UserControl object or by selecting appropriate procedure attributes. These special features represent additions to the control and do not affect the creation of properties, events, and methods as previously discussed.

Data Bound Controls

Perhaps the most complex of the special features to add is data binding. Data bound controls have the ability to connect to a Visual Basic data control to receive database information. Controls can be created to add, edit, update, and delete from an associated database.

Creating a bound control begins just as creating any other control does. You should, however, have in mind at the outset which properties will be data bound. Once you have built your control, you can set data binding features in the Procedure Attributes dialog. Simply select the property to

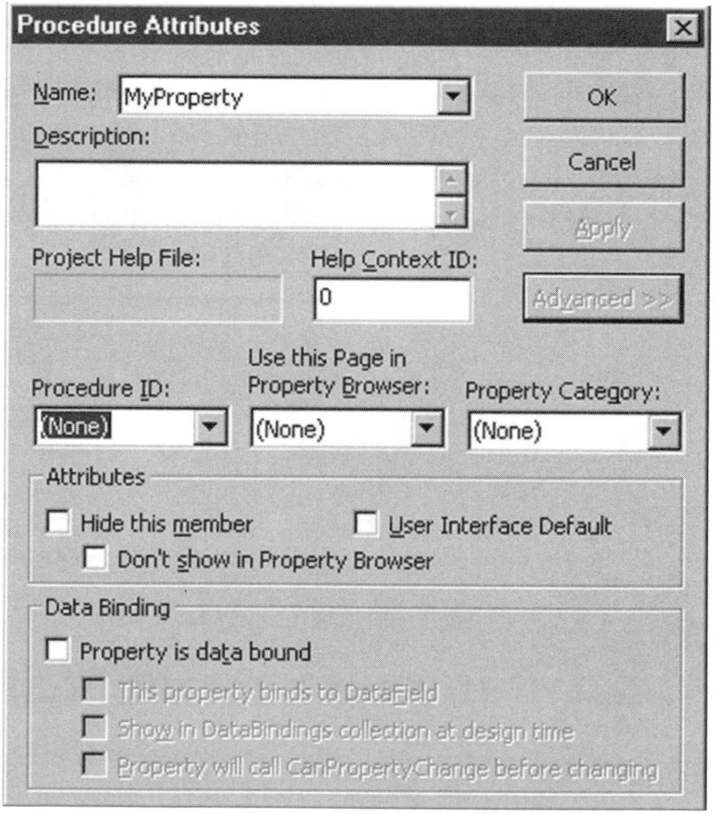

Figure 11-4 Data binding in the Procedure Attributes dialog.

bind from the Name list in the dialog and check the Property is Data Bound option.

Underneath the data binding option is a series of three suboptions that further specify the behavior of the data bound control. The first, titled This Property Binds to DataField is used to specify that this property should appear alone under the DataField property. The DataField property is generally used when a control has just a single bound property.

The second suboption is titled Show in DataBindings Collection At Design Time and determines whether the bound property appears in the Properties window under the DataBindings property. DataBindings normally contains a list of all bindable properties, if more than one exists.

The third suboption is titled Property will call CanPropertyChange before changing and specifies that the control will check with the container to see if the property can be changed. Bound properties may not be able to change if, for example, the database is opened for read-only access. Unfortunately, however, the current version of Visual Basic always returns True when this function is called, so it's of little use.

Container Controls

Visual Basic also supports creating controls that act as containers for other controls. The Frame control in Visual Basic is an example of a container control. When you place a new control in a container, it moves with the container and cannot be removed. Setting up your control as a container is a simple matter of setting the ControlContainer property of the UserControl object to True.

Once a control is designated as a container, you may want to be able to perform functions on the other controls that reside in the container. You may programmatically manipulate any hosted control through the ContainedControls collection that is accessible through the UserControl object. For example, if you wanted all of the controls in your container to resize proportionally when the container resizes, you could use the following code:

```
Private m_HeightScale As Integer
Private m_WidthScale As Integer

Private Sub UserControl_Resize()

  Dim objControl As Control

  For Each objControl In UserControl.ContainedControls
objControl.Left = Int(objControl.Left * (Width / m_WidthScale))
objControl.Top = Int(objControl.Top * (Height / m_HeightScale))
objControl.Width = Int(objControl.Width * (Width / m_WidthScale))
objControl.Height = Int(objControl.Height * (Height /
```

```
m_HeightScale))
  Next

  'Save New Dimensions
  m_HeightScale = Height
  m_WidthScale = Width
End Sub
```

Invisible Controls

Visual Basic also allows you to build controls that are invisible at run time. An example of such a control is the Timer control, which is used to fire code at a given interval. In the case of the timer, no user interface is necessary, so the control can be invisible at run time but provide an interface at design time for setting properties. To make your control invisible at run time, set the InvisibleAtRuntime property of the UserControl object to True.

Although you can create invisible controls in Visual Basic 5.0, you should probably not do it. The reason is that invisible controls are essentially ActiveX DLLs with more overhead. If you need the functionality of an ActiveX DLL, create one instead of an invisible control. You'll be happier with the performance.

Utilizing Property Pages

Although the Properties window offers an excellent way to set most properties, it is not always flexible enough to meet the needs of all controls. In many cases, controls have a special property called Custom that allows access to a special interface for setting properties. This interface is called a Property Page or Property Sheet. This is a feature that we can add to our own controls.

Just like the ActiveX Control project itself, Property Pages are created through a special object. The PropertyPage object can be easily added to your ActiveX Control project by selecting Project/Add Property Page... from the menu. When a PropertyPage object is added, it appears as a blank gray designer area similar to the UserControl object.

Each PropertyPage object that you include in your ActiveX Control project represents one tab in a tabbed dialog. The collection of these tabs makes the custom property dialog. Your job is to design the property pages so that common control features are grouped together. Typical tabs might include General, Fonts, and Colors.

Figure 11-5
A typical
PropertyPage.

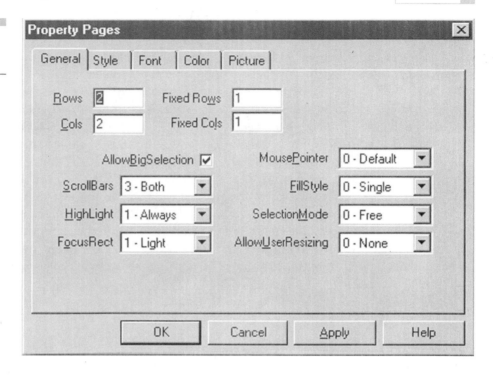

Figure 11-6
The PropertyPage
object.

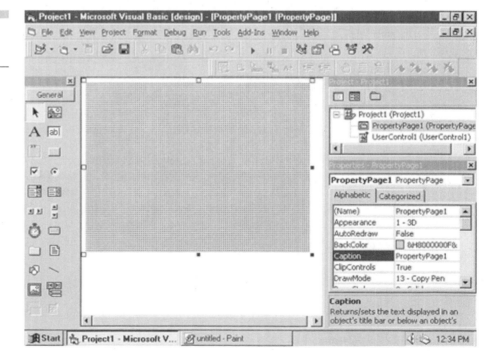

Showing Properties on the Page

Property Pages have key events that are used to manage control settings. In Visual Basic, if a PropertyPage object is associated with an ActiveX Control, then a Custom property will appear in the VB Properties window. This Custom property has an ellipsis associated with it. When the button is pushed, your PropertyPage object is notified through the SelectionChanged event. If fact, your page receives this event anytime, the group of selected controls are changed.

When the SelectionChanged event fires, you should populate your page with the current values of the control properties. Accessing the group of controls is done with the SelectedControls collection. By convention the first selected control is used to populate the page. If, for example, you had several Masked Edit controls selected, the following code would populate an option button called optMask using the Mask Property:

```
Private Sub PropertyPage_SelectionChanged()
  optMask(SelectedControls(0).Mask) = True

End Sub
```

Notice that the page is populated using SelectedControls(0). This is the typical technique and results in the expected control behavior in the Visual Basic design environment. When using the SelectedControls collection, you may always assume that it is populated solely with the class of controls handled by your page. Visual Basic will automatically disable property pages if controls of different classes are selected in the environment.

Applying Property Changes

When a PropertyPage object is open, users interact with the controls you have placed on it to set property values. When a property is changed, your code notifies the Visual Basic environment by calling the Changed method of the PropertyPage object. Calling this method causes the Apply button of the page to activate. Typically, you would call the Change method in the Changed or Clicked event of the GUI elements in your page.

Once the Apply button is active, users can click on the button to set all of the new property values into the group of selected controls. When the Apply button is pushed, your page is notified through the ApplyChanges event of the PropertyPage object. In this event, you apply all of the changes to each control in the SelectedControls collection. For example, the following code applies the changes from a page with an option button called optMask to all MaskEdit controls in the SelectedControls collection:

```
Private Sub PropertyPage_ApplyChanges()

  Dim objControl As MaskEdit

  For Each objControl In SelectedControls

    'Mask Property
    If optMask(0).Value Then objControl.Mask = 0
    If optMask(1).Value Then objControl.Mask = 1
    If optMask(2).Value Then objControl.Mask = 2

  Next

End Sub
```

Attaching Pages to Controls

Once the PropertyPage objects are constructed, they may be attached to the ActiveX Control. Attaching pages is done with the PropertyPages property of the UserControl object. This property opens a dialog you can use to attach pages to the Custom property. Attaching any PropertyPage object is done by simply checking the associated checkbox.

Figure 11-7
PropertyPages
dialog.

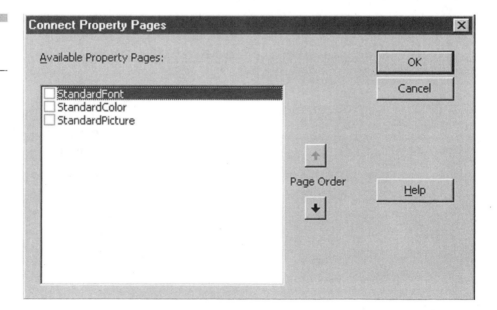

Pages can also be attached to individual properties. This is typically done when a property is complicated enough to warrant its own interface. The Procedure Attributes dialog allows you to specify a page for any property you create. Additionally, you can also get standard pages for properties such as pictures, colors, or fonts. Visual Basic sets these for you automatically whenever you create a property with a return value of StdFont, OLE_COLOR, or StdPicture.

EXERCISE 1: A COMPLETE ACTIVEX CONTROL

This example builds a Picture Button control from two image controls. In this exercise, you will learn to manage the initializing, saving, and retrieving of properties for your ActiveX control. When finished, you will have a fully functional control.

Requirements

1. Visual Basic 5.0

2. LIGHTON.ICO, LIGHTOFF.ICO, PBUTTON.BMP

Concepts Learned

1. Initialize, Read, and Write Properties

2. About Boxes

Control Fundamentals

Step 1

Create a new directory for this project called \VBCLASS\CONTROLS\PBUTTON. Start Visual Basic 5.0 and begin a new ActiveX Control project. Name the new UserControl object Pbutton, and place two image controls on the User Control object. Change the properties as follows:

```
Control     Name          Index
Image1      imgPButton    0
Image2      imgPButton    1
```

Step 2

Add a picture to each image control as shown next. These images should be available in your Visual Basic directories if you have installed the graphic files.

```
Control        Picture
imgPButton(0)  lightoff.ico
imgPButton(1)  lighton.ico
```

Step 3

Move the images to the upper left corner of the User Control and resize the User Control around the images. The image controls should be placed directly on top of each other. The images represent the two states of our Picture Button control.

Step 4

Set the properties for the images to establish the initial appearance of the button as follows:

```
Name           Visible    BorderStyle
imgPButton(0)  True       0-None
imgPButton(1)  False      1-Fixed Single
```

Step 5

Close all the open code windows and designers, and save the ActiveX Control project as PBUTTON.VBP.

Step 6

Add a new Standard EXE to Visual Basic using the File/Add Project menu. Save the Standard EXE project as TEST.VBP and the project group as PBUTTON.VBG.

Using the toolbox, place a new instance of the PButton Control on Form1 in the TEST.VBP project. You should see the control with the off light bulb image visible. Now switch to the ActiveX Control project.

Our button will not be sizable-it will always look like a toolbar button. In the code window for the User Control, select the Resize event. Add the following code to the event to prevent the control from resizing:

```
Private Sub UserControl_Resize()
Height = imgPButton(0).Height
Width = imgPButton(0).Width
End Sub
```

Control Properties

Step 7

For this control, we will create a MousePointer property and use this property to completely investigate the process of initializing, saving, and restoring property values in the design environment. Add the following code to expose the MousePointer property of the Pbutton Control:

```
Public Property Get MousePointer() As Integer
  MousePointer = imgPButton(0).MousePointer
End Property

Public Property Let MousePointer(ByVal intNewMouse As Integer)
  imgPButton(0).MousePointer = intNewMouse
  imgPButton(1).MousePointer = intNewMouse
End Property
```

The Property Procedures allow you to read and write the property, but what determines the initial value of the property when the control is first placed on a form? Visual Basic provides several events that are important to the life cycle of a control. These events control the initialization, saving, and restoring of properties when projects are opened and closed inside Visual Basic. These are the key events:

```
UserControl_InitProperties()
```

called when a control is first dropped on a form.

```
UserControl_WriteProperties()
```

called when the user saves the VB project.

```
UserControl_ReadProperties()
```

called when an existing VB project is opened.

InitProperties, WriteProperties, and ReadProperties allow you to control the initialization, saving, and restoring of properties to your ActiveX Control.

Step 8

When a user of your ActiveX Control first places an instance on a form in a new VB project, UserControl_InitProperties() is called. In this event, you can set the initial value for your property. Add the following code to UserControl_InitProperties() to set the initial value for the MousePointer property:

```
Private Sub UserControl_InitProperties()
  MousePointer = vbDefault
End Sub
```

Once the ActiveX Control is instantiated, Visual Basic must know if any of the control properties are changed. Calling the PropertyChanged method of the UserControl does this. The PropertyChanged method tells the VB environment that the form is dirty. This is vital because VB will not save the changed values of your control unless you tell it. To do this add the following line of code to the end of the Property Let MousePointer procedure:

```
PropertyChanged "MousePointer"
```

Step 9

Whenever a user saves changes to a project, Visual Basic must save any properties for the control. Control properties are saved as part of the FRM file that defines the form hosting the control. To save space in the FRM file, VB only saves properties for a control if the saved value is different from the default value. You can tell VB to save properties by using the WriteProperties event that fires whenever the user of your control saves changes to a project. In order to write the properties to the FRM file, you use the WriteProperty method of the PropBag object. Add the following code to the WriteProperties event to save the MousePointer property:

```
Private Sub UserControl_WriteProperties(PropBag As PropertyBag)
  PropBag.WriteProperty "MousePointer", MousePointer, vbDefault
End Sub
```

The syntax for writing properties with the PropBag object is this:

```
PropBag.WriteProperty name, value, default
```

Step 10

When a project that uses your control is opened for editing, the properties for the control must be read from the FRM file and set into the control. Visual Basic does this by using the ReadProperties event and the PropBag object. Add the following code to the ReadProperties event to restore the properties to the ActiveX control:

```
Private Sub UserControl_ReadProperties(PropBag As PropertyBag)
  MousePointer = PropBag.ReadProperty("MousePointer", vbDefault)
End Sub
```

The syntax for reading properties with the PropBag object is this:

```
property=PropBag.ReadProperty name, default
```

Control Events

Step 11

For this control, we want to add a Click event that will fire when the button is pushed. We must also track the state of the button so that the image can be changed accordingly. Add the following code to the [general][declarations] section to declare the Click event:

```
Event Click
```

Fire the Click event and change the images by adding the following code to the Click event of imgPButton:

```
Private Sub imgPButton_Click(Index As Integer)
  'Swap imagesIf Index = 0 Then
    imgPButton(0).Visible = False
    imgPButton(1).Visible = True
  Else
    imgPButton(0).Visible = True
    imgPButton(1).Visible = False
  End If

  RaiseEvent Click
End Sub
```

Step 12

Save the project group. Run the project group and test the button. You should have a functioning two-state button.

The Toolbox Icon

Step 13

Stop the project and switch to the ActiveX Control project. Add a toolbox icon by selecting the UserControl properties and setting the Tool-BoxBitmap property to PBUTTON.BMP, which is found on the companion CDROM. You can use any custom bitmap for the toolbox as long as it is 15 pixels wide and 16 pixels high.

The About Box

Step 14

Creating an About box for your control is a simple matter of creating a form for the box. Insert a new About dialog into your ActiveX Control project by selecting Project/Add Form from the menu. When prompted, select to add a new About dialog.

To display the About box, you will need to add a custom method to the ActiveX Control project to display the form. Using the Tools/Add Procedure dialog, add a new Sub called ShowAbout and place code in it to create the following:

```
Public Sub ShowAbout()
  frmAbout.Show vbModal
End Sub
```

This procedure must be called by VB whenever a user selects the About property from the Properties window. This connection is made through the Procedure Attributes dialog. Select Tools/Procedure Attributes from the menu, and then select ShowAbout from the procedure list.

Click on the Advanced button and select AboutBox in the ProcedureID box. Check "Hide this member" so that the subroutine cannot be called externally. This is important because the About box is not functionality we want called from VB code. Now close the dialog box.

Step 15

Save the project group and switch to TEST.VBP. Test the About box by clicking on the About property in the properties window.

Extra Credit

1. Finish the About box by adding any missing information.
2. Create a new read-only property for your control called State with enumerated values Up and Down, and display the value each time the button is clicked in TEST.VBP.

EXERCISE 2: CREATING PROPERTY PAGES

When creating an ActiveX Control, you can choose to allow the user to manipulate the control properties through a custom interface known as a Property Page. These pages can be found on many commercially available controls under the Custom property. In this exercise, you will create a rotating Moon control with custom property pages.

Requirements

1. Visual Basic 5.0
2. Files MOON01 through MOON08

Concepts Learned

1. Property Pages

Control Fundamentals

Step 1

Create a new directory for this project called \VBCLASS\CON-TROLS\MOON. Start Visual Basic 5.0 and begin a new ActiveX Control project. Name the new UserControl object Moon.

Step 2

Place eight image controls on the User Control object. Change the properties as follows:

Control	Name	Index
Image1	imgMoon	0
Image2	imgMoon	1
Image3	imgMoon	2
Image4	imgMoon	3
Image5	imgMoon	4
Image6	imgMoon	5
Image7	imgMoon	6
Image8	imgMoon	7

Step 3

Add a picture to each control using the eight moon images that ship with Visual Basic or the moon files for this exercise from the CDROM. Change the Visible property of each control to False.

Step 4

Place a Timer Control on the form and name it tmrMoon. Set the Interval property to 500.

Step 5

Move the images to the upper left corner of the User Control and resize the User Control around the images. The image controls should be placed directly on top of each other with the Top and Left properties set to zero. We do not want the user to be able to resize this control, so add the following code to the UserControl_Resize() event to prevent resizing:

```
Private Sub UserControl_Resize()
Height = imgMoon(0).Height
Width = imgMoon(0).Width
End Sub
```

Step 6

When the control runs, we want it to cycle the moon images. Add the following code to the Timer event of tmrMoon to cycle the images:

```
Private Sub tmrMoon_Timer()
Static intPhases As Integer

imgMoon(intPhases).Visible = False

intPhases = intPhases + 1
If intPhases > 7 Then intPhases = 0

imgMoon(intPhases).Visible = True

End Sub
```

Step 7

Save the project.

Add a Standard EXE to the project group by selecting File/Add Project from the menu. In the Standard EXE, place a moon control on Form1. Immediately, the moon should by cycling through its phases. Save the Standard EXE as TEST.VBP and the project group as MOON.VBG using the File/Save Project Group menu.

Control Properties

Throughout this section, the control behavior may be erratic until all of the code is in the project. If necessary, try dropping new copies of your control into the test project whenever you make a significant change to the property code.

Step 8

In this exercise, you will implement several properties for the moon control. We will start by creating them normally, and then add the custom property pages. To control the moon, we will create Enabled, Speed, and Direction properties.

Switch to the ActiveX Control project. Create the Enabled property by using the Tools/Add Procedure dialog and add a Property procedure called Enabled. The Enabled property is a special property implemented by the Extender object; however, the Extender will not provide an Enabled property unless you specifically create one. Very strange! Add the following code to create the Enabled property:

```
Public Property Get Enabled() As Boolean
  Enabled = UserControl.Enabled
End Property

Public Property Let Enabled(ByVal blnEnabled As Boolean)
  'Implement Enabled
  UserControl.Enabled = blnEnabled
```

```
'Affect animation
tmrMoon.Enabled = blnEnabled
PropertyChanged "Enabled"

End Property
```

Although provided by the Extender object, the Enabled property is not implemented unless you create a property procedure by that name and set the Enabled property of the UserControl object in code.

Step 9

In order to make the Enabled property work correctly, you must assign it to the Enabled ID using the Tools/Procedure Attributes dialog. In the dialog, select the procedure and push the Advanced button to locate the Procedure ID box. In the Procedure ID box, select Enabled. Save the project and try changing the Enabled property of the control in TEST.VBP.

Step 10

The Speed property determines the time interval for the animation sequence. This value is sent to the Interval property of the Timer control. Add a Property Procedure called Speed and the following code to create the property:

```
Public Property Get Speed() As Integer
Speed = tmrMoon.Interval
End Property

Public Property Let Speed(ByVal intSpeed As Integer)
If intSpeed < 0 Then intSpeed = 0

tmrMoon.Interval = intSpeed
PropertyChanged "Speed"

End Property
```

Save the project and switch to TEST.VBP. Change the Speed property and watch the results.

Step 11

The Direction property determines the rotation direction for the moon. This is an enumerated property with values ClockWise and Counter-ClockWise. To use the enumerated values, add the following code to the [general][declarations] section of the UserControl object:

```
Public Enum enmDirection
ClockWise
CounterClockWise
```

```
End Enum
Private m_Direction As enmDirection
```

Create the Direction property with a Property Procedure and the following code:

```
Public Property Get Direction() As enmDirection
  Direction = m_Direction
End Property
Public Property Let Direction(ByVal enmNewDir As enmDirection)
  m_Direction = enmNewDir
  PropertyChanged "Direction"
End Property
```

To use the Direction property, change the code in the Timer event of tmrMoon so that it appears as follows:

```
Private Sub tmrMoon_Timer()
  Static intPhases As Integer

  imgMoon(intPhases).Visible = False

  If m_Direction = CounterClockWise Then
    intPhases = intPhases + 1
    If intPhases > 7 Then intPhases = 0
  Else
    intPhases = intPhases - 1
    If intPhases < 0 Then intPhases = 7
  End If

  imgMoon(intPhases).Visible = True

End Sub
```

Save the project and try out the new properties.

Step 12

In order to properly manage the properties across development sessions, you must add code to the InitProperties, WriteProperties, and ReadProperties events. The following code will allow the properties to be saved and restored:

```
Private Sub UserControl_InitProperties()
imgMoon(0).Visible = True
Enabled = True
Direction = ClockWise
Speed = 500
End Sub
Private Sub UserControl_ReadProperties(PropBag As PropertyBag)
Enabled = PropBag.ReadProperty("Enabled", True)
```

```
Direction = PropBag.ReadProperty("Direction", ClockWise)
Speed = PropBag.ReadProperty("Speed", 500)
End Sub

Private Sub UserControl_WriteProperties(PropBag As PropertyBag)
PropBag.WriteProperty "Enabled",Enabled, True
PropBag.WriteProperty "Direction", Direction, ClockWise
PropBag.WriteProperty "Speed", Speed, 500
End Sub
```

Save the project and switch to the TEST.VBP project. Delete the moon control on Form1 and drop a new one. Try changing the Enabled, Speed, and Direction properties.

Adding Property Pages

Step 13

Now that the Moon control behavior is well defined. We can create a custom property page for graphically controlling the property values at design time. Insert a Property Page into your ActiveX Control project by activating the project and selecting Project/Add Property Page. Be sure to add a new blank page and do not use any of the Property Page wizards.

Add the following controls to the property page:

1 Checkbox

2 Option Buttons

2 Labels

1 Horizontal Scroll Bar

Step 14

Change the Properties of the Controls as follows:

PROPERTY PAGE

Caption	Moon
Name	ppgMoon

SCROLLBAR

Name	scrSpeed
Height	285
Left	225
Max	1000

Top	2970
Value	500
Width	4245

OPTIONBUTTON

Caption	CounterClockWise
Height	330
Name	optDirection
Left	180
Index	1
Top	2070
Width	2265

OPTIONBUTTON

Caption	Clockwise
Height	330
Name	optDirection
Left	180
Index	0
Top	1620
Width	2265

CHECKBOX

Caption	Enabled
Height	330
Left	180
Name	chkEnabled
Top	540
Width	2265

LABEL

Caption	Speed
Height	240
Left	225

Name	lblSpeed
Top	2655
Width	1905

LABEL

Caption	Direction
Height	285
Left	135
Name	lblDirection
Top	1215
Width	2310

Step 15

Save the Property Page as MOON.PAG.

Now that the property page interface is created, you can attach the page to the Moon control. In the ActiveX Control project, select the User-Control object and display its properties. Locate the property called Property Pages and open the Property Pages dialog.

In the dialog, place a check mark next to the ppgMoon Property Page. Save the project and examine the Custom property of the Moon control in TEST.VBP. When you press the ellipsis, the new Property Page should be visible, but not yet functional.

Step 16

Switch to the ActiveX Control project and open the code window for your Property Page. When a user of your control selects a control, the Property Page must reflect the values of the selected control. In order to do this, the Property Page supports an events called SelectionChanged. When this event fires, you must update the page to the new values. The following code reads the properties from the first control in the selection and populates the page:

```
Private Sub PropertyPage_SelectionChanged()

  'Enabled
  chkEnabled.Value = -1 * SelectedControls(0).Enabled

  'Direction
  If SelectedControls(0).Direction = ClockWise Then
    optDirection(0).Value = True
  Else
    optDirection(1).Value = True
```

```
      End If

      'Speed
      If SelectedControls(0).Speed > 1000 _
      Then SelectedControls(0).Speed = 1000
      scrSpeed.Value = SelectedControls(0).Speed

   End Sub
```

Step 17

Visual Basic needs to know whenever a user changes the control properties by using the Property Page. You let VB know about changes by calling the Changed method for the Property Page. This is analogous to calling the PropertyChanged method for a UserControl object. When you call the Changed method, VB automatically enables an Apply button for the page. Add the following code to the Click event for the checkbox and option buttons and the Changed event for the scroll bar:

```
      PropertyPage.Changed = True
```

Step 18

Updating the properties is done whenever the user presses the Apply button or selects OK. Pressing these buttons fires the ApplyChanges event in the Property Page. In this event, you add code to update all of the controls in the selection. Add the following code to the Property Page to update the selected controls:

```
   Private Sub PropertyPage_ApplyChanges()

     Dim ctlMoon As Moon

     For Each ctlMoon In SelectedControls

       'Enabled
       ctlMoon.Enabled = -1 * chkEnabled.Value

       'Direction
       If optDirection(0) Then
         ctlMoon.Direction = ClockWise
       Else
         ctlMoon.Direction = CounterClockWise
       End If

       'Speed
       ctlMoon.Speed = scrSpeed.Value

     Next

   End Sub
```

Step 19

Save the project and try out your new Property Page.

Increasing Your Knowledge

Help File Search Strings

- UserControl
- PropertyPage

Books OnLine References

- "Creating ActiveX Controls"
- "Building ActiveX Controls"
- "Creating Property Pages for ActiveX Controls"

Mastery Questions

1. Which object provides properties to ActiveX Controls hosted in a container?
   - **a.** UserMode object
   - **b.** Extender object
   - **c.** Ambient object
   - **d.** PropertyPage object

2. Name all the types of ActiveX Controls you can create in Visual Basic.
   - **a.** Owner drawn
   - **b.** User created
   - **c.** Single subclass
   - **d.** Multiple constituent

3. When a constituent control is used in an ActiveX Control project, what mechanism is used to expose properties?
   - **a.** Inheritance
   - **b.** Polymorphism
   - **c.** Encapsulation
   - **d.** Containment

4. What type of procedure is used to create a control method?
   - **a.** Subroutines
   - **b.** Functions
   - **c.** Property procedures
   - **d.** Friendly procedures

5. What object allows a control to synchronize with its container?
   - **a.** Extender
   - **b.** PropertyPage

    c. UserMode

    d. Ambient

6. What property is used to determine whether an ActiveX control is in the design environment?

    a. DesignMode

    b. RunMode

    c. UserMode

    d. OwnerMode

7. Where do you designate a control as data bound?

    a. The Procedure Attributes dialog

    b. The UserControl Property sheet

    c. The Project options

    d. The Project window

8. Where do you designate a control as a container?

    a. The Procedure Attributes dialog

    b. The UserControl Property sheet

    c. The Project options

    d. The Project window

9. Where do you declare new control events?

    a. Project options dialog

    b. In a Standard Module added to the control project

    c. In the RaiseEvent routine of the UserForm object

    d. The [general][declarations] section of the UserControl object

10. What event fires when a control is first placed on a form?

    a. ReadProperties

    b. InitProperties

    c. WriteProperties

    d. Save Properties

11. What event fires when a project is saved?

    a. ReadProperties

    b. InitProperties

    c. WriteProperties

    d. Save Properties

12. What event fires when a project is opened?

    a. ReadProperties

    b. InitProperties

    c. WriteProperties

    d. Save Properties

13. What collection is used to set the properties for controls from a Property page?
    a. GroupedControls
    b. ChosenControls
    c. PickedControls
    d. SelectedControls

14. What event fires when the Apply button of a PropertyPage object is pressed?
    a. FixChanges
    b. SetChanges
    c. ApplyChanges
    d. GetChanges

15. What happens to a Property Page if controls from different classes are selected?
    a. The page is disabled.
    b. Only controls of the same class are used.
    c. A trappable runtime error occurs.
    d. Nothing

Answers to Mastery Questions

1. b

2. a, c, d

3. d

4. a, b

5. d

6. c

7. a

8. b

9. d

10. b

11. c

12. a

13. d

14. c

15. a

12

Creating ActiveX Documents

What You Need to Know

You need to be familiar with how to use the Internet Explorer.

What You Will Learn

- ActiveX Documents
- VBD Files
- CAB Files
- Authenticode
- ActiveX Document distribution

An ActiveX Document, like just about every project type in Visual Basic, is a COM object. This project type, however, creates a special kind of COM object that can only be hosted in certain containers. The containers that host ActiveX Documents are known appropriately as ActiveX Document Container applications. Currently, only three applications have the ability to host ActiveX Documents. These are the Office Binder, a Visual Basic Tool Window, and the Internet Explorer version 3.0 and later. The Internet Explorer is the primary container for these applications and is the focus of this chapter.

ActiveX Documents Fundamentals

When you first set out to create an ActiveX Document project, you'll notice that you may select to create either an ActiveX Document DLL or an ActiveX Document EXE. The difference between these two projects is the same as when you choose between an ActiveX DLL or ActiveX EXE project. The ActiveX Document DLL runs in the process space of the container while the ActiveX Document EXE runs in its own separate memory space. Just like regular ActiveX components, ActiveX Document DLLs can run significantly faster than their out-of-process counterparts. However, ActiveX Document DLLs are incapable of showing modeless VB forms from within the browser. Therefore, the selection of in-process versus out-of-process is a choice of speed versus features.

When you have selected to create an ActiveX Document, Visual Basic begins the project with another special designer object. The foundation for every ActiveX Document is the UserDocument object. The UserDocument object contains all of the plumbing necessary for your project to be an ActiveX Document. All you have to do is focus on the functionality.

The UserDocument object is similar to a form in Visual Basic. You simply build a GUI on it and write code behind the events. When you are done, you can run the Document, but it will not run standalone as a regular VB application will. Instead, Visual Basic creates a special Internet page that can show in Internet Explorer. This page is called a Visual Basic Document or VBD. The VBD file is used by Internet Explorer to access the functionality of your project.

During the debugging process, Visual Basic creates a temporary VBD for you to use. This temporary VBD is placed in the root directory of Visual Basic and exists as long as the document is run inside the VB design environment. You must then go to Internet Explorer and open the file to view the results.

▨ Check It Out: Creating a Simple ActiveX Document

1. Start a new ActiveX Document DLL project in Visual Basic.
2. On UserDocument1, place a command button and change the Caption to Push Me!.

Figure 12-1
The UserDocument
object.

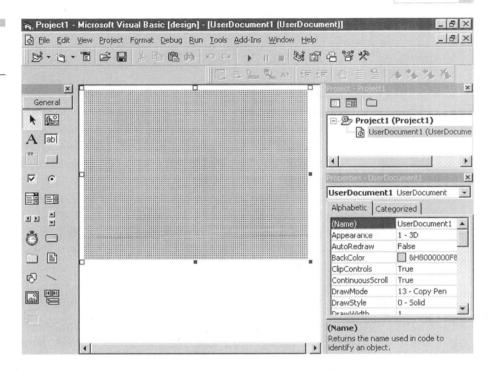

3. In the Click event of the button, place a Message Box statement
   with the classic "Hello, World!" announcement.

```
Public Sub Command1_Click()
  MsgBox "Hello, World!"
End Sub
```

4. Run the project. You'll notice that when the project is run, nothing
   happens. This is because you must use the Internet Explorer to view
   the results.

5. Minimize Visual Basic, but keep the project running.

6. Start Internet Explorer.

7. Using the File/Open menu item, browse to the root directory of Visual Basic and look for any files with a .VBD extension. You should see
   a file called USERDOCUMENT1.VBD.

8. Open the file in Internet Explorer.

9. Push the button and see what happens!

Data Persistence in ActiveX Documents

Developing an application for the Internet can be painstaking regardless of the technology you choose to implement the solution. The primary reason that Internet applications are difficult is because the function of the Internet is fundamentally different from that of an enterprise network. In particular, the relationship between client and server is handled differently.

Imagine utilizing any web server from within a browser. When you use a browser, you type the Internet address, or Uniform Resource Locator (URL), into the browser and a page appears. From the perspective of the server, each client request is a separate and distinct transaction, and the server maintains no history of the transactions. Each web page you request is unrelated to any that were requested before. The server simply hands out web pages to clients in response to their requests, but takes no responsibility for remembering the state of any transaction. This is the basic difficulty of web design.

If you build an application for the web, in any technology, users may come to your site to use this application, but how long will they stay? Suppose you have created an online catalog that allows a virtual shopping cart to be used. The user goes through the departments of the catalog, which are represented by different web pages, and then finally expects to go to one last page to pay the bill. Imagine that the user has viewed three different catalog pages, but before checking out, gets bored and decides to go to the Microsoft site for a while. At this point, should your application remember any items that were selected before the user left the site? When users come back to your site, should their shopping list still exist or do they have to now start over? What if they are gone for 5 minutes, 5 hours, or 5 days? When do you stop remembering the information for them?

The problem gets even worse. Imagine that someone is three pages through a ten-page application you have created. On page three, they simply use the browser to set a bookmark. Now, they turn off their machine and go on vacation. One week later, they start up the browser and immediately use the bookmark to jump right into the middle of your application. Now what do you do? The inherent statelessness of the Internet makes application design difficult.

Overcoming the limitations of the Internet is done by creating various means for persisting data in ActiveX Documents. The techniques used to remember state fall into two categories: managing state for an individual document and managing state for an entire set of documents. The techniques are not always pretty, but they are functional.

The Life Cycle of an ActiveX Document

In order to understand how state is maintained for an individual document, you must understand the life cycle of a document. The life of a User-Document is marked by several key events that are identical to the events received by an ActiveX Control created in Visual Basic. If you have read the chapter on ActiveX Control creation, these events will be familiar.

The InitProperties Event

The InitProperties event of the UserDocument object fires for your ActiveX Document the first time it is rendered in the browser. This property is used to initialize any data your document needs. For example, you might choose to clear the fields of a data input form or provide some default values for a list. The InitProperties event is similar to the Form_Load event for a regular VB form.

While the definition of InitProperties seems to make sense at first, it actually falls quite short. When, after all, is the "first time" a web page is viewed. Is it the first time any user views it? The first time for each user? And what if a user returns to the document later? How much time must go by before InitProperties fires again? Once again, the questions surrounding the development process all concern the management of state.

The answer to these issue lies in understanding how the Internet Explorer manages web pages. Internet Explorer actually remembers the last five web pages it viewed in RAM. This RAM cache should not be confused with disk cache. The last five web pages viewed by Internet Explorer are in memory. Therefore, the definition of the first time a web page is viewed is any time it is not in RAM cache. InitProperties fires whenever your UserDocument is requested by the browser and the document is not currently in the RAM cache.

The WriteProperties Event

The WriteProperties event of the UserDocument object is an event that fires whenever someone leaves your document. This event is designed to let your document know to save data to persistent storage because the user is leaving. When you save data to persistent storage from a User-

Document object, you utilize a special object in Visual Basic known as the Property Bag. The Property Bag is used by both UserDocuments and UserControls to save data to persistent storage just before the instance is destroyed. Saving the data is done with the WriteProperty method of the PropBag object. The following code writes some data called MyData to the Property Bag:

```
PropBag.WriteProperty "MyData","This is my data"
```

The Property Bag normally saves data to RAM until the user elects to save a document. If you want to prompt the user to save a document, then you need to let the Internet Explorer know that the particular document is dirty. Notifying the container that the current ActiveX Document is dirty is accomplished through the PropertyChanged method of the User-Document object. Using this method will force a dialog prompting the user to save the current document to the hard drive when the page is exited. The PropertyChanged method is generally used in any Property Let procedure.

Figure 12-2

Prompting to save an ActiveX Document.

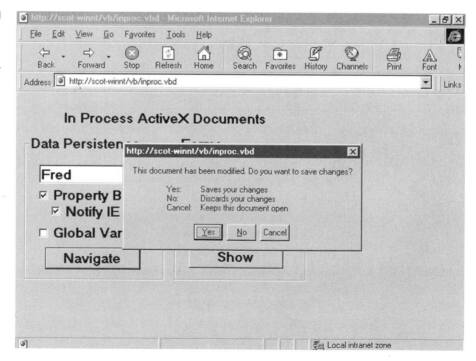

The ReadProperties Event

The ReadProperties event of the UserDocument object fires whenever the user returns to an ActiveX Document that was previously saved. This event provides notification to your document that it should read the Property Bag for data. Reading the Property Bag is accomplished through the ReadProperty method of the PropBag object. The following code reads some data called "MyData" from PropBag:

```
MyVar = PropBag.ReadProperties("MyData")
```

Utilizing Asynchronous Downloading

Whenever you are creating content for the Internet, size is an important factor. Often, Internet sites are filled with large files that must be downloaded before the application can function. ActiveX Documents provide a mechanism for minimizing the impact of downloads known as asynchronous downloads. Asynchronous downloads provide the ability for your application to function while downloading files over the Internet.

Asynchronous downloads are fairly simple to initiate and manage. The key to success is knowing when a download can be initiated. In ActiveX Documents, asynchronous downloading is available after the ActiveX Document is fully sited in the Internet Explorer. When a document is fully sited, it is visible in the browser and its Show event fires. Therefore, asynchronous downloads are not available until the Show event fires.

Once the document is sited, you may initiate an asynchronous download through the AsyncRead method of the UserDocument object. This methods takes as arguments the URL of the file to download, a file type specifier, and an alias for the data being loaded. The alias is used later to identify when the download is complete. Because the download is asynchronous, the alias is used in conjunction with the AsyncReadComplete event to notify the document when the data download is finished. In the AsyncReadComplete event, you may identify the download by the alias and then use the data wherever you need it.

Distributing ActiveX Documents

Perhaps one of the most compelling reasons to use ActiveX Documents is the ease with which they may be distributed to a client. ActiveX Documents, by definition, are distributed using the Internet to download the required components. The setup can be created easily using the Setup Wizard.

Once the ActiveX Documents are created and compiled, you can use the Internet Setup option of the Setup Wizard to create the distribution files. The Setup Wizard walks through the process with little fanfare. There are only a couple of areas where special attention is required. Specifically, the Setup Wizard uses a cabinet (extension .CAB) file as the container for all downloaded components, and requires some information about the project design before it can build the CAB file correctly. The Setup Wizard Internet Package step has a special button marked Safety where key options must be selected.

In this dialog, the ActiveX Documents may be marked as Safe for Initialization and Safe for Scripting. These two checkboxes are a verification on the part of the designer that the ActiveX Documents cannot be used to damage a client computer. This is not a certification on your part that the software is free of viruses, but rather that the design of the software will

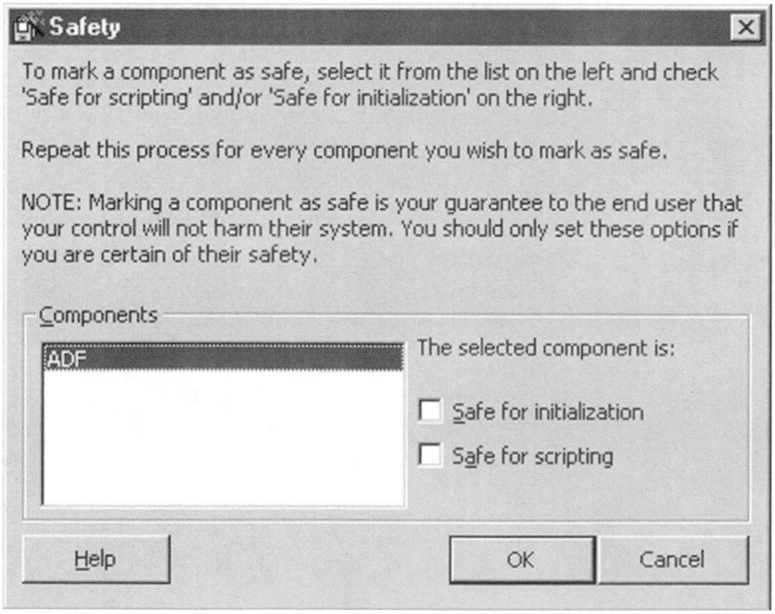

Figure 12-3 *Safety options.*

not allow someone else to use it for nefarious purposes. For example, if you create an ActiveX Document that performs file access, there is nothing inherently damaging about the component. However, someone else might use the file access capabilities to overwrite the AUTOEXEC.BAT, doing irreparable harm to the client. Therefore, checking the safety boxes requires a significant amount of thought on the part of the designer to ensure that the active content cannot be misused.

The second part of safety, verifying the software virus free, is done through a process known as Authenticode. Authenticode is an administrative process of verifying or "signing" active content to ensure it is safe to operate. Authenticode works on a dual-key encryption scheme in which Internet Explorer holds the public key and the software author holds the private key. When the active content is downloaded, the Internet Explorer can interrogate the component and determine if it is properly signed by the manufacturer. If it is, then the user sees the verification certificate. If not, the user is presented with a warning dialog cautioning that the active content is not known to be safe.

The tools required to sign active content are contained in the Internet Explorer Client SDK, or the InetSDK. The InetSDK is available as a free download from the Microsoft site at *www.microsoft.com/ie*. In order to

Figure 12-4

An Authenticode certificate.

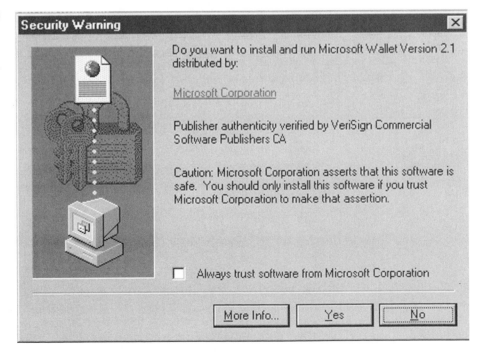

actually sign the software, developers need to have not only the code signing tools, but also a private key to use with the tools. Digital certificate keys are not assigned by Microsoft, but rather by an independent company known as Verisign. Verisign issues digital keys through its website at *www.verisign.com*.

Once the Setup Wizard has completed creating the Internet download, you can examine the resulting files. The output of the wizard consists of three files: a CAB file, a VBD file, and an HTML file. The process of downloading the components requires that a user navigate to the HTML file created by the Setup Wizard. The HTML file references the CAB file, causing Internet Explorer to download all of the components designated by the CAB file. Once the supporting components are downloaded and installed, Internet Explorer is redirected to the VBD file to display the ActiveX Document.

EXERCISE: ACTIVEX DOCUMENT FUNDAMENTALS

ActiveX Documents are special components that run inside ActiveX Document Containers. ActiveX Document Containers include the Internet Explorer, the Office Document Binder, and Visual Basic dockable windows. Although other containers can be used, ActiveX Documents are primarily used inside the Internet Explorer to create rich Visual Basic interfaces that can run on the Intranet.

Requirements

1. Visual Basic 5.0
2. Internet Explorer 3.01

Prerequisites

1. BFLY1.BMP and BFLY2.BMP installed in the VB directory

Concepts Learned

1. ActiveX Documents
2. Asynchronous Downloads

Graphic User Interface

Step 1

Create a new directory called BFLY using the File Explorer. Start a new ActiveX Document DLL project in Visual Basic. When the project starts, you will see an ActiveX Document designer known as a UserDocument object. The UserDocument object is like a Visual Basic form that can be hosted by the Internet Explorer.

ActiveX Documents can only be hosted by Internet Explorer 3.01 and higher. Make sure you have the correct version of IE running on your machine.

Step 2

On the UserDocument, place two image controls, a label control, and a timer control. Change the properties of the controls as follows:

USERDOCUMENT

BackColor	White
Name	Bfly
Height	3915
Width	6990

LABEL

BackStyle	Transparent
Name	lblBFly
Alignment	2-Centered
Caption	ActiveX Documents
Left	225
Top	180
Height	420
Width	6360
FontName	Arial
FontSize	14
FontBold	True

IMAGE CONTROLS

Name	imgBFly(0), imgBFly(1)
Visible	False
Left	0
Top	2745
Height	1155
Width	1155

TIME CONTROL

Name	tmrBFly
Interval	250

Step 3

Set the Project Properties by selecting Project/Project1 Properties...
Change the name of the project to ActDoc.

Asynchronous Downloads

Step 4

ActiveX Documents used inside the Internet Explorer have the ability to
asynchronously download content from a web site. In this exercise, you
will use the async download technique to place images in your image controls. Asynchronous downloading is available after the ActiveX Document
has initialized. Therefore, we will place the download code in the Show
event of the UserDocument. The Show event fires after the document is
visible in the browser. When Show fires, the ActiveX Document is sited
and asynchronous downloading is available. Add the following code to the
Show event of the UserDocument object to download the image files:

```
Dim strURL1 As String
Dim strURL2 As String

'Async data can be read from a file or URL
strURL1 = "[Your Required Files Root Directory]\Required
Files\BFly1.bmp"
strURL2 = "[Your Required Files Root Directory]\Required
Files\BFly2.bmp"

'The AsyncRead method starts the download
'NOTE: The Name you give the data in CASE SENSITIVE
UserDocument.AsyncRead strURL1, vbAsyncTypeFile, "BFly1"
UserDocument.AsyncRead strURL2, vbAsyncTypeFile, "BFly2"
```

Step 5

The AsyncReadComplete event of the UserDocument object fires when
the file download is complete. In this event, you can read the file into the
appropriate image control. Add the following code to the AsyncReadComplete event of the UserDocument object:

```
'Load the images
'Again, be careful to note the case sensitivity of the names!!
If AsyncProp.PropertyName = "BFly1" Then
  imgBFly(0).Picture = LoadPicture(AsyncProp.Value)
```

```
ElseIf AsyncProp.PropertyName = "BFly2" Then
  imgBFly(1).Picture = LoadPicture(AsyncProp.Value)
End If
```

Creating the Animation

Step 6

The animation is controlled by the timer. The position of the images is tracked by two variables. Add the following variable definitions to the [general][declarations] section to track the image position:

```
Private m_Top As Integer
Private m_Left As Integer
```

Step 7

Before the animation begins, the initial placement of the images is calculated. Add the following code to the Initialize event of the UserDocument to set the initial values:

```
'Set the initial values for Top and Left
m_Top = Height - imgBFly(0).Height
m_Left = 0
lblBFly.ZOrder
imgBFly(0).Visible = True
```

Step 8

When the timer fires, the images are alternated to generate the animation. Add the following code to the Timer event to create the animation:

```
'Calculate new positions
m_Top = m_Top - 100
m_Left = m_Left + 100

If m_Top < 0 Then
  m_Top = Height - imgBFly(0).Height
  m_Left = 0
End If

'Move images
imgBFly(0).Top = m_Top
imgBFly(0).Left = m_Left
imgBFly(1).Top = m_Top
imgBFly(1).Left = m_Left

'Animate the Butterfly
If imgBFly(0).Visible Then
  imgBFly(0).Visible = False
  imgBFly(1).Visible = True
Else
```

```
        imgBFly(0).Visible = True
        imgBFly(1).Visible = False
    End If
```

Viewing the Document

Step 9

Start the ActiveX Document by selecting Run/Start from the menu. When you start the ActiveX Document, Visual Basic creates a special file that can be access by the Internet Explorer. This document has a .VBD extension and is created in the root directory of your Visual Basic installation. Start the File Explorer and navigate to the Visual Basic directory. In that directory, you will find the file BFLY.VBD.

Now start the Internet Explorer. When started, select File/Open from the menu and navigate to the BFLY.VBD file. Open the file in Internet Explorer. The Internet Explorer will download the butterfly images asynchronously and start your animation. This technique is excellent for large files that take time to download across the Internet. You may allow users to start working in your web page while data is downloading in the background.

Increasing Your Knowledge

Help File Search strings

- ActiveX Documents
- Internet

Books OnLine References

- "Creating an ActiveX Document"
- "Building ActiveX Documents"

Mastery Questions

1. What types of ActiveX Documents can you create?
   a. ActiveX DLL
   b. ActiveX EXE
   c. ActiveX Document DLL
   d. ActiveX Document EXE

2. What features are supported by an ActiveX Document DLL?
   a. Modeless forms

    b. Modal forms

    c. Property bags

    d. Asynchronous download

3. What features are supported by an ActiveX Document EXE?

    a. Modeless forms

    b. Modal forms

    c. Property bags

    d. Asynchronous download

4. How many documents are kept in memory by Internet Explorer?

    a. 10

    b. 15

    c. 7

    d. 5

5. What method is used to mark an ActiveX Document as dirty?

    a. DocumentChanged

    b. BrowserChanged

    c. PropertyChanged

    d. ValueChanged

6. What events fire when an ActiveX Document is first viewed?

    a. ReadProperties

    b. WriteProperties

    c. Initialize

    d. InitProperties

7. What events fire when an ActiveX Document is dismissed?

    a. ReadProperties

    b. WriteProperties

    c. Initialize

    d. InitProperties

8. What features tell Internet Explorer that an ActiveX Document cannot be used to harm a client?

    a. Safe for Scripting

    b. Safe for Startup

    c. Safe for Initialization

    d. Safe for Properties

9. What program is used to mark active components as virus free?

    a. Supercode

    b. Genuinecode

    c. Authenticode

    d. Vericode

10. Name all the files created by an Internet setup.
- **a.** HTML
- **b.** CAB
- **c.** DLL
- **d.** VBD

Answers to Mastery Questions

1. c, d
2. b, c, d
3. a, b, c, d
4. d
5. c
6. c, d
7. b
8. a, c
9. c
10. a, b, c, d

CHAPTER **13**

Advanced Windows API

What You Need to Know

You should be familiar with declaring and calling Windows API functions. You should understand the role of KERNEL, USER, and GDI in the Windows operating system. You should know how to use the API Text Viewer in Visual Basic.

What You Will Learn

- Sending Messages to Windows
- Subclassing Windows
- Implementing Callbacks

Visual Basic 5.0 is tightly integrated with the Windows operating system. In fact, the events recognized by Visual Basic map directly to messages sent between windows. Normally, the features of Visual Basic are sufficient for many types of applications; however, VB does not support all of the functionality provided by the Windows operating system. In fact, Visual Basic 5.0 adds some new and powerful features for accessing the functionality present on any Windows client.

Understanding Messages

Messages are the primary communication mechanism in the Windows environment. Messages are sent between windows to request activity, set attributes, and announce events. Messages are also integral to the functionality of Visual Basic itself. Visual Basic has the ability to listen to the messages sent from various components, such as buttons and textboxes, and respond when a key message is heard. For example, double clicking on a list box in a Visual Basic application causes a Windows message to fire, which in turn causes the DblClick event code to execute. The event model in Visual Basic is one of listening to the message stream and responding by executing event code.

The Windows API allows us not only to listen to the various messages, but also to send our own messages. Sending our own messages can be useful in accessing functionality that has not been exposed in Visual Basic itself. Sending messages is accomplished through the SendMessage and PostMessage functions.

Before you can understand how the SendMessage and PostMessage functions work, you must understand how a window processes a message. All windows have a special function for processing messages known as the Window Function. The Window Function is responsible for receiving and processing messages sent to a window. The messages themselves are little more than long integer values that are processed by the Window Function. Think of it as a large Select Case statement that takes a different branch for each unique message received.

Because a window may receive more than one message at a time, every window supports a message queue for keeping messages that have not yet been processed by the Window Function. Messages sent to a window may be sent either directly to the Window Function or to the message queue. The SendMessage function sends a message by calling the Window function directly. The call to the window does not return to the client application until the message has been processed by the Window function. The PostMessage, on the other hand, sends a message to the message queue for later processing by the window. The call to PostMessage returns to the calling client immediately after the message is placed in the queue but before the message is actually processed.

The SendMessage Function

The SendMessage function is part of USER.EXE and can be declared using the following syntax:

```
Declare Function SendMessage Lib "user32" Alias "SendMessageA" _
(ByVal hwnd As Long, ByVal wMsg As Long, ByVal wParam As Long, _
lParam As Any) As Long
```

The hWnd argument for the SendMessage function is the window handle for the window that will receive the message. The window handle is an identifier used by Windows to uniquely represent every window in the operating system. This argument can be accessed for any form or control in Visual Basic by simply referencing the hWnd property.

The wMsg argument for the SendMessage function is the actual message to send. The available messages are many and varied. A complete description of every message you can send is beyond the scope of this book; however, this chapter will show examples for some of the more common messages.

The wParam and lParam arguments essentially provide a mechanism for passing additional information with the message. The type of information sent using these parameters varies based on the message being sent. Again, the possible values for these arguments are many, but once the function is declared, it's a simple matter of calling the function from Visual Basic to send the appropriate message. As an example, consider the message LB_SELECTSTRING, which can be sent to a Visual Basic ListBox. This message forces a ListBox to search for and highlight an entry matching a text string you provide. The declaration for the message constant is this:

```
Public Const LB_SELECTSTRING = &H18C
```

Searching a ListBox named List1 requires making a call to SendMessage with the appropriate parameters. The following code shows how to search List1 for the text string "New Technology Solutions":

```
Dim lngReturn As Long
lngResult = SendMessage(List1.hwnd, LB_SELECTSTRING, ByVal -1, _
ByVal "New Technology Solutions")
```

The PostMessage Function

The PostMessage function is similar to the SendMessage function and has the following syntax:

```
Declare Function PostMessage Lib "user32" Alias "PostMessageA" _
(ByVal hwnd As Long, ByVal wMsg As Long, ByVal wParam As Long, _
ByVal lParam As Any) As Long
```

Note that the arguments for the PostMessage function are identical to those for the SendMessage function. The only difference is when the message is processed. The PostMessage function should be used in cases where the return value of the function is not important. This is because PostMessage cannot return a meaningful value since the message was not processed by the Window Function before the call returned.

Subclassing Windows

Although every window provides a Window Function for processing messages, sometimes you may want to provide custom message processing. In this case, Visual Basic allows you to substitute a function that you write for the default Window Function. The process of creating a substitute Windows Function is known as subclassing a window. Subclassing relies on SetWindowLong function to specify the function that will act as the Window Function. The SetWindowLong function is declared as follows:

```
Declare Function SetWindowLong Lib "user32" Alias _
"SetWindowLongA" (ByVal hwnd As Long, _
ByVal nIndex As Long, ByVal dwNewLong As Long) As Long
```

The SetWindowLong function is used for changing the attributes of any particular window to which you have a handle. The hWnd argument is the window handle for the window you want to subclass. The nIndex argument specifies the type of window information to set. The constant GWL_WNDPROC tells the function to set the address for the Windows Function, and dwNewLong is the address of the new Window Function.

Notice that SetWindowLong relies on knowing the address in memory of the function to be the new Window Function. Until version 5.0, function addresses could not be retrieved for Visual Basic functions. However, in VB 5.0, the AddressOf keyword has been added to return the address in memory of a function that you create. The only restriction is that your new Window Function should be declared inside a Standard Module and not a form or Class Module.

As a simple example of subclassing, we have created a spy tool that displays all of the messages received by a Visual Basic form. This example simply shows messages sent to a form inside of a TextBox. The new Windows function is declared inside of a Standard Module using the following code:

```
Function WindowProc(ByVal hw As Long, ByVal uMsg As Long, _
ByVal wParam As Long, ByVal lParam As Long) As Long
  'This is the CallBack function that will intercept
  'all of the messages from the subclassed
  'window.

  frmSpy.txtMessages.Text = _
  "Message: hWnd(" & hw & "), Msg(" & uMsg & _
  "), wParam(" & wParam & "), lParam(" & lParam & ")" _
  & vbCrLf & frmSpy.txtMessages.Text
End Function
```

The arguments of the new procedure are strictly defined and match the arguments passed by the SendMessage function. This makes perfect sense since the new function will be called directly by any SendMessage call. Inside the function, the messages are simply printed in a textbox and the function exits. In most cases, however, you will not want to simply exit after examining a message, but rather pass the message along to the previous Window Function. This allows the window to process the message after your code is complete. Passing the message to the original Window Function is done with the CallWindowProc function, which is declared as follows:

```
Declare Function CallWindowProc Lib "user32" Alias _
"CallWindowProcA" (ByVal lpPrevWndFunc As Long, _
ByVal hwnd As Long, ByVal Msg As Long, _
ByVal wParam As Long, ByVal lParam As Long) As Long
```

This function takes as arguments the same parameters as were passed by SendMessage with the addition of the lpPrevWndFunc argument. The lpPrevWndFunc argument is the address of the original Window Function, which is returned by the call to SetWindowLong. Therefore, in order to subclass a window, you call SetWindowLong and save the return value that refers to the original function. This is called "hooking" the window. The following code hooks the window and sets the Window Function to WindowProc using the AddressOf keyword:

```
lpPrevWndProc = SetWindowLong(Form1.hWnd, GWL_WNDPROC, _
AddressOf WindowProc)
```

When the application is closed, we also want to unhook the window, that is, set the Window Function back to the original function because our application is no longer available. Unhooking the window is done with another call to SetWindowLong, but this time we pass the original function address.

```
Dim lngTemp As Long
lngTemp = SetWindowLong(Form1.hWnd, GWL_WNDPROC, lpPrevWndProc)
```

Subclassing windows allows you to essentially create events in Visual Basic that are not normally supported. You may hook a window and then examine the message stream that enters you function. When a given message is detected, you can execute code much as Visual Basic fires an event. You might even choose to use the RaiseEvent keyword to fire a custom event in code.

Receiving Callbacks

Callbacks are another technique that utilizes the AddressOf keyword. Callbacks allow you to send a pointer to a custom function and have Windows call that function periodically. Windows provides many functions that utilize callbacks. Some of the most common are the "enumerated" functions.

Enumerated functions are designed to make a callback and provide data. For example, the function EnumChildWindows makes a call to the callback function once for each child window of a parent window you specify. In this way, callbacks are used to provide a list of data to your program. To utilize an enumerated function, you declare it just as you would for any API call. The following is the declaration for EnumChildWindows:

```
Declare Function EnumChildWindows _
Lib "user32" (ByVal hWndParent As Long, _
ByVal lpEnumFunc As Long, ByVal lParam As Long) As Long
```

This function accepts as arguments the window handle of the parent window, a pointer to the callback function, and a user-defined parameter that holds some data you care about. Once the function is declared, you must create the callback function.

Enumerated functions have specific requirements for the arguments provided in a callback. The argument list varies according to each enumerated function. For the EnumChildWindows, the callback function must provide three long data types. These data types return the window handle of each child window and some associated data. The following is the declaration for a callback used with EnumChildWindows:

```
Public Sub Enumerate(lngHandle As Long, lngData As Long, _
lngReturn As Long)
```

EXERCISE 1: USING MESSAGES

The Windows Application Programming Interface (API) is a collection of C Language function calls housed in several Dynamic Link Libraries (DLLs). These libraries provide all of the functionality associated with the Windows Operating System. The primary libraries inside Windows are KERNEL, USER, and GDI, which may be accessed directly from your Visual Basic projects. In this exercise, you will use the functions of the Windows API to add functionality to a standard VB list box. In particular, you will add an incremental search feature and a horizontal scroll bar.

When you make calls using the Windows API, you do not get any error handling! This means that incorrect code will cause your system to crash. Save this project before you run any part of it or you may lose all your work!

The API Text Viewer

Step 1
Create a new directory in the File Explorer called VBCLASS\LISTAPI. Start a new Visual Basic Standard EXE project. Add a Standard Module to the project by selecting Project/Add Module from the menu. The Standard Module is used to keep the declaration statements for API calls.

Step 2
Add a Sub Main to your project by selecting Tools/Add Procedure from the menu. Set your project up to start from Sub Main by changing the Start Form option under the Project/Properties... menu.

Step 3
Minimize Visual Basic and find the icon for the API Text Viewer, which is located in the Visual Basic program group. Start the API Text Viewer. When you see the API Text Viewer, load the declarations by selecting File/Load Text File... and selecting WIN32API.TXT as the file to load. When the file is loaded, all of the API declarations are available to you.

In the API Text Viewer, click on any item under Available Items and then press the "s" key. The Viewer drops down to the entries beginning with S. Now find SendMessage and double click on it. Press the Copy button to copy it to the clipboard. This function allows you to send a message to any component in the Windows Operating System.

The types of messages that you can send to the list box are defined as constants in the API Text Viewer. You can copy these message constants to the clipboard as well. In the API Type box, select Constants. The message constants will load and you can select them from the list. Select the following constants:

```
LB_SELECTSTRING
LB_SETHORIZONTALEXTENT
```

Step 4

Return to Visual Basic and paste the declaration in the [general][declarations] section of the Standard Module. The declarations tell VB about the function you want to call, which is part of the Windows API. The following code is pasted in:

```
Declare Function SendMessage Lib "user32" Alias "SendMessageA" _
(ByVal hwnd As Long, ByVal wMsg As Long, ByVal wParam As Long, _
lParam As Long) As Long

Public Const LB_SETHORIZONTALEXTENT = &H194
Public Const LB_SELECTSTRING = &H18C
```

The constants you defined represent the actual messages being sent using this API. These messages will actually cause the incremental searching behavior and allow a horizontal scroll bar to appear.

Incremental Search

Step 5

On Form 1 place a textbox and a list box. Call the textbox txtAPI and the list box lstAPI. In the Sub Main routine, display this form and add some text entries by coding.

```
Form1.Show
Form1.lstApi.AddItem "Dan"
Form1.lstApi.AddItem "Dave"
Form1.lstApi.AddItem "David"
Form1.lstApi.AddItem "Davidson"
```

Step 6

Implementing the incremental search is now as simple as calling the SendMessage function. Add the following code to the txtAPI_Change event so that the incremental search occurs while we type:

```
'Make sure something is in the box
If txtAPI.Text = "" Then Exit Sub

Dim lngResult As Long

'Make the call to search the List Box
lngResult = SendMessage(lstAPI.hwnd, CLng(LB_SELECTSTRING), _
ByVal -1, ByVal txtAPI.Text)
```

This code uses the window handle for the list box (given by the hWnd Property) to send a message to search for the given text in the list box. The window handle is the identifier used by the operating system to track windows. Accessing the handle is a simple matter of using the hWnd property of the list box. Many components in VB have an hWnd property that allows messages to be sent to them. Check out the hWnd property in your VB Help file.

Horizontal Scroll Bar

Step 7
The standard Visual Basic list box provides a vertical scroll bar whenever the list of items is too long to be seen in the box. However, the list box does *not* provide a horizontal scroll bar if the entries are too wide to be read in the box. Adding a horizontal scroll bar is possible, however, if we send the list box a message telling it to produce one. In this section, you will add a horizontal scroll bar to your list box.

Step 8
First, you need to add some items that are particularly wide so that you can test the horizontal scroll bar. In your Sub Main procedure, add the following code to place some wide entries in your list box:

```
Form1.lstApi.AddItem "Microsoft Access"
Form1.lstApi.AddItem "MicroSoft Exchange Server"
Form1.lstApi.AddItem "Microsoft FoxPro"
Form1.lstApi.AddItem "Microsoft SQL Server"
Form1.lstApi.AddItem "Microsoft Systems Management Server"
Form1.lstApi.AddItem "Microsoft Visual Basic"
Form1.lstApi.AddItem "Microsoft Visual Basic Script"
Form1.lstApi.AddItem "Microsoft Visual C++"
```

Step 9
Now save and run your application. Look at your list box and make sure that the entries you just placed in the box are wider than the list box. If

not, shut down the application and make the list box smaller until the new entries are clipped.

Step 10

Add a button to your form and call it cmdAPI. Change the Caption of the button to Scrollbar. In the Click event of the button, add the following code to send a message to the list box:

```
'This routine makes a call to set the virtual width of
'the list box. This is known as the "Horizontal Extent".
'In this case, we set it to 3 times the physical width.

Dim lngReturnValue As Long
Dim lngExtent As Long

lngExtent = 3 * (lstAPI.Width / Screen.TwipsPerPixelX)
lngReturnValue = SendMessage _
(lstAPI.hwnd, LB_SETHORIZONTALEXTENT, lngExtent, 0&)
```

Step 11

Save and run your application. Pressing the button now sends a message to the list box telling the box to set a virtual width that is three times larger than the actual physical width. When the virtual width is increased, the list box automatically creates a horizontal scroll bar.

EXERCISE 2: SUBCLASSING WINDOWS

Subclassing windows allows you to replace the default Window Function with one of your own. This allows you to examine the messages entering a window and take action when you see one of interest. This exercise uses subclassing to create a status bar that shows a menu definition when the mouse passes over a menu selection.

Adding the API Calls

Step 1

Create a subdirectory called \SUBCLASS using the File explorer and then start a new Standard EXE project in Visual Basic 5.0. Add a Standard Module to the project.

Step 2

Minimize Visual Basic and start the API Text Viewer from the VB program group. In the API Text Viewer, locate, select, and copy the following functions and constants:

```
SetWindowLong
CallWindowProc
Const GWL_WNDPROC
Const WM_MENUSELECT
```

Step 3

Return to Visual Basic and paste the function and constant declarations into the Standard Module. Add the comments and variables shown to produce the following code:

```
Option Explicit

'This API function is used to change the structure
'of a window. To subclass, we must change the
'address of the "windows function" for the window
'we want to subclass.
Declare Function SetWindowLong Lib "user32" Alias _
"SetWindowLongA" (ByVal hwnd As Long, _
ByVal nIndex As Long, ByVal dwNewLong As Long) As Long

'This API function allows us to pass along the
'intercepted messages to the default "windows
'function" after we are through with them.
Declare Function CallWindowProc Lib "user32" Alias _
"CallWindowProcA" (ByVal lpPrevWndFunc As Long, _ByVal hwnd As
Long, ByVal Msg As Long, _ByVal wParam As Long, ByVal lParam As
Long) As Long

'API Constant used for nIndex argument of SetWindowLong.
'This tells SetWindowLong that we are changing the information
'regarding the address of the "windows function".
Public Const GWL_WNDPROC = -4

'This Message is used to detect a menu selection
Public Const WM_MENUSELECT = &H11F

'A variable to store the address of the default
'windows function we are subclassing
Public lpPrevWndProc As Long
```

Building the Interface

Step 4

Select Form1 to begin building the user interface. Add a menu to the form by selecting Tools/Menu Editor... from the menu. Add the following menu structure to Form1:

Step 5

Add a label control called lblStatus to Form1. Position the label so that it occupies the bottom of the form as a status bar. Add two command buttons to Form1. Name them cmdHook and cmdUnhook. Give them captions Hook and Unhook respectively.

Hooking and Processing Messages

Step 6

Open the Standard Module and define a new function called WindowProc. Add arguments to the function such that the final definition appears in the module as follows:

```
Function WindowProc(ByVal hw As Long, ByVal uMsg As Long, _
ByVal wParam As Long, ByVal lParam As Long) As Long

End Function
```

Step 7

The new function will be used as the Window Function to subclass Form1. When Form1 is hooked, all messages will be sent to this function. In this function, we want to detect a message that indicates the mouse is over a menu item in the Form1 menu. This message is WM_MENUSELECT. When we detect this message, a description of the menu item is written to the status bar of Form1. Add the following code inside of WindowProc to detect the message and print the status description:

```
'This is the function that will intercept
'all of the messages from the subclassed
'window.

'We look for the WM_MENUSELECT message which
'fires when the user puts the mouse over the message
If uMsg = WM_MENUSELECT Then
  Select Case Right$(Hex$(wParam), 1)
    Case 0
      Form1.lblStatus.Caption = ""
    Case 2
      Form1.lblStatus.Caption = "Menu One"
    Case 3
      Form1.lblStatus.Caption = "Menu Two"
    Case 4
      Form1.lblStatus.Caption = "Menu Three"
  End Select
End If
```

TABLE 13-1

Menu structure

Menu Name	Menu Caption	Indent
MnuMenus	Menu	0 spaces
MnuOne	One	1 space
MnuTwo	Two	1 space
MnuThree	Three	1 space

```
'After the messages are intercepted, we pass them
'along to the original default function so that it can
'actually handle them.
WindowProc = CallWindowProc _
(lpPrevWndProc, hw, uMsg, wParam, _
lParam)
```

Step 8

Hooking the window is done under the Click event of cmdHook. Add the following code to the click event:

```
lpPrevWndProc = SetWindowLong(Form1.hWnd, GWL_WNDPROC, _
AddressOf WindowProc)
```

Unhooking is done under the Click event of cmdUnhook using the following code:

```
Dim lngTemp As Long
lngTemp = SetWindowLong(Form1.hWnd, GWL_WNDPROC, _
lpPrevWndProc)
```

Stop 9

Save and run the project. Try passing your mouse over the menus.

Increasing Your Knowledge

Help File Search Strings

- AddressOf
- API Functions

Books OnLine References

- "Accessing DLLs and the Windows API"

Mastery Questions

1. Messages are processed in a window by the:
   - **a.** Window Procedure.
   - **b.** Window Function.
   - **c.** Window Handle.
   - **d.** Window Caption.

2. Windows messages are:
   a. long integers.
   b. function pointers.
   c. window handles.
   d. Window Functions.

3. What API function places a message in the message queue?
   a. EnumChildWindows
   b. Send Message
   c. PostMessage
   d. WM_MENUSELECT

4. Subclassing refers to:
   a. sending messages to a window
   b. firing VB events.
   c. replacing the default Window Function.
   d. AddressOf.

5. Callbacks are used by windows to:
   a. periodically call a function.
   b. call a function after an event.
   c. call a function when data is available.
   d. call a function based on time.

Answers to Mastery Questions

1. b

2. a

3. c

4. c

5. a

14

ODBC and the Remote Data Objects

What You Need to Know

You need to have an understanding of the fundamentals of relational databases and of using ActiveX servers.

What You Will Learn

- ODBC Data Sources
- The RDO Object Model
- Connecting to Data Sources
- Adding, Editing, and Deleting Records
- Cursors

Understanding data access principles is probably the single most important skill a Visual Basic developer can possess. The bulk of all programs written today concern data access in one form or another. This statement is particularly true for a business application running on an enterprise network. Databases are the backbone of business applications, and your success as a VB programmer depends on this knowledge. In this chapter, we will examine the high-performance data access tools available to Visual Basic developers.

Open Database Connectivity

Understanding data access begins with understanding the Open Database Connectivity (ODBC) specification. ODBC is a database communication standard that aims to provide database independence. Database independence is a concept that allows access to any database, regardless of manufacturer, through a single common data manipulation language known as Structured Query Language (SQL). A single unified language for data access provides enormous benefits to developers because they can easily migrate data from one source to another without recoding the applications that are accessing the data.

Achieving database independence is the job of the ODBC driver. This is a middleware component that translates the standard SQL statements of database front ends to the proprietary language of the particular back end. In this way, a single SQL statement can be run on any proprietary server that supports SQL.

The reality is, of course, slightly less perfect than the theory. Although SQL is widely supported, not all databases have the same level of support. The most fundamental SQL support is known as Core SQL. All data sources that claim to support ODBC must support Core SQL. However, SQL has several different additional levels that may or may not be supported. You should know to what degree a target data source supports SQL prior to beginning any project.

In order to write Visual Basic code against an ODBC-compliant database, you must first establish an ODBC data source, or DSN. A data source is simply the definition of a database and ODBC driver under a single alias that can be accessed through code. Establishing the ODBC data source is the job of the ODBC Administrator. The ODBC Administrator is an application found under the Control Panel as the ODBC icon.

When you start the ODBC Administrator, you can view all of the ODBC data sources on your machine as well as establish new ones. ODBC sources may be built against any database for which you have a driver such as SQL Server, Microsoft Access, or Oracle. When you build a data source, you provide an alias name that can be used to access the source later from code. This name has no significance and can be nearly anything you want. Typically, the name is closely related to the actual database name.

Figure 14-1
The ODBC Administrator.

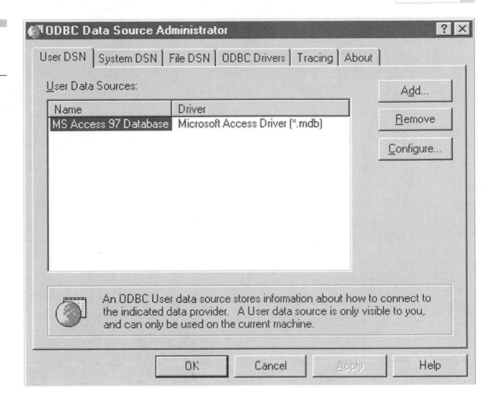

ODBC data sources may be created with varying levels of scope. A DSN may be either a User, System, or File DSN. User data sources are available only to the user who created the source. System data sources are available to all of the users on a machine and the Windows NT services that run on the machine. File data sources are available to any machine that has security access to the DSN and the appropriate drivers available. In a nutshell, User, System, and File data sources provide individual, machine, and enterprise scope respectively. Additionally, Visual Basic supports the concept of DSN-less connections that allow you to specify a data source without using the ODBC Administrator.

Check It Out: Creating a DSN

1. Start the ODBC Administrator by selecting the ODBC icon from the Control Panel.

2. In the ODBC Administrator, click on the System DSN tab.

3. Click on the Add button.

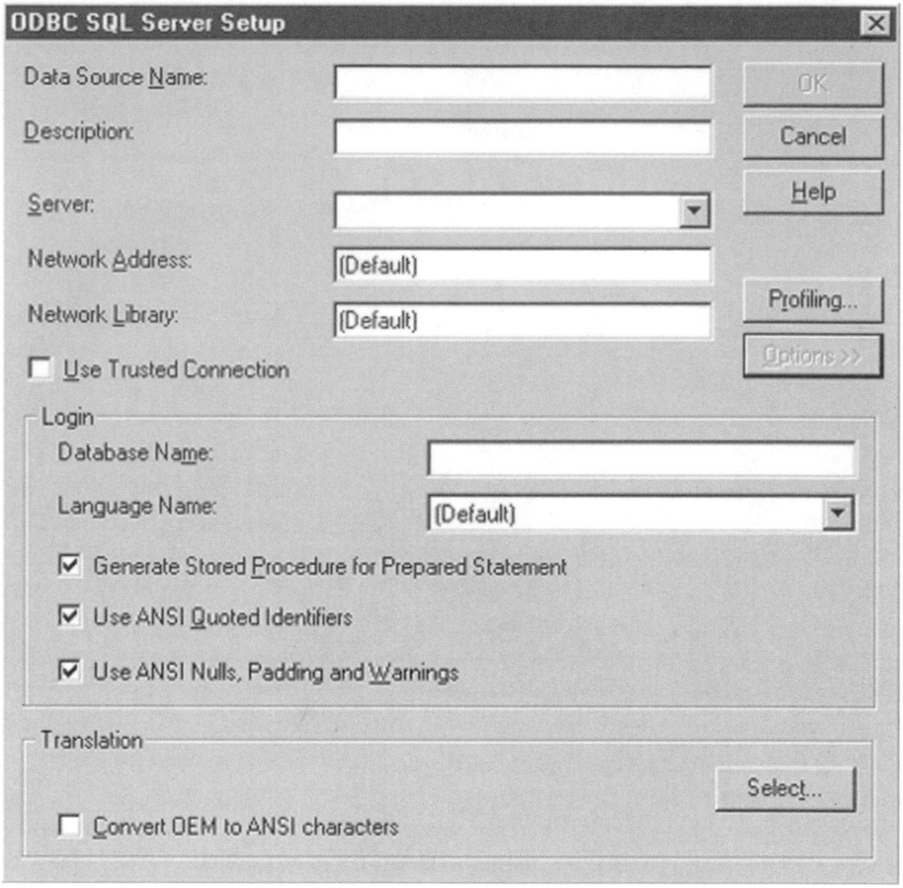

Figure 14-2
Setting up a new
DSN.

4. In the Create New Data Source dialog, select to create a new SQL Server DSN. Push the Finish button.

5. Follow the dialog instructions to set up a new data source named Publications utilizing the "pubs" database that ships with SQL Server.

6. Save the DSN for later.

The Remote Data Objects Model

Visual Basic 5.0 provides a fast and powerful mechanism for accessing ODBC data sources through code known as the Remote Data Objects (RDO). This is an ActiveX Component that consists of several classes that support all of the features of ODBC. Using this component you can easily

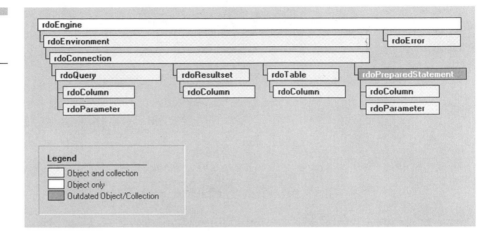

Figure 14-3
The RDO Object
Model.

query and update ODBC sources with the speed and control required for enterprise applications.

Understanding RDO comes from understanding the objects contained within it. These objects together form the RDO Object Model. This model is used to graphically represent the features of RDO and is similar in use to any object model you have seen before such as Excel or even Visual Basic itself. The objects contained in RDO are specialized for data access using ODBC.

The rdoEngine Object

The top-level object in the RDO model is the rdoEngine object. The rdo-Engine object is an object that represents the entire RDO component. The rdoEngine object contains information such as the version of RDO you are using. Generally, you do not have to explicitly create an rdoEngine object because RDO gives you the first one automatically.

The rdoError Object

The rdoError object is an object that you can examine to determine whether an operation on a data source has resulted in an error. The errors trapped by the rdoError object are not Visual Basic errors, but rather errors that occur at the database server. Therefore, you may not receive a trappable runtime error for these errors. Instead, you can examine the rdoErrors collection after an operation to determine success or failure. The following code shows how:

```
Dim objError As RDO.rdoError
Dim strError As String

'Simple Error Handling
'Collect Errors
  For Each objError In rdoEngine.rdoErrors
    strError = strError & vbCrLf
  Next

'Display Errors
MsgBox "The following errors occurred: " _
& vbCrLf & strError, vbOKOnly, "User Connection Object"
```

The rdoEnvironment Object

The rdoEnvironment object is an object that represents a security context inside which you can access a database. Using this object, you might specify User IDs and passwords or perform database transactions. Like the rdo-Engine, you generally do not create an rdoEnvironment directly, since you get the first one automatically. If needed, however, you could create additional environments that are automatically managed through a collection.

The rdoConnection Object

Most of the real functionality of RDO begins with the rdoConnection object. The rdoConnection object is used to establish a connection with an ODBC data source. Once the data source is defined, you may use the properties and methods of the rdoConnection communicate with the database server. The simplest access requires only that you instantiate the object, set the name of the data source as it appears in the ODBC Administrator, and establish the connection:

```
Set m_Connection = New RDO.rdoConnection
m_Connection.Connect = "DSN=Biblio;"
m_Connection.EstablishConnection
```

The rdoQuery Object

The rdoQuery object is an object that can be used to execute SQL queries on an open ODBC data source. Executing queries can be done with simple SQL statements in your code or from precompiled stored procedures that are supported by a database like SQL Server. Parameters for stored procedures are managed using the rdoParameter object. At its most fundamental, the rdoQuery object needs only to know what connection you want to query and the SQL statement to use in the query:

```
'Create SQL Statement
Dim strSQL As String
strSQL = "SELECT * FROM Publishers"

'Create Query
Dim m_Query As RDO.rdoQuery
Set m_Query = New RDO.rdoQuery
Set m_Query.ActiveConnection = m_Connection
m_Query.SQL = strSQL

'Run Query
m_Query.Execute
```

The rdoResultset Object

The rdoResultset object is used to manage records returned from an open ODBC connection. Resultsets can be returned from an rdoQuery object or the OpenResultset method of an rdoConnection object. In any case, you can access the most recently created record set through the Last-QueryResults property of the rdoConnection object:

```
Set m_Resultset = m_Connection.LastQueryResults
```

Once the rdoResultset is obtained, you can navigate the records using the MoveFirst, MoveLast, MoveNext, and MovePrevious methods. These methods allow you to scroll through the records for display and update purposes. Once the record pointer is moved to the correct position, you may read the data from the record and show it in a Visual Basic form using the rdoColumn object.

The rdoTable Object

The rdoTable object is an object that represents a table in an ODBC data source. Its use, however, is quite limited. This object is not particularly useful for normal database applications, but it can be helpful if you are trying to determine the structure of a data source. Generally, you will work with rdoResultset objects to retrieve data from the data source.

RDO Events

Along with the properties and methods of RDO, the object model also supports several useful events. These events can notify you when an asynchronous activity completes or allow you to customize the application behavior during database operations. Accessing any of the supported events is accomplished by declaring variables against the model using the WithEvents keyword. When declared this way, supported RDO events automatically appear in the code window:

```
Private WithEvents m_Connection As RDO.rdoConnection
```

Check It Out: Viewing the RDO Object Model

1. Start Visual Basic.

2. From the menu select Help/Microsoft Visual Basic Help Topics.

3. Click on the Index tab and type Remote Data Objects.

4. Push the Display button.

5. Select Remote Data Objects and Collections.

6. Push the Display button.

7. You should see the complete RDO object model. Scroll down and see the definitions for the various objects and collections in RDO.

8. Click on rdoConnection.

9. A Help topic for the rdoConnection object should appear. Check out the EstablishConnection method.

10. This Help file gives you information on all of the properties, events, and methods of RDO.

Connecting to Data Sources

Using RDO from Visual Basic begins by properly referencing the RDO object model and connecting to a data source. Referencing the RDO object model is accomplished using the References dialog just as it is for any COM object. The reference for RDO is listed as Microsoft Remote Data Object 2.0.

Once a reference is established, you may connect to a data source using the rdoConnection object. The process of using this object is the same as that for using any COM object. First you must declare a variable as an rdoConnection. Next, you must instantiate the object, and finally, you can call the properties and methods to accomplish the connection. When making the DSN connection, you must manage several key aspects, which include the Connect string and the cursor driver

Figure 14-4
The References dia-
log.

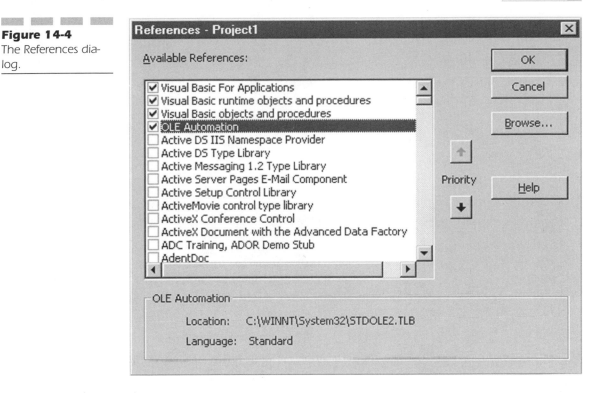

ODBC Connect Strings

ODBC Connect strings are used during the connection process to pass key information to the ODBC driver for accessing the data source. This information can contain the DSN name, User ID, and Password as well as other key information. Table 14-1 shows all of the possible arguments for a Connect string.

Connect string arguments are optional, and you can even cause ODBC to automatically request missing parameters that are required for a connection. Typically, you pass the ODBC Connect string through the Connect property of the rdoConnection object. The following code shows a sample string for a SQL Server data source called Publications:

```
Set m_Connection = New RDO.rdoConnection
m_Connection.Connect = "DSN=Publications;UID=sa;PWD=;"
```

In addition to the standard DSN connection using the ODBC Administrator, RDO also supports DSN-less connections. These connections are made without defining the DSN in the ODBC Administrator. Instead, all

TABLE 14-1

ODBC Connect
String Arguments

Argument	Description
DSN	The ODBC data source name
UID	The database User ID
PWD	The database password
DRIVER	The ODBC driver
DATABASE	The name of the database to connect with
SERVER	The machine name of the database server
WSID	The machine name of the database client
APP	The name of the EXE File

of the data necessary to connect to the database is provided as a part of the Connect string. The following code connects to a SQL Server database named "pubs" without using a DSN:

```
Set m_Connection = New RDO.rdoConnection
m_Connection.Connect = _
"DRIVER={SQL
Server};SERVER=(local);DATABASE=pubs;UID=sa;PWD=;DSN=;"
```

Note that the Connect string provides additional information such as the ODBC driver to use and the server where the database can be found. Normally, this information is entered into the ODBC Administrator, but when provided as part of the Connect string, it removes the requirement for a DSN.

Cursor Drivers

Once the DSN is identified, you should also specify the cursor driver to use with your connection. The cursor driver determines where and how records returned from a database query are maintained. In general, the records may be maintained on either the client or the server. You may also select to retrieve the actual record information or simply a set of record identifiers, known as keys, that map to the results of a query. The cursor driver is selected by setting the CursorDriver property of the rdoConnection object to one of the following constants: rdUseIfNeeded, rdUseODBC, rdUseServer, rdUseClientBatch, or rdUseNone. The following code shows a typical example of establishing the cursor driver:

```
Set m_Connection = New RDO.rdoConnection
m_Connection.Connect = "DSN=Publications;UID=sa;PWD=;"
m_Connection.CursorDriver=rdUseClientBatch
```

The CursorDriver property affects all of the queries that are run on a connection and can only be set before the connection is actually made. After you make the connection, the property is read-only. Therefore, if you want to change the cursor driver, you will have to open a new connection to the database. The following sections describe the cursor driver options in detail.

rdUseIfNeeded

This option allows the ODBC driver to select the cursor to use. The driver may select client-side cursors or server-side cursors. If available, RDO will preferentially select a server-side cursor. Server-side cursors store records returned from queries on the server where the database is located. The records are placed in temporary storage and returned to the client in groups as needed.

rdUseODBC

This option uses the standard ODBC cursor driver that places all retrieved records on the client machine. If the results are too large to fit in RAM, they are placed on the disk until all records are retrieved. Because all records are returned directly to the client, scrolling the records can be faster if the records all exist in memory. However, if you return large results sets to the client, you may quickly exhaust resources and cause delays while records are written and read from disk.

rdUseServer

This option forces the use of server-side cursors for all result sets. Server-side cursors are excellent for use with larger result sets that may exhaust client resources if returned through a client-side cursor. However, you must make sure that the machine running the database has the capability to handle large data sets from multiple users. This can be an issue if you have hundreds of users accessing the same database and they are all returning thousands of records. The best strategy is to use server-side cursors, but strictly limit the number of records returned from a query by narrowing searches requested by the client.

rdUseClientBatch

This cursor driver causes records to be returned to the client just like the rdUseODBC option, but it has more capabilities than the rdUseODBC option. This cursor driver is designed to support features such as multi-table updates. It is really an upgrade to the rdUseODBC option, and you should employ this option preferentially when using client-side cursors.

rdUseNone

This option prevents the use of cursors and simply returns the records to the client one at a time when requested. This cursor is the lightest and fastest of all cursors and should be used whenever possible. In order to maintain speed, however, this cursor driver only allows records to be navigated in one direction-forward. This cursor driver is ideal for populating list boxes from lookup lists.

Establishing the Connection

Once the Connect string and cursor driver are set, you are ready to make the connection to the data source. Making the connection is done through the EstablishConnection method of the rdoConnection object. The EstablishConnection method has the arguments prompt, read-only, and options.

The prompt argument determines whether or not you want ODBC to prompt users for arguments not provided in the Connect string. The prompting consists of dialog boxes that users may fill out to specify missing arguments from the Connect string. Table 14-2 shows a complete list of the prompting arguments.

The read-only argument is a simple Boolean value that indicates whether the connection is opened for read-only access. If opened as read-only, then no updates are permitted. This argument is optional, and the default is False.

The options argument allows you to specify whether the connection should be opened asynchronously. If you want the connection opened

TABLE 14-2	Constant	Description
ODBC Prompt Options	rdDriverPrompt	Always shows the prompt dialog using the arguments of the connect string and the dialog
	rdDriverNoPrompt	Never shows the prompt dialog. Incomplete connect information raises an error
	rdDriverComplete	Raises the ODBC dialog if the connect string does not contain sufficient information
	rdDriverCompleteRequired	Raises the ODBC dialog if the connect string does not contain sufficient information. Also disables controls in the dialog where information entry is not required.

asynchronously so that programmatic control returns to your application immediately, simply specify the constant rdAsyncEnable and RDO will then start opening the connection and notify you when the connection is complete through its event-handling structure, described later.

Establishing a complete connection requires setting the Connect string and cursor driver before calling the EstablishConnection method. The following code shows a typical connection example that incorporates all of the features discussed:

```
Set m_Connection = New RDO.rdoConnection
m_Connection.Connect = _
"DRIVER={SQL
Server};SERVER=(local);DATABASE=pubs;UID=sa;PWD=;DSN=;"
m_Connection.CursorDriver = rdUseClientBatch
m_Connection.EstablishConnection rdAsyncEnable
```

rdoConnection Events

RDO supports a number of events related to the rdoConnection object. These events provide your application with notification (e.g., when your application connects or disconnects from a data source). These events can be particularly useful when your application engages in asynchronous connection. Under asynchronous connections, programmatic control returns immediately to your application even before the actual connection is made. This feature allows your application to continue processing, but you cannot run any queries until the connection is complete. The Connect event provides notification that the connection has been established so that you may run queries. Table 14-3 shows a complete list of events supported by the connection object.

In order to use the event model of RDO, you must declare the rdoConnection variable WithEvents. This keyword was covered in detail earlier in

TABLE 14-3

rdoConnection Object Events

Event	Description
BeforeConnect	Fires just before connecting to a DSN.
Connect	Fires after connecting to a DSN
Disconnect	Fires after disconnecting from a DSN.
QueryComplete	Fires immediately after a query completes.
QueryTimeout	Fires when elapsed time exceeds the value set in the QueryTimeout property.
WillExecute	Fires just before a query executes.

the book when you created custom events for your Class Modules. When used with RDO, it enables all of the events for RDO in the VB code window.

■ Check It Out: Using RDO Events

1. Start a new Standard EXE project in Visual Basic.

2. Set a reference to RDO by selecting Project/References... from the menu.

3. In the References dialog, select Microsoft Remote Data Object 2.0.

4. Close the dialog.

5. Open the code window for Form1. In the [general][declarations] section, declare a variable as an rdoConnection object using the following code:

```
Private WithEvents m_Connection As RDO.rdoConnection
```

6. Examine the Object box in the code window. You should see the object m_Connection listed in the drop-down combo. Select this object.

7. Examine the Procedure box. You should see all of the events associated with the rdoConnection. You may select any of these events and write code in them.

Creating Queries

Queries are the heart of any database application. Even if you understand Visual Basic and RDO, you cannot successfully create a database application without knowledge of the Structured Query Language. Creating and executing queries on a DSN can have a tremendous impact on the overall performance of an enterprise system. Developers need knowledge of stored procedures, queries, and cursors to be successful.

Structured Query Language

SQL is the language of database applications. A complete treatment of SQL is far beyond the scope of this book, but if you want to write enterprise database applications, you must commit yourself to mastering the language. Many good books are available for learning SQL, but you can just as easily begin with the *Transact-SQL Reference for SQL Server 6.5*, which contains all of the SQL constructs supported by SQL Server.

Fundamentally, database applications concern themselves with adding, deleting, and editing data in the database. In this section, we will examine the basics of SQL and show how to write queries that perform these functions. The idea is to get you started. From here, you can go on to learn more complex SQL structures.

Selecting Records

One of the first things any application needs to do is select records from the database for display. Selecting records is done with a SQL SELECT statement. A SELECT statement specifies which data to return from tables in the database. Although the SELECT statement has many parts, most of them are optional (as is the case for most SQL constructs). A minimal SELECT statement contains the columns to return and the table names where the data can be found.

```
SELECT LastName FROM Customers
```

A statement like this is likely to bring back many, many records. In most cases your application is not interested in every name of every customer. Therefore, in addition to specifying the name of the column to return and the table, you might also want to specify a conditional statement that narrows the set of records returned. Conditional statements are created with the WHERE clause.

```
SELECT LastName FROM Customers WHERE ZipCode=22030
```

This statement is better because it limits the returned records to only customers in a certain zip code. WHERE clauses are useful in creating manageable result sets that do not waste system resources. You may use a WHERE clause against any information in the database. You may also perform partial string searches with wildcards using the LIKE statement.

```
SELECT LastName FROM Customers WHERE LastName LIKE 'A%'
```

This statement only returns customers whose last name begins with 'A'. Notice the use of the wildcard character (%) which specifies any string following the letter A. You can use wildcards like this to narrow any search. You can also combine criteria in the WHERE clause.

```
SELECT LastName FROM Customers WHERE LastName LIKE 'A%' AND Zip-
Code=22030
```

This statement returns only customers whose last names begin with A and who live in zip code 22030. This strategy is sure to limit the size of records returned from the database and ease the overall load on your system. The only problem left to solve is the order of the records. Even though you may have returned the records you want, the order will probably not be alphabetical. You can fix this using the ORDER BY clause.

```
SELECT LastName FROM Customers WHERE LastName LIKE 'A%' AND Zip-
Code=22030 ORDER BY LastName
```

Editing Records

Once records are selected and presented, users will begin to make changes to the data in your application. At some point in the process, the user will expect to be able to save the changes to the database. This is done through a SQL UPDATE statement. For example, the following statement updates the address of a customer:

```
UPDATE Customers SET Address = '444B Washington Ave' WHERE Cus-
tomerID = 1234567
```

Note the use of the WHERE clause in the statement, which strictly limits the changes to a single record using the CustomerID field. The WHERE clause is a vital part of the UPDATE statement, because without it, changes can be global. The following statement would change the last name of every customer in the database to Smith:

```
UPDATE Customers SET LastName = 'Smith'
```

What a disaster! You just lost all of your customer data! Remember, SQL statements directly affect the database, so test them on noncritical data before placing them in production. Global changes are not always bad, however. Sometimes you need to change all the records. The following statement raises all product prices by 10%:

```
UPDATE Products SET Prices = Prices * 1.1
```

Adding Records

Adding records to a database is done with a SQL INSERT statement. An INSERT statement can be used to add a new record using specific values or simply default values. Adding a new blank Customer record with default values is done with the following statement:

```
INSERT INTO Customers DEFAULT VALUES
```

If you want to insert exact information, you may specify the values of the fields. When you specify field values, you can give a value for every field, or just the ones you want. The following example shows how to specify an insert with selected fields:

```
INSERT INTO Customers (LastName, ZipCode) VALUES ('Smith',22030)
```

Deleting Records
Records are removed from a table using a SQL DELETE statement. DELETE statements specify the conditions for deletion using a WHERE clause. The following example deletes a single customer record based on CustomerID:

```
DELETE FROM Customers WHERE CustomerID = 1234567
```

The Microsoft Data Tools

Creating SQL statements can be tedious at best. Fortunately for VB programmers, we have tools to help. Visual Basic 5.0 ships with an add-in known as the "Microsoft Data Tools." These tools are designed specifically to help create SQL statements for use in your applications. In order to use them, you will have to have the tools installed (they are actually part of the Visual Studio '97 package) and select the add-in from the Add-In Manager.

To activate the Data Tools, simply select them from the Add-Ins menu in Visual Basic. When you do, the Developer Studio shell will launch and present you with the Data Tools. This assumes, of course, that you have Developer Studio installed. The point is, the Data Tools are available from Visual Basic, but they are not actually part of Visual Basic. Once the tools are launched, however, you can easily view every table and all the data from any ODBC data source.

In order to create a SQL statement, you must select the SQL button from the toolbar in Developer Studio. Then, simply drag the tables you want to query from the project window to the data area. As you drag tables, you will see a list of all fields in the table and relationships to other tables. Select the fields you want to return and the Data Tools will build an SQL query for you. When you are happy with the query, copy it to the clipboard and paste it in your code.

Figure 14-5

The Microsoft Data Tools.

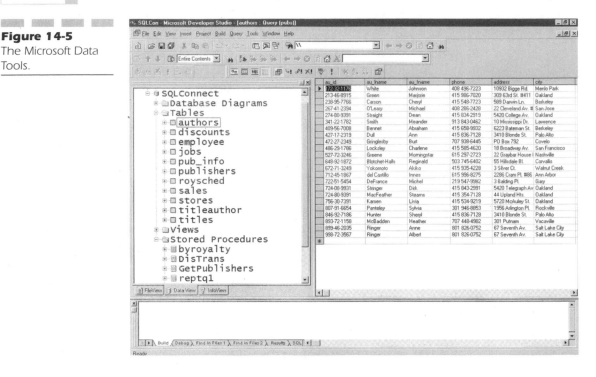

Figure 14-6

Building queries with the Data Tools.

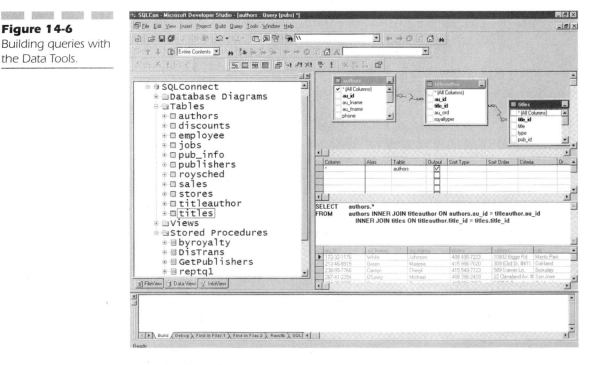

In addition to SQL statements created on the fly, you may also choose to utilize stored procedures. Stored procedures are precompiled queries that you can execute from RDO. Stored procedures are much faster than straight SQL queries because they are already compiled in the database itself. The Data Tools show all stored procedures that exist in the data source.

Check It Out: Building Queries

1. Start a new Standard EXE project in Visual Basic.

2. Save the project in a temporary directory. Projects must be saved to use the Data Tools.

3. Open the Add-In Manager by selecting Add-Ins/Add-In Manager… from the menu.

4. In the Add-In Manager, select Microsoft Data Tools.

5. Close the Add-In Manager.

6. Activate the Data Tools by selecting Add-Ins/Visual Database Tools… from the menu. Visual Studio '97 should start.

7. Inside Developer Studio, select the Publications DSN that you created earlier in this chapter.

8. Expand the table view for the Publications DSN.

9. Double click on the Authors table. You should see all of the data in the table.

10. Locate the Data Tools toolbar and push the button marked SQL. A window with the following SQL statement should appear:

```
SELECT  "authors".*
FROM  "authors"
```

11. Now push the Show Diagram Pane button on the same toolbar.

12. Drag the TitleAuthor and Titles tables from the project window to the work area. You should see the tables and their relationships to other tables.

13. Use the checkboxes in the table and select Au_Lname from Authors and Title from Titles. No other fields should be selected. The following SQL statement should be visible:

```
SELECT  authors.au_lname, titles.title
FROM  authors INNER JOIN titleauthor ON authors.au_id = titleau-
thor.au_id INNER JOIN titles
ON titleauthor.title_id = titles.title_id
```

14. Run the query by pushing the Run button that appears as an exclamation point.

15. Note the resulting records that appear at the bottom of the work area.

Understanding Cursors

Once you have created the SQL statement or stored procedure you want to execute, then you can use the rdoQuery object to run the statement against an ODBC data source. To use an rdoQuery object, you simply instantiate it and set the SQL property to the statement you want to run. The Execute method then runs the statement against the DSN specified by the ActiveConnection property.

```
Set m_Query = New RDO.rdoQuery
Set m_Query.ActiveConnection = m_Connection
m_Query.SQL = "SELECT * FROM Customers"
m_Query.CursorTye = rdOpenStatic
m_Query.Execute
```

When you run a query and return records, you are bound by the limitations of the cursor driver you selected when the connection was made. Within these limitations, rdoQuery objects can then specify the type of cursor to employ. A cursor is different from a cursor driver. The driver sets some boundaries inside which the rdoQuery object must operate, but the CursorType property refines the features of the resulting rdoResultset object. These features include scrolling and update capabilities. The CursorType is defined by the constants rdOpenForwardOnly, rdOpenStatic, rdOpenKeyset, and rdOpenDynamic.

rdOpenForwardOnly

The rdOpenForwardOnly option produces a set of records that can only be scrolled in the forward direction using the MoveNext method of the rdoResultset object. This cursor is the lightest and fastest of all the available cursors, but it is also the most limiting. This cursor is always generated in conjunction with the rdUseNone cursor driver. This cursor is ideal for populating lookup lists or read-only data.

rdOpenStatic

The rdOpenStatic option produces a set of records that is fully scrollable, but has a static membership. This option generates a snapshot of the data at a moment in time. As updates are made to the database, they are not reflected in the recordset. The returned records may exist on the client or the server depending on the cursor driver selected.

rdOpenKeyset

The rdOpenKeyset option does not actually produce records, but rather a set of keys that reference records. The membership of the keyset is fixed, but the set is fully scrollable. The keyset will accept updates to the records it references through methods of the rdoResultset object such as Edit and Delete. These cursors may be built on either the client or the server depending on the cursor driver selected.

rdOpenDynamic

The rdOpenDynamic option produces a keyset just like the rdOpenKeyset, but the rdOpenDynamic option produces a result that dynamically reflects all changes made to the records either through methods of the rdoResultset or through direct changes via SQL UPDATE statements. This is the heaviest and slowest of all the cursors and is generally not used in an enterprise application.

rdoQuery Events

Just like the rdoConnection object, the rdoQuery object supports events. The events may be used to perform asynchronous queries and receive notification when the query completes. Running an asynchronous query is accomplished by using the rdAsyncEnable option in the Execute method. When the query finishes, it fires the QueryComplete event.

Managing Resultsets

When a query is executed, you will want to be able to access the resulting records. RDO provides an object that allows you to access the results of the last query. This result set is referenced through the LastQueryResults property of the rdoConnection object.

```
Dim m_Connection As RDO.rdoConection
Dim m_Query As RDO.rdoQuery
Dim m_Resultset As RDO.rdoResultset

m_Connection.Connect = "DSN=BIBLIO"
m_Connection.CursorDriver = rdUseClientBatch
m_Connection.EstablishConnection
Set m_Query = New RDO.rdoQuery
Set m_Query.ActiveConnection = m_Connection
m_Query.SQL = "SELECT * FROM Customers"
m_Query.CursorTye = rdOpenStatic
m_Query.Execute

Set m_Resultset = m_Connection.LastQueryResults
```

Once the rdoResultset is captured, you can directly manipulate the records through methods of the rdoResultset object. Depending on the cursor type you create, you can use the MoveFirst, MoveLast, MoveNext, and MovePrevious methods to scroll through the records. Once you point to the desired record, you can read the data out and place it in ActiveX controls for display.

Whenever you want to populate a GUI with the results of a query, you may choose to display all of the results at once, say in a grid, or you may want to display the results one at a time in a detail view. In either case, there are accepted routines that can be used to produce the desired output. If, for example, you have a set of records that represent customer data, displaying all at once is done by looping through the rdoResultset object and populating the grid. The following code shows how you might add customer data to a grid from a resultset named m_Resultset:

```
Do While Not m_Resultset.EOF
  Grid1.AddItem m_Resultset("LastName").Value
  M_Resultset.MoveNext
Loop
```

Displaying records one at a time requires the combination of Move methods and a common routine used for display. This display routine populates an individual control based on the current record pointer. Whenever you read data from the database, be careful of NULL values. NULL values can cause run-time errors if you try to assign them to a TextBox. Instead, concatenate an empty string on the end of every field you read to prevent NULL values. The following code shows a routine that populates a series of TextBoxes in a control array from a resultset called m_Resultset. The bang operator (!) is used to specify the name of the database field to place in the TextBox.

```
txtFields(0).Text = m_Resultset!Title & ""
txtFields(1).Text = m_Resultset!Au_lname & ""
txtFields(2).Text = m_Resultset!Pub_Name & ""
txtFields(3).Text = m_Resultset!PubDate & ""
txtFields(4).Text = m_Resultset!Title_ID & ""
```

The UserConnection Object

While understanding the complete RDO model is essential, managing the code and queries can be difficult. In an effort to ease the management and

Figure 14-7 The UserConnection object.

coding requirements to use RDO effectively, Visual Basic 5.0 introduces the UserConnection object. The UserConnection object is a special interface to the RDO model that you use at design time to set up connections and queries rather than writing code to set them up at run time.

In order to make the UserConnection object available, you must select it in the Components dialog. This is done on the Designers tab after selecting Project/Components... from the menu. After you select it, the object appears on the VB menu under Project/Add ActiveX Designer/Microsoft User Connection.

Once inserted, you can use the UserConnection to establish a connection to a DSN and set up query objects. Accessing the objects is accomplished through property sheets. In these sheets, you may specify the DSN and cursor driver as well as SQL statements and stored procedures.

EXERCISE: USING REMOTE DATA OBJECTS

Visual Basic offers many different data access techniques. For creating client/server applications in the 32-bit operating systems, RDO is the tool of choice. In this exercise, you will investigate RDO features through the UserConnection object.

Setting Up the ODBC Data Source

Step 1

Before you can use the ODBC API, you must set up an ODBC data source for use. In this exercise, we will be interacting with the Microsoft Access database BIBLIO.MDB that ships with Visual Basic.

Establishing the data source is not a Visual Basic function, but rather a Windows Operating System function. Setup is done using the ODBC Administrator utility found in Windows. Locate the ODBC Administrator in the Main Program Group under the Control Panel icon. In the Control Panel, you will see an icon labeled ODBC. Double click on this icon to start the ODBC Administrator.

Step 2

When you start the ODBC Administrator, you will see the Data Sources dialog box. This dialog box lists all of the ODBC data sources currently available on your machine. In order to create a new data source, push the Add button.

When you push the Add button, you will see the Add Data Source dialog. This dialog lists all of the ODBC drivers available on your machine. For this exercise, you will need the Microsoft Access ODBC drivers. Select this driver from the list and press OK.

Step 3

When you press OK, you will see the ODBC Setup dialog. In this dialog, you can define the new data source. First, you must provide a name for the data source. This can be any name, but for this exercise, name the data source BIBLIO.

Next, select the BIBLIO.MDB database by pushing the Select button and navigating the file system until you find the BIBLIO.MDB file. Select this file. After you have selected the file, push the OK button and back out of the ODBC Administrator. You have now defined the data source.

Creating the GUI

Step 4

Create a new subdirectory called /RDO for this Visual Basic project using the File Manager. Start a new Visual Basic Standard EXE project. Construct a graphic user interface on Form1 using the following information:

```
frmUserConnection
  BorderStyle                        3 'Fixed Dialog
  Caption                            Remote Data Objects
  Font Name                          MS Sans Serif
  Font Size                          13.5
cmdMovePrevious
  Caption                            Prev
  Height                             510
  Left                               7875
  Top                                3240
  Width                              1545
cmdMoveNext
  Caption                            Next
  Height                             510
  Left                               5385
  Top                                3240
  Width                              1545
cmdMoveLast
  Caption                            Last
```

```
                    Height              510
                    Left                2895
                    Top                 3240
                    Width               1545
               cmdMoveFirst
                    Caption             First
                    Height              510
                    Left                405
                    Top                 3240
                    Width               1545
               txtFields
                    Height              480
                    Index               4
                    Left                3495
                    Top                 2355
                    Width               6120
               txtFields
                    Height              480
                    Index               3
                    Left                3495
                    Top                 1815
                    Width               4680
               txtFields
                    Height              480
                    Index               2
                    Left                3495
                    Top                 1260
                    Width               6120
               txtFields
                    Height              480
                    Index               1
                    Left                3495
                    Top                 720
                    Width               6120
               txtFields
                    Height              480
                    Index               0
                    Left                3495
                    Top                 180
                    Width               6120
               lblLabels
                    Height              345
                    Index               4
                    Left                180
                    Top                 2415
                    Width               2985
               lblLabels
                    Caption             Year Published:
                    Height              345
                    Index               3
                    Left                180
                    Top                 1875
                    Width               2985
               lblLabels
                    Caption             Company Name:
                    Height              345
                    Index               2
                    Left                180
                    Top                 1335
                    Width               2985
```

```
lblLabels
  Caption                          Author:
  Height                           345
  Index                            1
  Left                             180
  Top                              780
  Width                            2985
lblLabels
  Caption                          Title:
  Height                           345
  Index                            0
  Left                             180
  Top                              240
  Width                            2985
```

The User Connection Object

Step 5

In this section, you will use the UserConnection Object to access the functionality of RDO. Select Project/Components... from the menu to display the Component dialog. In the Component dialog, select the Designers tab. In the Designers tab, ensure the Microsoft UserConnection object is selected. Add a new UserConnection to the project by selecting Project/Add ActiveX Designer/Microsoft UserConnection from the menu. A new UserConnection object should appear.

Step 6

Name the new UserConnection object conData. The UserConnection object is a new object in VB5 that gives you graphical access to an RDO connection. The UserConnection allows you to define the ODBC connection and queries/stored procedures without writing extra code. Microsoft estimates that the UserConnection can reduce code by up to a factor of 10 for data connections with RDO.

Step 7

In the Properties of the UserConnection, connect to the BIBLIO ODBC data source by selecting it under the Use ODBC Datasource option. When the connection is defined, close the properties dialog for the User-Connection.

Step 8

Right click on the UserConnection icon in the workspace area to reveal a popup menu. Select Insert Query from this menu to add a new query to the UserConnection. The new query dialog will appear and allow you to define the query. Name the new query AllTitles.

Select the option Based on User-Defined SQL and add the following SQL statement to the query designer:

```
SELECT Titles.Title, Titles.ISBN, Authors.Author, Titles.[Year Pub-
lished], Publishers.[Company Name] FROM Publishers, Titles,
Authors, [title author] WHERE((((Authors.Au_ID = [title
author].Au_ID) AND ([title author].ISBN = Titles.ISBN)) AND
(Titles.PubID = Publishers.PubID))) ORDER BY Titles.Title
```

Step 9

Close the Query designer to save the new query definition in the User-Connection. Saving the SQL this way prevents you from having to create it in several lines of code. It is simply stored with the UserConnection object.

Coding the Application

Step 10

Open the code window for frmUserConnection. In the [general][declarations] section, define the following variables for the RDO Connection and Resultset objects:

```
'This variable is for the User Connection
Private m_Connection As conData

'This variable is for the Recordset
Private m_Resultset As RDO.rdoResultset
```

Step 11

The database connection is made in the Form_Load event. After the connection is made, we run the predefined SQL query contained in the User-Connection. Add the following code to the Form_Load event to access the data source:

```
'Establish the Connection
Set m_Connection = New conData
m_Connection.EstablishConnection

'Get Results
Set m_Resultset = m_Connection.rdoQueries("AllTitles"). _
OpenResultset(rdOpenDynamic)

'Fill Form
ShowData
```

The ShowData routine is used to populate the textboxes with data. Create a new procedure as follows to show the data:

```
Public Sub ShowData()

'This routine shows data in the Text Boxes
txtFields(0).Text = m_Resultset!Title
txtFields(1).Text = m_Resultset!Author
txtFields(2).Text = m_Resultset![Company Name]
txtFields(3).Text = m_Resultset![Year Published]
txtFields(4).Text = m_Resultset!ISBN

End Sub
```

Step 12

The remainder of the code provides the functionality for the buttons. Add the following code to activate the buttons:

```
Private Sub cmdMoveFirst_Click()
  m_Resultset.MoveFirst
  ShowData
End Sub

Private Sub cmdMoveLast_Click()
  m_Resultset.MoveLast
  ShowData
End Sub

Private Sub cmdMoveNext_Click()
  m_Resultset.MoveNext
  If m_Resultset.EOF Then
    Beep
    m_Resultset.MoveLast
  Else
    ShowData
  End If
End Sub

Private Sub cmdMovePrevious_Click()
  m_Resultset.MovePrevious

  If m_Resultset.BOF Then
    Beep
    m_Resultset.MoveFirst
  Else
    ShowData
  End If
End Sub
```

Step 13

Save the project and run it.

Increasing Your Knowledge

Help File Search Strings

■ Remote Data Objects

- ODBC
- Cursors

Books OnLine References

- "Understanding Remote Data Access Options"
- "Using Remote Data Objects and Remote Data Control"

Mastery Questions

1. Name all of the different kinds of ODBC data sources.
   a. User DSN
   b. Folder DSN
   c. File DSN
   d. System DSN

2. Which type of data source is available to all users of a single machine?
   a. User DSN
   b. Folder DSN
   c. File DSN
   d. System DSN

3. What methods can create an rdoResultset?
   a. rdoConnection.OpenResultset
   b. rdoConnection.LastQueryResults
   c. rdoQuery.OpenResultset
   d. rdoQuery.LastQueryResults

4. What keyword is used to include RDO events in your code?
   a. HaveEvents
   b. RaiseEvents
   c. WithEvents
   d. Event

5. What arguments are valid parts of an ODBC connect string?
   a. TIME
   b. PWD
   c. DSN
   d. UID

6. Which constants are valid cursor drivers in RDO?
   a. rdDoNotUse
   b. rdUseIfNeeded
   c. rdUseClientBatch
   d. rdUseOnly

7. Which is a valid DSN-less Connect string?
   a. "DRIVER={SQL Server};
   DSN=;SERVER=(local);DATABASE=pubs;UID=sa;PWD=;"
   b. "DRIVER={SQL Server};SERVER=(local);
   DSN=;DATABASE=pubs;UID=sa;PWD=;"
   c. . "DRIVER={SQL Server};SERVER=(local);DATABASE=pubs;
   DSN=;UID=sa;PWD=;"
   d. . "DRIVER={SQL
   Server};SERVER=(local);DATABASE=pubs;UID=sa;PWD=;
   DSN=;"

8. What constants represent valid cursors in RDO?
   a. rdOpenForwardOnly
   b. rdOpenStatic
   c. rdOpenKeyset
   d. rdOpenDynamic

9. Which cursors are available for use if the cursor driver is selected as rdUseNone?
   a. rdOpenForwardOnly
   b. rdOpenStatic
   c. rdOpenKeyset
   d. rdOpenDynamic

10. Which of the following methods are available to a cursor designated as rdOpenForwardOnly?
    a. MoveFirst
    b. MoveLast
    c. MoveNext
    d. MovePrevious

11. If the cursor driver is set to rdUseIfNeeded, which type of cursor is preferred?
    a. Server
    b. ClientBatch
    c. ODBC
    d. ForwardOnly

12. Which cursor type is fully updatable and has dynamic membership?
    a. rdOpenForwardOnly
    b. rdOpenStatic
    c. rdOpenKeyset
    d. rdOpenDynamic

13. What types of queries can be used with the UserConnection object?
   a. SQL queries
   b. Stored procedures
   c. SQL parameter queries
   d. SQL UPDATE queries

Answers to Mastery Questions

1. a, c, d

2. d

3. a, b, c

4. c

5. b, c, d

6. b, c

7. d

8. a, b, c, d

9. a

10. c

11. a

12. d

13. a, b, c, d

15

Scaling Visual Basic Applications

What You Need to Know

You need to be familiar with ODBC concepts and the Remote Data Objects (RDO).

What You Will Learn

- Three-Tier Architecture
- Business Objects
- Unattended Execution
- Microsoft Transaction Server

Visual Basic 5.0 is built from the ground up to support three-tier architecture. In three-tier architecture, you create thin clients that access business objects distributed throughout the enterprise. These objects in turn access the data sources used in the application. The entire design is aimed at providing thin clients and robust, scalable applications. Figure 15-1 shows the fundamental architecture of a typical three-tier application.

Figure 15-1
Simplified three-tier
architecture.

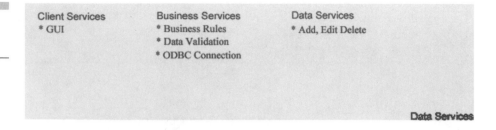

Visual Basic provides several features that allow business objects (typically ActiveX EXE projects) to scale. Many of these features are somewhat limited, but depending on the number of users, VB may handle the projected load easily. However, as user load increases from one to hundreds or even thousands, system resources such as threads, ODBC connections, and object instances play a major role in the overall performance of you enterprise.

Single-User Applications

Single-user applications are the simplest to build in Visual Basic, and they obviously have no scaling issues to worry about. In this type of application, developers might open a connection to a data source through RDO and leave the connection open throughout the session. Because the app has only one user, you do not have to be concerned with monopolizing resources like data connections.

Similarly, threading is not an issue. Visual Basic applications are single-threaded applications, which means that they can have only one execution point within them. Other languages (e.g., Visual C++) support multithreading, which allows an application to executed multiple lines of code nearly simultaneously. Multithreading means an application can perform more than one function at a time-such as printing while saving to a file. In Visual Basic, multiple functions are queued and executed in the order requested. This is not a problem for one user, but it can be for many users.

Memory usage is hardly a concern either. With only a single user, no one is competing for resources, so your application can be much less concerned with releasing objects that are not currently in use. Most of your business objects should be created as ActiveX DLLs to enhance speed, and they will run directly on the client. Perhaps the only real memory issue you will face is returning a record set from the database that is larger than the available memory, resulting in disk caching.

All in all, single-user apps are simple, and most Visual Basic programming reflects this style. Many VB developers simply do not understand scaling issues and continue to develop large applications as if they were being used by just a few people. Scaling requires breaking old habits.

Multi-User Applications

As soon as you leave the world of single-user applications, you must immediately concern yourself with conserving system resources. Immediately, you must take the business objects off the client and move them to the network. This means that you can no longer use ActiveX DLLs because they cannot be remotely accessed. Instead, you will use ActiveX EXEs and call to them through the Distributed Component Object Model (DCOM). At this point, several key features of ActiveX EXEs come in to play.

Class Modules in ActiveX EXEs have a special property known as the Instancing property that helps business objects scale. This property has several settings, but we are primarily concerned with two: Multi-Use and Single-Use. These settings determine how an ActiveX EXE behaves when its services are requested by many users at the same time.

Multi-Use ActiveX EXE

Setting the Instancing property to Multi-Use causes the ActiveX EXE to produce a new instance of the requested class module each time a new client calls. All of the created instances run together inside the same memory space occupied by the ActiveX EXE. This model is best described as "One Server-Many Instances."

The problem with a Multi-Use ActiveX EXE goes back to the single-threaded nature of Visual Basic. In the single-threaded business object, all of the instances share the same memory space. Therefore, they cannot execute simultaneously on the one thread. This means that although each client has its own instance of the business object, VB cannot service more than one client at a time.

Imagine that the business object is responsible for querying a database and performing a multitable join. Suppose further that the operation takes six seconds to complete. Since the object cannot multithread, it must make each client wait in the queue while requests are processed in turn. If you are the tenth person in the queue, you would wait 60 seconds for service! In fact, you may never get service, since remote requests typically time out in 10 seconds based on the default value of the OLE-ServerBusyTimeout property of the App object.

Single-Use ActiveX EXE

One solution to the problem of single-threaded business objects is to set the Instancing property to Single-Use. Unlike Multi-Use objects, Single-Use objects do not create multiple instances in the same memory space. Single-Use components create a whole new instance of the ActiveX EXE-one for each client that calls. Essentially, when you can't multithread, you multitask. This leads to a situation in which no client has to wait for service because a new instance of the ActiveX EXE is always created for each individual user.

The issue with Single-Use components, of course, is resource depletion. Each client is going to start a new copy of the component, which may rapidly consume system resources on the target machine. This is the same old story-performance versus resources.

Unattended Execution

Another solution to the scaling problem lies in declaring your ActiveX EXE as an unattended execution component. Components declared for unattended execution support multithreading through the apartment threading model. Declaring unattended execution is done in the properties dialog for the project. In this dialog, you may check the Unattended Execution option and declare whether you want to provide a thread for each client or utilize a thread pool.

Selecting unattended execution is not a panacea by any means. This choice has some significant consequences and design issues. Many people unfamiliar with multithreaded applications believe that multithreaded apps always run faster and better than single-threaded apps. This is not necessarily true. Multithreaded apps must share the microprocessor resources, and the constant swapping can actually lead to poorer performance for an individual process while providing improved performance on the aggregate.

Additionally, threading works best when the number of active threads roughly matches the number of processors in the machine. If you have a single processor, this means that you cannot have more than one active thread at a time. Any more will actually lead to performance degradation.

Performance is also greatly affected by the threading options you select. Selecting Thread per Object causes your component to spawn a

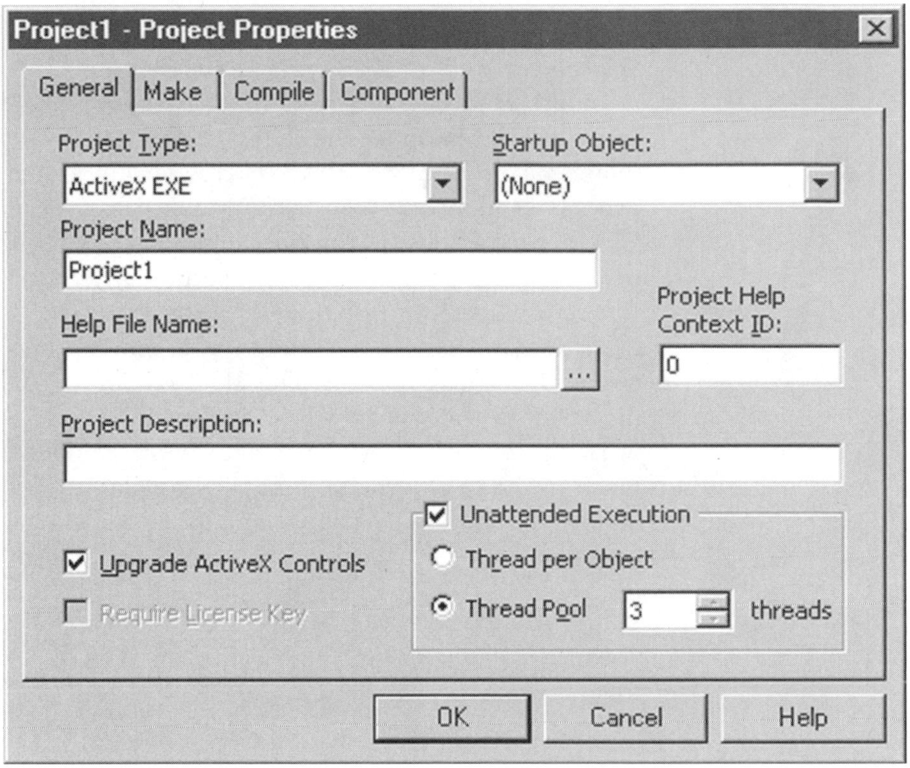

Figure 15-2
Selecting unattended execution.

new thread for each client that calls. This can cause problems quickly as dozens of threads are created on a machine with just a single processor. Better still is to select Thread Pool. Here you can limit the threads to just a few and they will be reused when they arc inactive. Thread pooling is one of the cornerstones of scalability.

The Microsoft Transaction Server

While the options built in to Visual Basic provide some nice scaling features, they generally are not powerful enough to scale into large-enterprise applications. When you begin to build applications that are used by dozens of people, you must be concerned with more than just the options you can set in VB. Specifically, you must concern yourself with pooling three major resources: threads, object instances, and ODBC connections.

Providing control over the major resources is the job of the Microsoft Transaction Server (MTS). This provides all of the resource and component management capabilities your application needs while requiring very few changes in you VB code. Best of all, you do not concern yourself with issues like Multi-Use versus Single-Use. You simply create VB business objects and let MTS manage the details.

In fact, MTS does not even utilize ActiveX EXEs. It uses ActiveX DLLS instead. The reason for the change is that once you place an ActiveX DLL under MTS control, MTS takes care of beahind like the surrogate out-of-process component. Microsoft Transaction Server provides all of the DCOM capabilities that allow clients to call your object while managing resources that promote scalability and fault tolerance.

The primary interface to MTS is the Transaction Server Explorer. The MTS Explorer provides a way to add, remove, and manage components under the control of MTS. The MTS Explorer is built on the same GUI metaphor as the Windows File Explorer.

In the MTS Explorer, you may view components that are on any machine in the enterprise. Components placed under MTS control reside inside a package. A package is an administrative grouping of components that allows them to exist in the same security context. Inside of a package, you install one or more components which can be nothing more than Visual Basic class modules compiled into an ActiveX DLL.

The process of placing an object under MTS control is relatively simple. First you must construct an ActiveX DLL project in Visual Basic. At the

Figure 15-3
The MTS Explorer.

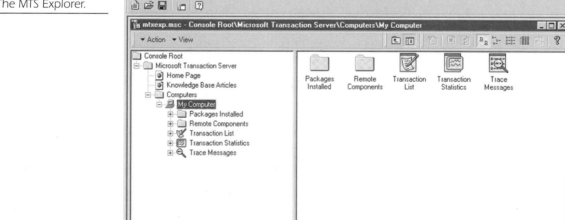

simplest level, the DLL you create need not have any special features associated with it. Simply create the DLL and provide at least one public method in the Class Module. As a simple example, we compiled an ActiveX DLL to test MTS. This DLL called MyObject has one Class Module called Test. This class has one method called CallMe. The following is the complete code for SimpleObject.Simple:

```
Public Function Test() As String
  Test = "I received your call!"
End Function
```

Once compiled into a DLL, it is placed under MTS control. In the MTS Explorer, we define a new package by clicking the Packages folder and selecting File/New... from the menu. In the Package Wizard we can define a new package for the ActiveX DLL. Then MTS prompts for a name and creates the new package.

Figure 15-4
Package Wizard

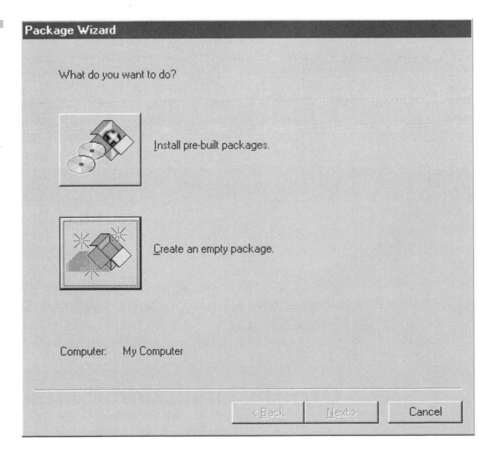

Figure 15-5
The Component
Wizard.

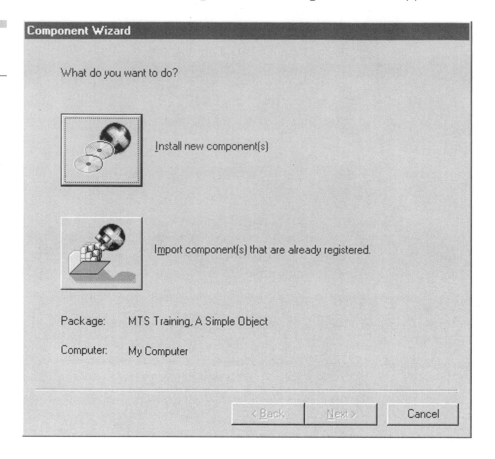

Once the package is created, we can easily install the component by clicking on the new package to open it. Inside the package, we must click on the Components folder and select File/New… from the menu. In the Component Wizard, we then select to install our new component by picking it from a list of all registered components on the machine.

At this point, the component is installed under MTS control. Now whenever the component is called by a VB application, MTS will intercept the request and process it using a pool of objects created from the MyObject ActiveX DLL. Now the object is available to all applications on the enterprise that support DCOM, and just for placing the object in MTS, you gain the benefits of automatic thread pooling.

ODBC Connection Pooling

ODBC Connection pooling is a requirement for scaling applications. In most Visual Basic applications, developers do not concern themselves with ODBC connection pooling. Front end components routinely open connections at the beginning of a session and leave them open until the application is terminated. Under MTS, ODBC connections are not opened by the front end or web page, but rather by the business object itself. Therefore, the ActiveX DLLs we create in Visual Basic must open a connection, read or write to the data source, and then close the connection. The clients never accesses the database directly.

If the client never has direct access to a record set, then you may wonder how the data can be retrieved. Clients calling MTS objects rely on the objects to return data from a data source in a predictable format. This format may be as simple as a delimited string or as complex as a collection of objects. In any case, the business object is responsible for selecting records from the database, packaging them and returning them to the client

As a simple example, consider a business object that reads entries from a data source and returns them as a delimited string. The client will then parse the returned string and place the entries in a list box. This is a typical technique for creating lookup lists. This object is an ActiveX DLL with one method. The business object uses the Remote Data Objects (RDO) to access an ODBC data source created from the data source. The component opens a connection to the data source and reads out all required data. The data are added to a string that is delimited by pipe characters (|). In order to allow MTS to pool the ODBC connection, the object opens a connection at the beginning of the function and closes the connection before returning the string of publishers.

```
'Open Connection
Dim objConnection As RDO.rdoConnection
Dim objResultset As RDO.rdoResultset

Set objConnection = New RDO.rdoConnection
objConnection.Connect = "DSN=Publications"
objConnection.EstablishConnection rdDriverNoPrompt, True

'Run Query
Set objResultset = objConnection.OpenResultset _
("SELECT Pub_Name FROM Publishers ORDER BY Name")

'Create Return String
Dim strTemp As String
strTemp = ""

Do While Not objResultset.EOF
  strTemp = strTemp & objResultset!Name & "|"
  objResultset.MoveNext
Loop

'Close Connection
objResultset.Close
objConnection.Close
Set objResultset = Nothing
Set objConnection = Nothing
```

The string is returned to a client through a call to the business object. The data are then parsed and added as entries to a list box. Other applications, such as Internet web pages, may use the same object to fill a lookup list as well.

```
'Variables
Dim objQuery As MyObject.MyData

'Create the business object
Set objQuery = New MyObject.MyData

'Run the Query
Dim strReturn As String
strReturn = objQuery.GetDelimData

'Fill the ListBox
Dim intStart As Integer
Dim intCurrent As Integer

intStart = 1
intCurrent = 1

'Parse the returned string
Do While intCurrent < Len(strReturn)
  intCurrent = InStr(intStart, strReturn, "|")
  If intCurrent = 0 Then Exit Do
  List1.AddItem Mid$(strReturn, intStart, intCurrent - intStart)
  intStart = intCurrent + 1
Loop
```

As your site gets more complicated, you will want to return more than just a simple list from a lookup table, you will want to return complete sets of records. In this case, we will use an array to retrieve records and return them to the client. This follows the same general technique as producing a delimited string, but instead you pass a Variant array. Since arrays cannot be passed directly, you must pass array data inside a Variant. Typically your business object would provide a Variant as an argument that is passed ByReference. Passing ByRef ensures that changes in the business object are reflected in the calling client.

```
Public Function GetData(ByRef vRecords As Variant) As Boolean
```

The business object fills the array using a special method of the rdoResultset object called GetRows. GetRows is specifically designed to fill a Variant array with data from a rdoResultset object. Once the Variant array is filled, all of the data is reflected back to the client since the original array was passed By Reference.

```
'Run A Stored Procedure Called "TitlesByCompany"
m_Connection.EstablishConnection
m_Connection.TitlesByCompany "'%Microsoft%'"
```

```
'Get Results
Dim m_Resultset As RDO.rdoResultset
Set m_Resultset = m_Connection.LastQueryResults

If m_Resultset.BOF And m_Resultset.EOF Then
  ReDim vArray(1, 0)
  vArray(0, 0) = "No Records!"
Else
  vArray = m_Resultset.GetRows(lngMaxRows)
End If
```

In the client, the array is used to populate a GUI element like a grid. In this way a complete set of data can be returned without having direct access to an ODBC connection in the web page. The page can also write data back by calling methods that perform SQL UPDATE functions inside of Visual Basic business objects. Thus add, edit, and delete functionality can be provided through objects under MTS. These objects are subsequently available to all applications on the enterprise without rewriting.

Object Instance Pooling

The final concern when scaling applications is object instance pooling. Object pooling allows for the creation of a group of objects that may be shared by many applications. Since a pool of objects is shared, the system has to create far fewer objects than if each user had his or her own copy. Therefore, resources are used more efficiently. MTS provides for object instance pooling, but you must make slight changes to your business object in order to take full advantage of MTS.

Whenever a business object is created under MTS, a special object is also created to monitor the process. This new object is called an Object-Context. The ObjectContext is responsible for observing the resource usage of a business object and notifying MTS when the object can be returned to the pool. Normally, objects are only pooled after the client releases any reference being held in a variable. However, if we change our code slightly, we can actually notify MTS directly that our object is ready to return to the pool. Cycling business objects in and out of the pool in MTS is called Just-In-Time (JIT) activation and As-Soon-As-Possible (ASAP) deactivation. JIT/ASAP pooling is the most efficient way to use system resources.

Enabling JIT/ASAP pooling requires that your business object communicate directly with the ObjectContext assigned by MTS. This is done by using a special function called GetObjectContext. The GetObjectContext function is a member of the Transaction Server API. To access these functions, you must set a reference to MTS in the References dialog of Visual Basic. Once the reference is set, you may use the following code to access your ObjectContext:

```
Dim objContext As MTxAS.ObjectContext
Set objContext = GetObjectContext()
```

When your business object has a reference to the ObjectContext, it can easily tell MTS to recycle the object by calling the SetComplete method. This method causes MTS to place the business object back in the pool for reuse. When objects are reused this way, MTS can use just a few objects to service many clients simultaneously. Of course, recycling objects means that any data retained in the object is lost. Therefore, you should not create business objects that persist data. Instead save any required data in the client that calls through Session variables.

Distributing Enterprise Applications

After you have mastered the principles of three-tier applications, you will need to distribute your software across the enterprise. Normally, applications should be developed on a single workstation for ease and distributed for testing. The process of distributing your components assumes you are deploying from your development machine.

Business Object Distribution

Whether MTS is the final target for a Visual Basic business object or the object will simply run distributed as an EXE, a remote distribution file must be created for all distributed COM objects. This special file, also called a VBR file, contains information required by the client software to access a distributed component. Creating a VBR file is a matter of checking the Remote Server Files option of the Component tab under the Project Properties dialog. When the business object is compiled, Visual Basic will create the VBR file.

If the application is an Active EXE, then a setup may be created by using the Visual Basic Setup Wizard. The Setup Wizard requires no special information except whether your component will be distributed using

Figure 15-6
The Project Properties
dialog.

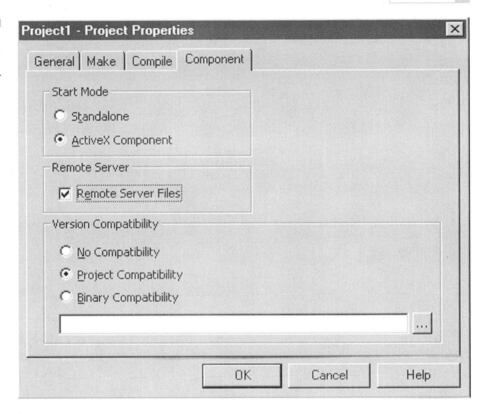

DCOM or Remote Automation-typically, distributed components should use DCOM. When the setup wizard is complete, the resulting setup disks may be used to distribute the server to another machine. On the target machine, the DCOM configuration utility DCOMCNFG may be run to set security properties for the component.

If the server component will be distributed as an ActiveX DLL running under MTS, then you can create a setup by simply using the MTS Explorer. From the MTS Explorer, packages can easily be moved to other machines by using the Export function. Export creates all of the files necessary to move a package from one Transaction Server environment to another. When exported, MTS will create a PAK file. The PAK file can be installed on any other machine running MTS. Once it is installed, any client can call the object.

Client Software Distribution

Client software is also distributed using the Setup Wizard. The difference between the distributed and normal setup has to do with the use of the

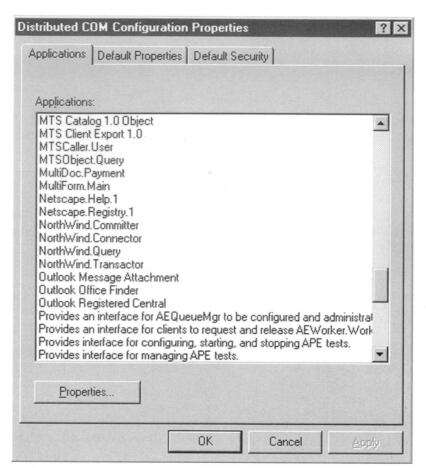

Figure 15-7 DCOMCNFG.

VBR file. During the setup, Visual Basic prompts you to specify any ActiveX servers that are required by the application. Normally, VB will select the local components in this screen. At this point, the distributed version of the ActiveX component can be added by selecting Add Remote... from the Wizard. Then, simply navigate the VBR file instead of the local DLL or EXE.

When software is distributed on the enterprise client, a special Registry setting is installed to reference remote components. The key is called RemoteServerName and refers to the remote server where the ActiveX component is installed. The VBR file is used to create this key. The value of the key is defined by the SetUp Wizard when you add the VBR file. After this, the setup program may be used on any client in the enterprise.

EXERCISE: USING MICROSOFT TRANSACTION SERVER

Microsoft Transaction Server (MTS) is a product designed to manage all of the overhead associated with scalable business applications. Using MTS, you can immediately gain the benefits of resource pooling and transaction processing. This exercise will show you how to use MTS to build fault-tolerant, scalable applications.

Setting up the ODBC Data Source

Step 1
Before you can use the ODBC API, you must set up an ODBC data source for use. In this exercise, we will be interacting with the Microsoft Access database NORTHWIND.MDB that ships with Visual Basic.

Establishing the data source is not a Visual Basic function, but rather a Windows Operating System function. Setup is done using the ODBC Administrator utility found in Windows. Locate the ODBC Administrator in the Main Program Group under the Control Panel icon. In the Control Panel, you will see an icon labeled ODBC. Double click on this icon to start the ODBC Administrator.

Step 2
When you start the ODBC Administrator, you will see the Data Sources dialog box. This dialog box lists all of the ODBC data sources currently available on your machine. In order to create a new data source, push the Add button.

When you push the Add button, you will see the Add Data Source dialog. This dialog lists all of the ODBC drivers available on your machine. For this exercise, you will need the Microsoft Access ODBC drivers. Select this driver from the list and press OK.

Step 3
When you press OK, you will see the ODBC Setup dialog. In this dialog, you can define the new data source. First, you must provide a name for the data source. This can be any name, but for this exercise, name the data source NorthWind.

Next, select the NORTHWIND.MDB database by pushing the Select button and navigating the file system until you find the NORTH-WIND.MDB file. Select this file.

Step 4

After you have selected the file, push the OK button and back out of the ODBC Administrator. You have now defined the data source.

Creating the Business Objects

Step 5

Create a new sub directory called \MTS for the Visual Basic project using the File Manager. Beneath this new directory, create two new folders: one called Client and one called Server.

Start a new Visual Basic ActiveX DLL project. Select Project/References... from the menu and set a reference to the Remote Data Objects 2.0 component and the Microsoft Transaction Server 1.0 Library.

Step 6

Change the properties of Project1 by selecting Project/Project1 Properties... from the menu. In the Project Properties dialog, change the name of the project to MTSObject and change the description to MTS Training Exercise Objects. Press OK.

Step 7

This DLL will use a single class to query the NorthWind data source and return records. In this exercise, you will use custom marshaling techniques to return data from a lookup table and records from a query. Change the name of the Class1 component to Query, and save the project in the \MTS\SERVER directory.

Step 8

Create a new method in the Query class by selecting Tools/Add Procedure... from the menu. In the Add Procedure dialog, set the following attributes:

```
Name: GetCategories
Type: Function:
Scope: Public
```

Step 9

After inserting the new function, modify the function to return a String data type. The complete function should now look like this:

```
Public Function GetCategories() As String
End Sub
```

The GetCategories method returns a delimited string with all of the category names. This technique is a quick way to return lookup table information to a list. Add the following code to the GetCategories method to read the category names from the data source and format the return string:

```
On Error GoTo GetCategoriesErr
  Dim objConnection As RDO.rdoConnection
  Dim objResultset As RDO.rdoResultset
  'Make Connection
  'NOTE: Connections made to MTS must
  'be made with the rdDriveNoPrompt option!!
  Set objConnection = New RDO.rdoConnection
  objConnection.Connect = "DSN=NorthWind"
  objConnection.EstablishConnection rdDriverNoPrompt, True

  'Run Query
  Set objResultset = objConnection.OpenResultset _
  ("SELECT CategoryName FROM Categories")

  'Build Return String
  Dim strReturn As String
  strReturn = ""

  Do While Not objResultset.EOF
    strReturn = strReturn & objResultset("CategoryName") & "|"
    objResultset.MoveNext
  Loop

  'Close connection to allow pooling
  objResultset.Close
  objConnection.Close
  Set objResultset = Nothing
  Set objConnection = Nothing
GetCategoriesExit:
  GetCategories = strReturn
  Exit Function

GetCategoriesErr:
  strReturn = Err.Description
  Resume GetCategoriesExit
```

Step 10

After all of the category names are returned to the front end, the user will select a category and want to see the associated products from the data source for that category. This is accomplished with a method called Get-Products. GetProducts will return all of the products in an array to the client. The client can then use the array to fill a grid. Add the GetProduct method by selecting Tools/Add Procedure... from the menu. In the Add Procedure dialog, set the following attributes:

```
Name: GetProducts
Type: Function
Scope: Public
```

Change the GetProducts function to accept two arguments and return a Boolean value to indicate success or failure. The resulting function should look like this:

```
Public Function GetProducts(strCategory As String, _
ByRef arrProducts As Variant) As Boolean

End Function
```

GetProducts uses the category to search for products and fill an array with results. It also uses Just-In-Time Activation to ensure that instances of the Query class are recycled as soon as possible. Add the following code to the GetProducts function to return an array of products to the client:

```
On Error GoTo GetProductsErr
  'Get Object Context
  Dim objContext As MTxAS.ObjectContext
  Set objContext = GetObjectContext()

  'Declare Database Objects
  Dim objConnection As RDO.rdoConnection
  Dim objResultset As RDO.rdoResultset

  'Make Connection
  Set objConnection = New RDO.rdoConnection
  objConnection.Connect = "DSN=NorthWind"
  objConnection.EstablishConnection rdDriverNoPrompt, True

  'Run Query
  Dim strSQL As String
  strSQL = "SELECT ProductName,CompanyName,UnitPrice "
  strSQL = strSQL & "FROM Products,Suppliers,Categories "

  strSQL = strSQL & "WHERE Products.SupplierID=Suppliers.SupplierID
"   strSQL = strSQL & "AND Products.CategoryID=Categories.CategoryID
"   strSQL = strSQL & "AND Categories.CategoryName=
  '" & strCategory & "
  '"
  Set objResultset = objConnection.OpenResultset(strSQL)

  'Build Return Array
  arrProducts = objResultset.GetRows(100)

  'Close connection to allow pooling
  objResultset.Close
  objConnection.Close
  Set objResultset = Nothing
  Set objConnection = Nothing

  'Tell MTS we are done
  GetProducts = True
  objContext.SetComplete

GetProductsExit:
  Exit Function

GetProductsErr:
  'Tell MTS we failed
  GetProducts = False
  objContext.SetAbort
  Resume GetProductsExit
```

Placing the Business Object under MTS control

Step 11
Now that the business object is complete, you can compile it by selecting File/Make MTSOBJECT.DLL from the menu. When the DLL is created, save your work and exit Visual Basic.

Step 12
From the Start menu, start the Microsoft Transaction Server Explorer. In the MTS Explorer, select My Computer and Packages Installed. Add a new package to MTS by selecting File/New from the menu. In the Package Wizard, select to install a new empty package. Push Next. Name the new package MTS NorthWind Exercise. Push Next. Set the package identity to Interactive User. Push Finish.

Step 13
Double click on the new package you installed to reveal the Components folder. Now double click on the Components folder itself. Select File/New from the menu to add a new component to your package. In the Component Wizard, select to import components that are already registered. Locate the component MTSObject.Query and select it. Push Finish. Your component is now registered and under control of the Microsoft Transaction Server.

Creating the Front End

Step 14
Start Visual Basic and select to create a new Standard EXE project. This project will be the front end to the business objects. Select Project1/Properties... from the menu and change the project name to FrontEnd.

Select Project/References... from the menu and set a reference to the business object you just created called MTS Training Exercise Objects. Save your project in the \MTS\CLIENT directory.

Step 15
Add a list box to Form1 in the new project. Name the list box lstCategories. This list box will be used to show a list of all of the product categories. Filling the list is done by calling to the business object and retrieving the category names as a delimited string. When the string is received, you can parse it and fill the list. Add the following code to the Form_Load event to fill the list:

```
'Show Form
Show
DoEvents

'Variables
Dim objQuery As MTSObject.Query

'Create the business object
Set objQuery = New MTSObject.Query

'Run the Query
Dim strReturn As String
strReturn = objQuery.GetCategories

'Fill the ListBox
Dim intStart As Integer
Dim intCurrent As Integer

intStart = 1
intCurrent = 1

'Parse the returned string
Do While intCurrent < Len(strReturn)
  intCurrent = InStr(intStart, strReturn, "|")
  If intCurrent = 0 Then Exit Do
  lstCategories.AddItem Mid$(strReturn,
  intStart, intCurrent - intStart)
  intStart = intCurrent + 1
Loop
```

Step 16

Save your project and run it. Verify that the list box fills with product categories. Stop your project and return to design mode.

Step 17

Add a grid control to your project by selecting Project/Components... from the menu. In the Components dialog, select the Microsoft Flex Grid 5.0 control. Push OK.

Place a grid control on your form and name it grdProducts. This grid will be used to see all of the products in a given category. The grid will fill when a user clicks on a category. Add the following code to the Click event of lstCategories to finish the project:

```
'Create Business Object
  Dim blnReturn As Boolean
  Dim objProducts As MTSObject.Query
  Set objProducts = New MTSObject.Query

  'Run Query
  Dim varReturn As Variant
  blnReturn = objProducts.GetProducts _
  (lstCategories.List(lstCategories.ListIndex), varReturn)

  'Handle return value If blnReturn Then
  'Set default grid appearance
  grdProducts.Clear
  grdProducts.FixedRows = 0
```

```
grdProducts.FixedCols = 0
grdProducts.Rows = 0
grdProducts.Cols = 3
grdProducts.AddItem "Product" & vbTab & "Company" & vbTab &
"Price"
 'Fill the Grid
 Dim i As Integer
 For i = 0 To UBound(varReturn, 2)
   grdProducts.AddItem varReturn(0, i) _
   & vbTab & varReturn(1, i) & vbTab _
   & varReturn(2, i)
 Next

 If grdProducts.Rows > 1 Then grdProducts.FixedRows = 1
Else

 'Show Error Message
 MsgBox "Error:" & varReturn(0, 1) & "; Source:" & varReturn(1, 1)
End If
```

Step 18
Save your project and run it. Try clicking on a product category in the list box!

Increasing Your Knowledge

Help File Search Strings

■ Remote Data Objects

■ Instancing

■ Unattended Execution

Books OnLine References

■ "Guide to Building Client Server Applications with Visual Basic"

Mastery Questions

1. Which are valid settings for the Instancing Property?
   a. Multi-Use
   b. Many-Use
   c. Single-Use
   d. Mono-Use

2. What are valid selections for the unattended execution feature?
   a. Threads per client
   b. Threads per object
   c. Thread pool
   d. Multi-thread

3. What threading model is supported by unattended execution?
   a. Free threading
   b. Trailer threading
   c. Home threading
   d. Apartment threading

4. What resources are pooled by the Microsoft Transaction Server (MTS)?
   a. Disk space
   b. Object instances
   c. ODBC connections
   d. Threads

5. What kind of components are used in MTS?
   a. ActiveX Document DLL
   b. ActiveX Document EXE
   c. ActiveX DLL
   d. ActiveX EXE

6. How is ODBC Connection pooling implemented in MTS?
   a. By using the function PoolConnection
   b. By managing connections in the business layer
   c. By managing connections in the client services layer
   d. By preventing data access with stored procedures

7. What method of the rdoResultset object puts records into an array?
   a. GetRecords
   b. GetResults
   c. GetColumns
   d. GetRows

8. What function is used to get a reference to the ObjectContext?
   a. GetRef
   b. GetRows
   c. GetObjectContext
   d. GetContext

9. What methods are used to tell MTS that an object can be pooled?
   a. SetAbort
   b. SetComplete
   c. SetDone
   d. SetFinish

10. What file is used to aid in remote client installs?
    a. VBR file
    b. VBD file
    c. MSC file
    d. MMM file

Answers to Mastery Questions

1. a, c
2. b, c
3. d
4. b, c, d
5. c
6. b
7. d
8. c
9. a, b
10. a

16

Creating
Visual Basic
Add-Ins

What You Need to Know

- Classes
- Interfaces
- ActiveX DLLs

What You Will Learn

- The VB5 Object Model
- Connecting and Disconnecting an Add-In
- Creating Menus and Toolbars on VB5
- Handling Design-Time events

Visual Basic Add-Ins are ActiveX components that you can use to extend the functionality of the Visual Basic Integrated Debugging Environment (VBIDE). Add-ins are built to add features that Visual Basic does not implement or to provide tools

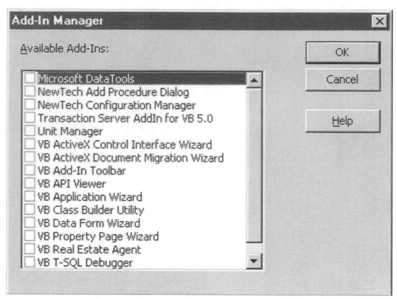

Figure 16-1   *The Add-In Manager.*

that are specific to your particular business. They are an excellent way to enhance your productivity.

Visual Basic 5.0 ships with several add-ins that were created for you by Microsoft. You can find a complete list of these by examining the contents of the Add-In Manager. The Add-In Manager is used to activate selected add-ins for use inside Visual Basic. You can access the Add-In Manager by selecting Add-Ins\Add-In Manager... from the menu.

The Visual Basic Object Model

Creating add-ins for Visual Basic requires an intimate knowledge of the VBIDE object model. The object model is a graphical representation of the objects contained inside Visual Basic. These objects help you communicate with and manipulate the components inside a Visual Basic project. Using these objects, you may add menu items to the VBIDE, change properties of forms and modules in a project, detect when projects are changed, and perform a host of other useful functions.

The VBIDE object model begins with the top-level object, named VBE. The VBE object is an object that represents the current Visual Basic envi-

Figure 16-2
The Visual Basic
object model.

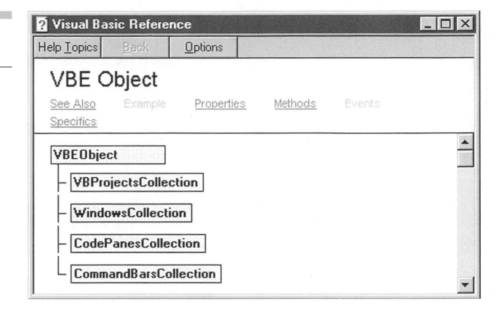

ronment. This is a reference to the running instance of VB itself. This object is the gateway to all of the other aspects of the VBIDE. All of your add-ins will begin by establishing a reference to this object.

After you establish a reference to the VBE object, you can access one of several collections that represent projects and components inside VB. The VBProjects collection is a collection of all of the currently open projects inside VB. Because VB can open more than one project at a time, the collection can be used to programmatically examine and manipulate components inside any open project.

The CodePanes collection is a collection of all of the currently visible code windows. Each time a new code window is open, VB adds it to the CodePanes collection. This collection allows you to access code in any open project so that you can create tools that generate code. The Code-Pane itself is considered to be the window that contains the code module. In the VB object model, you actually access individual lines of code by using the CodeModule property of any selected CodePane object.

The set of all open windows including dockable windows is represented by the Windows collection. The Windows collection allows you programmatic access to a Window object for the purpose of managing display and real estate. You can easily open and close windows from your add-in using this collection.

One of the key features of an add-in is the ability to place menu items and toolbars in the VBIDE. Establishing new menus and tools is done using CommandBar objects. CommandBar objects are generic selection objects that can represent either a button on a toolbar or a menu item. In fact, CommandBar objects allow you to create all kinds of menu controls. Using CommandBar objects you can place a ListBox control directly inside a menu or toolbar, for example. You are not limited to simple text and buttons. VB manages the set of all menu and toolbar controls through the CommandBars collection.

Once you have established new menu items or tools, you will want to know when a user selects your control. Detecting events inside VB is the function of the Events collection. The Events collection provides a number of interesting and useful events for your add-in. Visual Basic supports menu item and toolbar events through the CommandBarEvents object. You can detect File events such as File Save actions through the FileControlEvents object. When users change the references in the References dialog, you can detect the action through the ReferenceEvents object. Detecting modified controls and components is accomplished through the VBControlsEvents, SelectedVBControlsEvents, VBComponentsEvents, and VBProjectsEvents objects.

In addition to the VBIDE object model, you will also need to understand the Office object model. The Office object model provides all the features your VB add-in will use to create toolbars and menus. These items are actually part of a shared library that is used not only by Visual Basic, but by each member of the Office family as well. If you look closely at various Office products, you will notice that these applications have the same types of menu items and tools as VB supports. Office products, for example, can support lists and combo boxes just like VB. This is because the menus are driven from the same common library.

Accessing these two critical object models from your VB add-in requires that you set a reference to them in the References dialog. The VBIDE object model can be found in the References dialog listed as Microsoft Visual Basic 5.0 Extensibility. The Office model is listed as Microsoft Office 8.0 Object Library. To begin creating any add-in, you start with an ActiveX DLL project and set a reference to these two libraries.

Once you have set the references for these models, you should spend some time examining them in the Object Browser. The Object Browser is an excellent way to gain knowledge about the objects, properties, events, and methods of the two models. The VB extensibility library is listed as VBIDE in the Object Browser, whereas the Office 8.0 library is simply listed as Office. You will refer to the Object Browser often while developing add-ins.

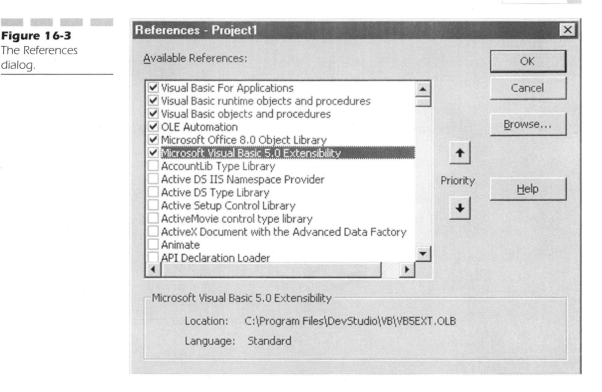

Figure 16-3
The References dialog.

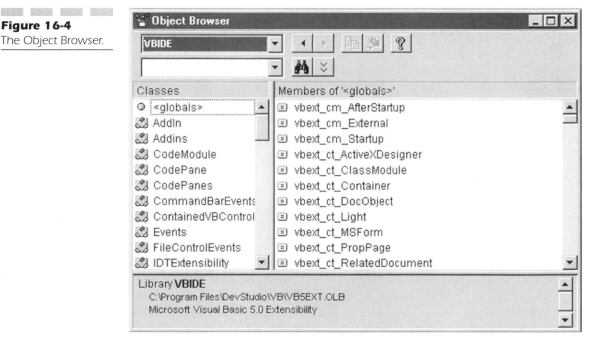

Figure 16-4
The Object Browser.

Connecting an Add-In to VB

Add-ins are made available to Visual Basic when they are listed in a special file used by the Add-In Manager. This file is called VBADDIN.INI. This file lists all of the add-ins that are available to VB and is the basis for the list generated when you select the Add-In Manager from Visual Basic. Before your add-in can appear in the list, it must make an appropriate entry in this file.

VBADDIN.INI is an initialization file that contains only one section, which is dedicated to maintaining the list of available add-ins. You can examine this file by opening it in Notepad. The file typically resides in the \WINDOWS directory on your system. The file contains an entry for an add-in based on the ProgID of the primary connection class in the add-in. The value of each entry is either 1 for True or 0 for False and specifies whether the add-in should be immediately loaded when VB starts. Those that are not loaded immediately may be loaded at any time by using the Add-In Manager. Any add-ins left running when VB shuts down are automatically restarted when VB is run the next time. Listing 16-1 shows a typical VBADDIN.INI file.

```
[Add-Ins32]
RealEstateAgent.Connect=1
AddProc.Connect=1
AttilaPro.Connector=0
WebDock.Connect=0
RVBAddInMenus.Connector=0
RVBAddIn.Connector=0
MTxAddIn2.RegRefresh=1
Dashboard.Connect=0
UnitMan.Connect=1
VBSDIAddIn.Connect=0
DataToolsAddIn.Connect=0
AppWizard.Wizard=0
WizMan.Connect=0
ClassBuilder.Wizard=0
AddInToolbar.Connect=0
ControlWiz.Wizard=0
DataFormWizard.Wizard=0
ActiveXDocumentWizard.Wizard=0
PropertyPageWizard.Wizard=0
APIDeclarationLoader.Connect=0
ResEdit.Connect=1
```

Normally, an entry for your add-in should be made during the installation process. However, you must also make an entry at design time so that the add-in can be tested. You could certainly make the entry by hand, but a simple way is to provide a helper function that you use only during debugging. This helper function is created in a standard module. It can be called anything, but we use AddToINI. This function makes an API call to add the necessary entry for an add-in to the VBADDINI.INI. Listing 16-2 shows a typical routine for a project titled MyAddIn.

```
Declare Function WritePrivateProfileString& _
Lib "kernel32" Alias "WritePrivateProfileStringA" _
(ByVal AppName$, ByVal KeyName$, _
ByVal keydefault$, ByVal FileName$)

Sub AddToINI()
  'Call this to update the INI in VBIDE
  Dim ErrCode As Long

  ErrCode = WritePrivateProfileString _
  ("Add-Ins32", "MyAddIn.Connect", "1", _
  "vbaddin.ini")

  Debug.Print "Added to VB INI file!"

End Sub
```

Once the routine is added to your project, you can execute it in Break mode to make the appropriate entry. Typically, you do this by starting the add-in and then immediately entering Break mode. In the Immediate window, you type AddToINI and the routine responds with "Added to VB INI file!" Once this is finished, you can continue running the add-in for testing.

When an add-in is connecting to the VBIDE, VB expects to find a certain set of methods available in a class module that it can call to provide information. Therefore, the next step in connecting is to create a class with the desired methods. Although you can create either an ActiveX EXE or ActiveX DLL project as an add-in, ActiveX DLLs are much faster and are therefore the preferred method. When you create an ActiveX DLL, VB provides your project with a default class named Class1. Typically, we change the name of this class to Connect or Connector and use it as the primary class to receive input from the VBIDE. Note that the name you give this class is arbitrary, but it must be properly reflected in the AddToINI routine or your add-in will not be properly registered with the VBADDIN.INI file.

Once you have your class module named, you must provide a set of four guaranteed methods that VB can call when your add-in connects. From our previous work, we know exactly how to guarantee that a class implements certain methods. This is accomplished through interfaces. In fact, the VBIDE object model provides an interface that your class must implement to be an add-in. This interface is called IDTExtensibility. Therefore, every add-in must have a class that contains the following code in the [general][declarations] section:

```
Implements IDTExtensibility
```

When your class implements this interface, you will be forced to provide four methods for your class: OnConnection, OnDisconnection,
OnStartupComplete, and OnAddinsUpdate. OnConnection is called by
the VBIDE when your add-in is first connected to VB. OnDisconnection is
called when your add-in is disconnected. Both of these methods may be
called if VB starts up or shuts down, or when your add-in is manipulated
in the Add-In Manager. The OnStartupComplete method is called when
the complete startup of the VBIDE is finished. OnAddinsUpdate is called
when changes to the VBADDINI.INI file are saved. These methods represent the starting point for much of what your add-in does.

Menus and Toolbars

When your add-in is connecting to the VBIDE, you will probably want to
add menu items and tools to the environment. The menu items and tools
are typically removed again when the add-in disconnects. You can add
and remove these items in the OnConnection and OnDisconnection methods respectively. Creating these items is done using the Office object
model.

All of the menu items and toolbars that you interact with in VB are
members of the CommandBars collection. This collection has members
that give you access to nearly everything you can select from in VB. The
main toolbars are accessed by the names Standard, Edit, Debug, and
Form Editor. The Main menu is accessed by the name Menu Bar. Through
these, you can easily add new items, delete items, and replace items.

Before you can place your own items, you must declare a variable to
represent the new item. This is done in the [general][declarations] section
of the primary Connect class. Typically objects are declared as Com-
mandBarControl objects, but you can also get special controls such as
CommandBarButton objects and CommandBarComboBox objects. When
you declare these special objects you can get additional properties,
events, and methods for the object. For example, declaring an object as
CommandBarComboBox allows you to use the AddItem method to place
items in a list on a toolbar. After the new items are declared, you must
add them to the toolbar or menu of your choice. The following code shows
how to add a new button to the Standard toolbar in Visual Basic.

```
'Declare Toolbar Buttons and Menus
Public objCommandBar As Office.CommandBarControl

Private Sub IDTExtensibility_OnConnection _
```

```
(ByVal VBInst As Object, _
ByVal ConnectMode As VBIDE.vbext_ConnectMode, _
ByVal AddInInst As VBIDE.AddIn, Custom() As Variant)

  'Save instance of VB
  Set objVBE = VBInst

  'Create CommandBar
  Set objCommandBar = objVBE.CommandBars("Standard").Controls.Add _
  (msoControlButton, , , , True)

End Sub
```

Once the new control is attached to a menu or toolbar, you will want to receive events from the control in your add-in. This is done by declaring a Handler object. Handlers are members of the VBIDE object model and are of type CommandBarEvents. The following code declares a Handler object for a new item.

```
Private WithEvents objMenuHandler As VBIDE.CommandBarEvents
```

Notice that the object is declared WithEvents. Creating the object this way enables your class module to receive events from the new control. All you have to do is connect the Handler object to the menu item of your choice. The following code shows how to add a new menu item to the View menu in VB and trap when the item is clicked.

```
'Menu Item and Handler Objects
Public WithEvents objMenuHandler As VBIDE.CommandBarEvents
Public objMenu As Office.CommandBarControl

Private Sub IDTExtensibility_OnConnection(ByVal VBInst As Object,
ByVal ConnectMode As VBIDE.vbext_ConnectMode, ByVal AddInInst As
VBIDE.AddIn, Custom() As Variant)

  'Save the vb instance
  Set m_VBIDE = VBInst

  'Build Menus
  Set objMenu = m_VBIDE.CommandBars("Menu Bar"). _
  Controls("View").CommandBar.Controls.Add(1, , , 1, True)

  objMenu.Caption = "&My Item"

  Set objMenuHandler = m_VBIDE.Events.CommandBarEvents(objMenu)
End Sub
```

With the Handler object in place, all you have to do is write the appropriate code in the Click event generated for the Handler object. You will find this event in the code module because you declared the Handler object WithEvents. After you receive the click, you can use the rest of the VB object model to manipulate the current project.

Manipulating Project Components

Menu items are not the only way to receive events in your add-in. Visual Basic 5.0 supports a wide range of events that are generated in response to user actions. These events can be very useful in creating tools that effectively manage the VB environment. When you create add-ins, it is often useful to know when a user is manipulating the current project. In particular, changes to the elements of a project may be of interest. Visual Basic allows you to discover, for example, when a user adds a new project, component, or control to the environment. You can also detect when one of these elements is removed or selected.

To set up your add-in to detect project events, you must declare an appropriate event-handling object using the WithEvents keyword. You may declare objects that handle project, component, and control events. You declare these handlers based on the design of your add-in. The following lines of code declare events for all the supported project events.

```
Private WithEvents m_ProjectHandler As VBIDE.VBProjectsEvents
Private WithEvents m_ComponentHandler As VBIDE.VBComponentsEvents
Private WithEvents m_ControlsHandler As VBIDE.VBControlsEvents
Private WithEvents m_SelectedHandler As VBIDE.SelectedVBControlsEv-
ents
```

Once the event handler is declared, you must connect it to the appropriate project in the VBIDE. Connecting the event handlers always begins by connecting the VBProjectsEvents object first. You can connect an event handler to the current project using the following code:

```
Set m_ProjectHandler = VBE.Events.VBProjectsEvents
```

This line of code is typically included in the OnConnection method of the IDTExtensibility interface. When you connect this event handler, your add-in can receive four project events: ItemAdded, ItemRemoved, ItemRenamed, and ItemActivated. You can use these events to take action in your add-in, but you must also use them to connect the event handler for the project components. For example, when your add-in receives the ItemAdded event, you know that a new project was added. Therefore, you must connect the component handler. The following code shows how this is done.

```
Private Sub m_ProjectHandler_ItemAdded _
(ByVal VBProject As VBIDE.VBProject)
```

```
    Set m_ComponentHandler = m_VBE.Events.VBComponentsEvents(VBPro-
ject)
End Sub
```

When you connect the component events handler, your add-in can receive six events related to forms and modules. These events are ItemAdded, ItemActivated, ItemReloaded, ItemRemoved, ItemRenamed, and ItemSelected. You can then use these events to take action when a user manipulates a form or module in the project. In addition, you should also connect the controls events handler. So when a new component is added, you receive the ItemAdded event just as you did for a new project. You could then connect controls events through the following code:

```
Set ControlsHandler = _
objVBE.Events.VBControlsEvents _
(m_VBE.ActiveVBProject, m_VBComponent.Designer)

Set SelectedHandler = _
objVBE.Events.SelectedVBControlsEvents _
(m_VBE.ActiveVBProject, m_VBComponent.Designer)
```

Once you are set up to receive events about the projects, components, and controls, your add-in can easily manipulate these same components. Suppose, for example, that you were creating an add-in that did nothing except change the BackColor of forms from the default color to white. Every time a user places a new form in the project, your add-in will automatically change the color. This could easily be done by detecting the ItemAdded event of the Component Handler object. Then, in the event, you manipulate the currently selected component as follows:

```
Private Sub objCompHandler_ItemActivated(ByVal VBComponent As
VBIDE.VBComponent)

  If TypeOf VBComponent Is vbext_ct_VBForm Then
    VBComponent.Properties("BackColor").Value = &HFFFFFF
  End If

End Sub
```

In addition to the project components, you can also modify the project code. The Visual Basic add-in model allows access to the code windows of any component through the CodePanes and VBComponents collections. To access these collections, you simply address the CodeModule property of any CodePane or VBComponent. As an example, the following code can be used to get a reference to the CodeModule object of the currently selected form or module:

```
Set objModule = m_VBE.ActiveCodePane.CodeModule
```

Once you have a reference to the CodeModule, VB provides a number of properties and methods that allow you to generate code. The AddFrom-File and AddFromString methods allow you to add code to a module from either a saved file or a string variable, respectively. You can delete code from a module using the DeleteLines method. Code replacement is accomplished by using the ReplaceLine method. VB also provides properties and methods to search a code module, count the lines in various sections, and manipulate the variable and procedure declarations. In general, you can accomplish a wide variety of coding tasks using the add-in architecture.

Distributing the Add-In

Distributing a Visual Basic Add-In is generally as easy as distributing any in-process ActiveX DLL project. Simply use the Setup Wizard that ships with Visual Basic to generate the distribution disks. The only rub in the entire procedure is that an add-in must have an entry in the VBADDIN.INI file before it can be seen inside the VB environment.

To make an entry in VBADDIN.INI, you must customize the setup created by the Setup Wizard. The Setup Wizard actually uses a Visual Basic project to perform the installation. This project already exists in your VB installation. You can find this project in the directory structure \PRO-GRAM FILES\DEVSTUDIO\VB\SETUPKIT\SETUP1. The project is SETUP1.VBP. Open this file directly in VB to modify the Setup Wizard.

The SETUP1.VBP project is specifically designed to be easily modified. Although the project has many pieces, everything you need to change is kept in the Form_Load event of the form called frmSetup1. In this section, you can add code to make the appropriate entry in VBADDIN.INI. Typically, we place the code right after the label marked ExitSetup. The code looks the same as the code you use for registering the add-in during debugging:

```
ExitSetup:
  '**********
  'Custom Code for Add-In installation
  '**********
  Dim ErrCode As Long

  ErrCode = WritePrivateProfileString _
  ("Add-Ins32", "MyAddIn.Connect", "1", _
  "vbaddin.ini")
```

Once the new code is added, you will need to save the project and recompile it. After recompiling, run the Setup Wizard and distribute the add-in as an ActiveX DLL project. Because the Setup Wizard uses the compiled version of SETUP1.VBP, your changes will become part of the installation. When the user sets up your add-in, the appropriate entries will be made to VBADDIN.INI. The only thing to remember is to comment out these lines of code before using the Setup Wizard to distribute any other application.

Existing Add-Ins

Although you can easily create add-ins yourself with Visual Basic, many useful ad-ins already ship with the product. In many cases, these add-ins help solve design problems or simplify repetitive coding. In this section, we'll briefly review the add-ins that ship with Visual Basic.

The Microsoft Data Tools

Visual Basic 5.0 is part of the Visual Studio suite of tools. Several of these tools, including Visual C++, Visual J++, and Visual InterDev, share a common development environment known as the Developer Studio shell. The Developer Studio hosts these tools as well as related productivity tools. In particular, the Developer Studio shell is home to the Microsoft Data Tools. The data tools provide a graphical editing environment for data contained in ODBC-compliant data sources.

Although Visual Basic is a member of the Visual Studio suite, it does not share the Developer Studio shell. Instead, as you know, VB has its own environment. However, the data tools add-in allows you to connect to the Developer Studio from Visual Basic so that you can use the data tools in your VB projects. Figure 16-5 shows the data tools in use with Visual Studio.

The data tools are most useful when you connect to data sources using the UserConnection object. The UserConnection object contains information about the ODBC data source your VB application is accessing. This information can be passed directly to the data tools for manipulating the data source. The data tools allow the creation of SQL queries and stored procedures as well as direct data manipulation.

Figure 16-5
The Microsoft
Data Tools.

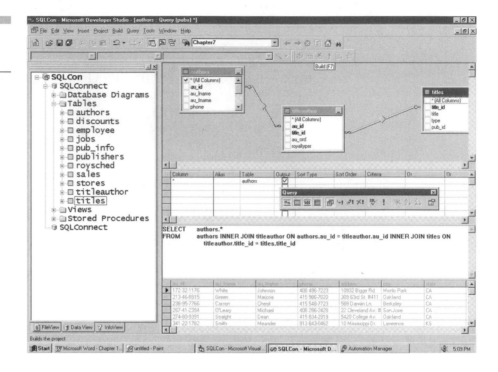

SQL Debugging

The SQL debugging add-in allows you to create and debug SQL stored procedures from within Visual Basic. When it is used in conjunction with the Microsoft Data Tools, you gain a powerful set of tools for manipulating database elements. Figure 16-6 shows the SQL debugging interface.

Establishing SQL debugging from Visual Basic can be tricky. You will find several of the necessary installation elements located on the \TOOLS directory of your Visual Basic CDROM. In order to set up the debugging, be sure to install these components on the server where the SQL Server databases reside. You must also be sure that the SQL Server service is not set to logon as a service. You can check this setting in the Services applet of the Control Panel.

With the Server components installed and the SQL Service properly set up, you can then access SQL debugging through the add-in. As with the Microsoft Data Tools, the best way to use the debugger is in conjunction with the UserConnection object. This allows you to move between your project, debugging, and the data tools easily.

Figure 16-6
SQL debugging.

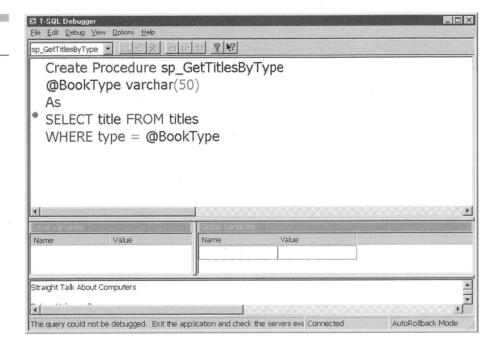

ActiveX Control Interface Wizard

The ActiveX Control Interface Wizard is an add-in designed to generate the laborious code required to subclass an ActiveX control. In the control creation portion of this book, we examined the required steps for subclassing and extending existing ActiveX controls. Because Visual Basic does not support inheritance, each property, event, and method of a subclassed control must be re-created by hand. This code is difficult to write accurately, which is why you should use the add-in to automatically generate the code. Figure 16-7 shows the ActiveX Control Interface Wizard.

ActiveX Document Migration Wizard

The ActiveX Document Migration Wizard is an add-in designed to create an ActiveX Document from an existing Visual Basic form. This is a helpful utility, but you should be aware that this tool actually changes the existing project. Therefore, be sure you back up any project before using this add-in. Figure 16-8 shows the ActiveX Document Migration Wizard.

Figure 16-7
The ActiveX Control
Interface Wizard.

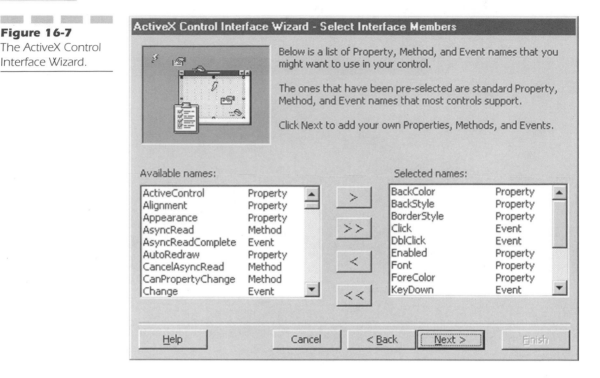

Figure 16-8
The ActiveX
Document Migration
Wizard.

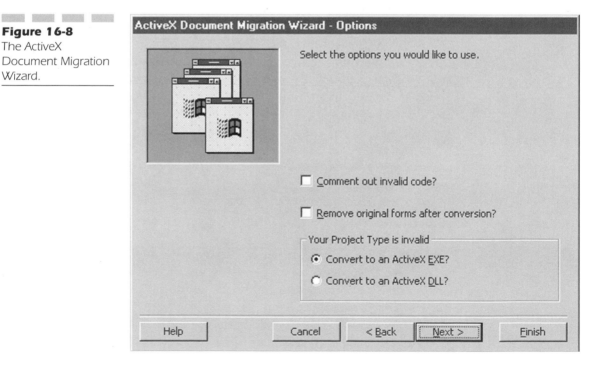

API Viewer

The API Viewer add-in is a utility for retrieving template function call definitions for the Windows API. Using this tool, you can search for, copy, and paste into code Declare Function statements, constants, and structures. This tool is the same as the tool found in the VB program group but is accessible from the IDE. Figure 16-9 shows the API Viewer.

Application Wizard

The Application Wizard is an add-in that generates a framework for an MDI, SDI, or Explorer application. Using this add-in, you can easily create the essentials of any application style including fundamental interfaces. Figure 16-10 shows the Application Wizard.

Class Builder Utility

The Class Builder Utility is perhaps one of the most useful add-ins that ships with Visual Basic. The Class Builder allows you to design a complete object hierarchy of collections and objects. You design the hierarchy using

Figure 16-9
The API Viewer.

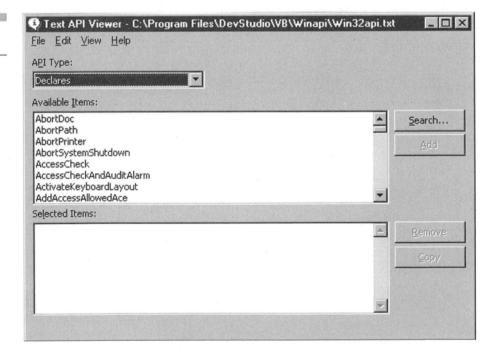

Figure 16-10
The Application
Wizard.

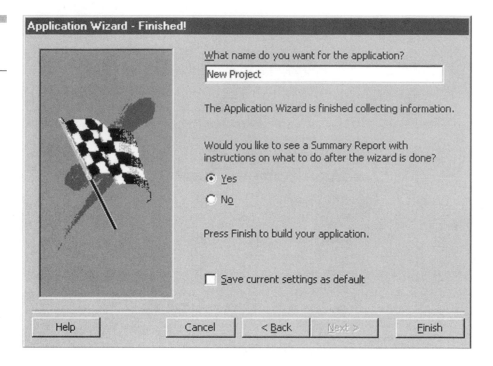

visual elements in a tree structure. For each element, you can specify
properties and methods. The utility also supports collections. When you
are finished defining the hierarchy, you can generate all of the class mod-

Figure 16-11
The Class Builder
Utility.

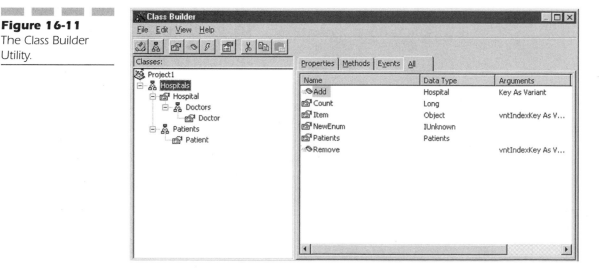

ules and code necessary to implement the structure. The Class Builder even takes care of setting key procedure attributes for default members as well as implementing the For Each… syntax for collections. Figure 16-11 shows the Class Builder.

Data Form Wizard

The Data Form Wizard builds a data input form from a database definition. This tool makes it easy to construct data input forms. The forms, however, are relatively plain and consist of textboxes as the input mechanism. The add-in is a good starting point, but probably not useful in a wide range of applications. Figure 16-12 shows the Data Form Wizard.

Property Page Wizard

The Property Page Wizard helps construct Property Pages for your ActiveX Control projects. These pages can then be attached to any prop-

Figure 16-12
The Data Form
Wizard

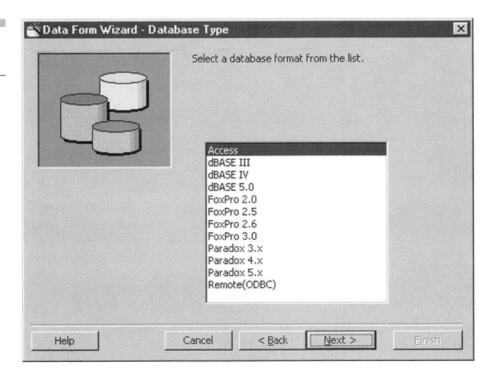

Figure 16-13
The Property Page
Wizard.

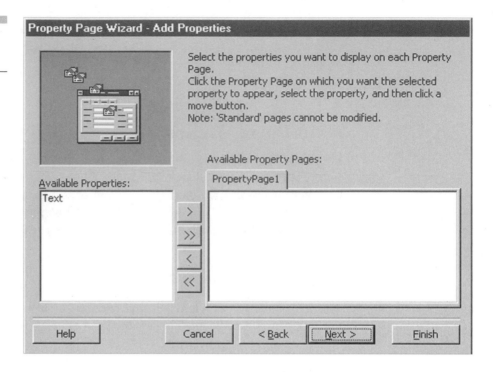

erty of your control and allow a graphical way to set design time properties. Property Pages are discussed in more detail in the ActiveX Control section of this book. Figure 16-13 shows the Property Page Wizard.

EXERCISE: THE VB REAL ESTATE AGENT

This exercise creates a useful add-in in Visual Basic 5.0 that manages the set of open windows. As you have used Visual Basic throughout this book, you have probably noticed that managing screen real estate can be time-consuming. One key reason that VB has a cluttered environment is because it has no "full-screen" feature to clear away all the open windows. This add-in creates a full-screen feature and adds it to the VB menu.

Add-ins are an advanced topic in Visual Basic. Before beginning this exercise, you should be well versed in topics such as interfaces and classes. You will need all of this knowledge to successfully create an add-in. This particular exercise also makes use of several API calls to read and write to an initialization (INI) file.

Requirements:

1. Visual Basic 5.0

Interfacing to the VBIDE

Step 1

Using the File Explorer, create a directory called \REALESTATE for your project. Start a new ActiveX DLL project in Visual Basic 5.0. All add-ins are ActiveX components and essentially operate exactly as any component you have ever created.

Step 2

To communicate with the Visual Basic Integrated Debugging Environment (VBIDE), you must set a reference to the object library that defines the available VBIDE objects. Set the reference by choosing Project/References... from the menu. The References dialog should appear. In the References dialog, select Microsoft Visual Basic 5.0 Extensibility. This is the library for the VBIDE.

In addition to the VBIDE reference, we also need to set a special reference to help us manage menu items and toolbars. Menus and tools are managed from a central component for all Microsoft Office and VB menus. This library is the Microsoft Office 8.0 Object Library. Set a reference to this item as well.

Close the References dialog and open the Object Browser by selecting View\Object Browser... from the menu. With the Object Browser visible, select VBIDE from the library list to view the set of all objects in the VBIDE model. Scroll to the bottom of the object list and select the object Windows. This is the collection of VB's windows. We will use this collection to manage the available screen real estate. After you are done, close the Object Browser.

Step 3

Select Class1 from the Project Window and view its properties in the Properties window. Change the name of Class1 to Connect. This is the class we will use to communicate with the VBIDE.

To communicate with VB, we must implement a special interface that allows the VBIDE to contact our add-in when it is loaded by the user. The name of this interface is IDTExtensibility. To implement this interface,

Figure 16-14
The References dialog.

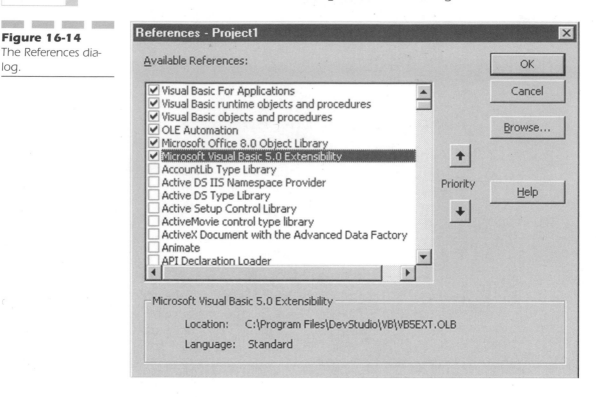

Figure 16-15
The VBIDE library.

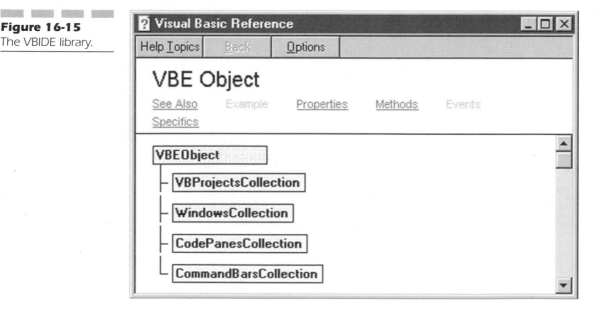

add the following code to the [general][declarations] section of the Connect class:

```
'The Add-In Interface
Implements IDTExtensibility
```

Once this interface is implemented, you can select it from the Object box in the code window. This interface enforces four methods: OnConnection, OnDisconnection, OnStartupComplete, and OnAddinsUpdate. Because this interface is implemented in our Connect class, we must have code for each of the four methods or the compile will fail. This add-in only uses two of the four methods, however, so we will simply place a comment in the unused methods. To add the unused methods, place the following code in the Connect class:

```
Private Sub IDTExtensibility_OnStartupComplete(Custom() As Variant)
  'Required
End Sub

Private Sub IDTExtensibility_OnAddInsUpdate(Custom() As Variant)
  'Required
End Sub
```

Connecting to the VBIDE

Step 4

The first thing any add-in must do is connect to the VBIDE. Connecting to the VBIDE allows our add-in to receive events from the environment, such as a user adding a new component or saving a project. Connecting is done with the OnConnection method. VB calls this method when the add-in is first loaded.

The primary purpose of the OnConnection method for this add-in is to set up the menu structure inside the VBIDE. Using objects found inside the VBIDE, we can easily add menu items to and delete menu items from the VB menu structure. For our add-in, we want to add a menu item to the View menu that a user can select to maximize the available work area. We also want to add a menu item to the Add-Ins menu that can be used to set some options for the add-in.

Before we can create the menus, we must declare some object variables to manage them. Menus require two different objects to do their work: the Menu object and the Menu Handler object. The Menu object represents the actual menu; the Menu Handler object is used to trap when the user selects the menu. Declare these variables in the [general][declarations] section of the Connect class as follows:

```
'Menus
Public WithEvents objMenuHandler As VBIDE.CommandBarEvents
Public objMenu As Office.CommandBarControl

Public WithEvents objFormMenuHandler As VBIDE.CommandBarEvents
Public objFormMenu As Office.CommandBarControl
```

Step 5

When the OnConnection method is called, we are passed an object reference to the running instance of Visual Basic. To preserve this reference, we store it in a Private variable. Declare that variable in the [general][declarations] section of the Connect class as follows:

```
'Variables
Private m_VBIDE As VBIDE.VBE
```

Step 6

Now that the variables are all declared, we can connect to the VBIDE and create our menus. In the OnConnection method, we create the menus and attach the handlers. Add the following code to the OnConnection method to initialize the add-in:

```
'Save the vb instance
Set m_VBIDE = VBInst

'Build Menus
Set objMenu = m_VBIDE.CommandBars("Menu Bar"). _
Controls("View").CommandBar.Controls.Add(1, , , 1, True)

objMenu.Caption = "&Max Work Area"

Set objMenuHandler = m_VBIDE.Events.CommandBarEvents(objMenu)

Set objFormMenu = m_VBIDE.CommandBars("Menu Bar"). _
Controls("Add-Ins").CommandBar.Controls.Add(1, , , 3, True)

objFormMenu.Caption = "VB Real &Estate Agent..."

Set objFormMenuHandler = m_VBIDE.Events.CommandBarEvents(objForm-
Menu)
```

Step 7

The IDTExtensibility interface also provides the OnDisconnection method, which is called when our add-in is unloaded. In this method, we must restore the menu structure and undo the changes we made in the OnConnection method. Add the following code to the OnDisconnection method to reset the VBIDE:

```
'Remove Menu Items
  objMenu.Delete
  objFormMenu.Delete
```

Specifying Options

Step 8

When the add-in is in use, we want to allow users to select options that will specify which windows to close when the work area is maximized and which windows to open when the work area is restored. In this way, users can select, for example, to leave the toolbox visible while the screen is maximized working on a form.

All of the options are selected as checkboxes in a special options form. When the add-in is unloaded, these options are written via API calls to an INI file. The INI file is subsequently read at startup to preserve the user's options. In this step, you will create the options form. The form consists of two frames with nine checkboxes in each frame. Add a new form to the project by selecting Project\Add Form... from the menu. Name the new form frmOptions.

As far as the add-in is concerned, only the names of the checkboxes and form are important. Each of the checkboxes is a member of a two-element control array. Index zero is the checkbox associated with the Close frame,

Figure 16-16   The Options form.

and Index one is associated with the Open frame. You may construct the form in any way as long as the following controls are available:

fraOptions(0)

fraOptions(1)

chkImmediate(0)

chkImmediate(1)

chkLocals(0)

chkLocals(1)

chkWatch(0)

chkWatch(1)

chkProject(0)

chkProject(1)

chkProperties(0)

chkProperties(1)

chkBrowser(0)

chkBrowser(1)

chkLayout(0)

chkLayout(1)

chkToolbox(0)

chkToolbox(1)

chkColor(0)

chkColor(1)

cmdCancel

cmdOK

Step 9

The options are managed by reading from and writing to an INI file. This file is accessed through API calls. These calls are defined inside a standard module. Add a new module to the project by selecting Project\Add Module from the menu. Name the new module basRealEstate.

Open the module. In this module you will define the routines necessary to read from and write to the INI file. Reading from and writing to the file are done with two API calls: WritePrivateProfileString and GetPrivateProfileString. You can obtain the function declarations for these calls by using the API text viewer that ships with Visual Basic. When you have identified the function declarations, paste them into the [general][declarations] section of the module as follows:

```
Declare Function WritePrivateProfileString& _
Lib "kernel32" Alias "WritePrivateProfileStringA" _
(ByVal AppName$, ByVal KeyName$, _
ByVal keydefault$, ByVal FileName$)
Declare Function GetPrivateProfileString Lib "kernel32" _
Alias "GetPrivateProfileStringA" (ByVal lpApplicationName As
String, _
ByVal lpKeyName As Any, ByVal lpDefault As String, _
ByVal lpReturnedString As String, ByVal nSize As Long, _
ByVal lpFileName As String) As Long
```

Step 10

The module contains two functions that are used to obtain information from the INI file. These functions are called CloseWindow and OpenWindow. These two functions are used to determine if a particular window should be opened or closed on a maximize or restore operation. They take as an argument the type of window you want information about and return a Boolean value indicating whether the window should open or close. Add these two functions to the module using the following code:

```
Public Function CloseWindow(intType As Integer) As Boolean

  Dim strReturn As String * 255
  Dim lngErr As Long
  Dim strWindow As String

  CloseWindow = False

  strReturn = String$(255, Chr$(0))

  'Get Window Type
  Select Case intType
    Case vbext_wt_Watch
      strWindow = "Watch"
    Case vbext_wt_ToolWindow
      strWindow = "Toolbox"
    Case vbext_wt_PropertyWindow
      strWindow = "Property"
    Case vbext_wt_ProjectWindow
      strWindow = "Project"
    Case vbext_wt_Locals
      strWindow = "Locals"
    Case vbext_wt_Immediate
      strWindow = "Immediate"
    Case vbext_wt_ColorPalette
      strWindow = "Color"
    Case vbext_wt_Browser
      strWindow = "Browser"
    Case vbext_wt_Preview
      strWindow = "Layout"
  End Select
```

```
'Read Flag
lngErr = GetPrivateProfileString(strWindow, "MaxClose", _
"NotFound", strReturn, Len(strReturn) + 1, App.Path &
"\RealEstate.ini")

'Return Value
If UCase$(Left$(strReturn, InStr(strReturn, Chr$(0)) - 1)) = "-1"
Then
    CloseWindow = True
End If

End Function

Public Function OpenWindow(intType As Integer) As Boolean

Dim strReturn As String * 255
Dim lngErr As Long
Dim strWindow As String

OpenWindow = False

strReturn = String$(255, Chr$(0))

'Get Window Type
Select Case intType
  Case vbext_wt_Watch
    strWindow = "Watch"
  Case vbext_wt_ToolWindow
    strWindow = "Toolbox"
  Case vbext_wt_PropertyWindow
    strWindow = "Property"
  Case vbext_wt_ProjectWindow
    strWindow = "Project"
  Case vbext_wt_Locals
    strWindow = "Locals"
  Case vbext_wt_Immediate
    strWindow = "Immediate"
  Case vbext_wt_ColorPalette
    strWindow = "Color"
  Case vbext_wt_Browser
    strWindow = "Browser"
  Case vbext_wt_Preview
    strWindow = "Layout"
End Select

'Read Flag
lngErr = GetPrivateProfileString(strWindow, "RestoreOpen", _
"NotFound", strReturn, Len(strReturn) + 1, App.Path &
"\RealEstate.ini")

'Return Value
If UCase$(Left$(strReturn, InStr(strReturn, Chr$(0)) - 1)) = "-1"
Then
    OpenWindow = True
End If

End Function
```

Step 11

Although not related to managing options, the module also contains an important routine that helps us test the add-in at design time. This routine is called AddToINI and is used to register our add-in with the Visual Basic Add-In Manager. The Add-In Manager is responsible for displaying a list of all available add-ins. To be shown on the list, our add-in must make an entry in a special file called VBADDIN.INI. Add the following code to the module to make this entry:

```
Sub AddToINI()
   'Call this to update the INI in VBIDE
   Dim ErrCode As Long

   ErrCode = WritePrivateProfileString _
   ("Add-Ins32", "RealEstateAgent.Connect", "1", _
   "vbaddin.ini")

   Debug.Print "Added to VB INI file!"
End Sub
```

Step 12

Close the module and save your work. Now open the code window for the options form. In this form, you must write routines used to display the current options in the checkbox groups. The data is saved in the INI file for each window and specifies its state when a maximize or restore operation is performed. Listing 16-3 shows a typical INI file for this add-in.

```
[Immediate]
MaxClose=-1
RestoreOpen=0

[Locals]
MaxClose=-1
RestoreOpen=0

[Watch]
MaxClose=-1
RestoreOpen=0

[Project]
MaxClose=-1
RestoreOpen=-1

[Property]
MaxClose=-1
RestoreOpen=-1

[Browser]
MaxClose=-1
RestoreOpen=0

[Toolbox]
MaxClose=-1
RestoreOpen=-1

[Color]
MaxClose=-1
RestoreOpen=0
```

```
[Layout]
MaxClose=-1
RestoreOpen=0
```

Step 13

Before we can code the routines to update the options form, we must add some constants to the form. These constants are simply used to track the mode for which we are saving options. Add the following constants to the [general][declarations] section of frmOptions:

```
Private Const ntsMaxClose% = 0
Private Const ntsRestoreOpen% = 1
```

Step 14

Options are read from the INI file when the options form is first loaded. The options are read for each of the window by making the appropriate API call. Read these options by adding the following code to the Form_Load event of frmOptions:

```
'Immediate
chkImmediate(ntsMaxClose).Value = _
CloseWindow(vbext_wt_Immediate) * -1
chkImmediate(ntsRestoreOpen).Value = _
OpenWindow(vbext_wt_Immediate) * -1

'Locals
chkLocals(ntsMaxClose).Value = _
CloseWindow(vbext_wt_Locals) * -1
chkLocals(ntsRestoreOpen).Value = _
OpenWindow(vbext_wt_Locals) * -1

'Watch
chkWatch(ntsMaxClose).Value = _
CloseWindow(vbext_wt_Watch) * -1
chkWatch(ntsRestoreOpen).Value = _
OpenWindow(vbext_wt_Watch) * -1

'Project
chkProject(ntsMaxClose).Value = _
CloseWindow(vbext_wt_ProjectWindow) * -1
chkProject(ntsRestoreOpen).Value = _
OpenWindow(vbext_wt_ProjectWindow) * -1

'Property
chkProperties(ntsMaxClose).Value = _
CloseWindow(vbext_wt_PropertyWindow) * -1
chkProperties(ntsRestoreOpen).Value = _
OpenWindow(vbext_wt_PropertyWindow) * -1

'Browser
chkBrowser(ntsMaxClose).Value = _
CloseWindow(vbext_wt_Browser) * -1
chkBrowser(ntsRestoreOpen).Value = _
OpenWindow(vbext_wt_Browser) * -1
```

```
'Layout
chkLayout(ntsMaxClose).Value = _
CloseWindow(vbext_wt_Preview) * -1
chkLayout(ntsRestoreOpen).Value = _
OpenWindow(vbext_wt_Preview) * -1
'Toolbox
chkToolbox(ntsMaxClose).Value = _
CloseWindow(vbext_wt_ToolWindow) * -1
chkToolbox(ntsRestoreOpen).Value = _
OpenWindow(vbext_wt_ToolWindow) * -1

'Color
chkColor(ntsMaxClose).Value = _
CloseWindow(vbext_wt_ColorPalette) * -1
chkColor(ntsRestoreOpen).Value = _
OpenWindow(vbext_wt_ColorPalette) * -1
```

Step 15

Options are written to the INI file when the user presses the OK button. If the user presses the Cancel button, options are not written. Add the following code to the Click event of the Cancel button to bypass the INI file:

```
Unload Me
```

Writing the options requires calling the API function and is essentially the reverse of the read operation. Add the following code to the Click event of the OK button to write the options:

```
Dim lngErr As Long
'Immediate
lngErr = WritePrivateProfileString _
("Immediate", "MaxClose", _
CStr(chkImmediate(ntsMaxClose).Value * -1), _
App.Path & "\RealEstate.ini")
lngErr = WritePrivateProfileString _
("Immediate", "RestoreOpen", _
CStr(chkImmediate(ntsRestoreOpen).Value * -1), _
App.Path & "\RealEstate.ini")

'Locals
lngErr = WritePrivateProfileString _
("Locals", "MaxClose", _
CStr(chkLocals(ntsMaxClose).Value * -1), _
App.Path & "\RealEstate.ini")
lngErr = WritePrivateProfileString _
("Locals", "RestoreOpen", _
CStr(chkLocals(ntsRestoreOpen).Value * -1), _
App.Path & "\RealEstate.ini")

'Watch
lngErr = WritePrivateProfileString _
("Watch", "MaxClose", _
CStr(chkWatch(ntsMaxClose).Value * -1), _
```

```
App.Path & "\RealEstate.ini")
lngErr = WritePrivateProfileString _
("Watch", "RestoreOpen", _
CStr(chkWatch(ntsRestoreOpen).Value * -1), _
App.Path & "\RealEstate.ini")

'Project
lngErr = WritePrivateProfileString _
("Project", "MaxClose", _
CStr(chkProject(ntsMaxClose).Value * -1), _
App.Path & "\RealEstate.ini")
lngErr = WritePrivateProfileString _
("Project", "RestoreOpen", _
CStr(chkProject(ntsRestoreOpen).Value * -1), _
App.Path & "\RealEstate.ini")

'Property
lngErr = WritePrivateProfileString _
("Property", "MaxClose", _
CStr(chkProperties(ntsMaxClose).Value * -1), _
App.Path & "\RealEstate.ini")
lngErr = WritePrivateProfileString _
("Property", "RestoreOpen", _
CStr(chkProperties(ntsRestoreOpen).Value * -1), _
App.Path & "\RealEstate.ini")

'Browser
lngErr = WritePrivateProfileString _
("Browser", "MaxClose", _
CStr(chkBrowser(ntsMaxClose).Value * -1), _
App.Path & "\RealEstate.ini")
lngErr = WritePrivateProfileString _
("Browser", "RestoreOpen", _
CStr(chkBrowser(ntsRestoreOpen).Value * -1), _
App.Path & "\RealEstate.ini")

'Layout
lngErr = WritePrivateProfileString _
("Layout", "MaxClose", _
CStr(chkLayout(ntsMaxClose).Value * -1), _
App.Path & "\RealEstate.ini")
lngErr = WritePrivateProfileString _
("Layout", "RestoreOpen", _
CStr(chkLayout(ntsRestoreOpen).Value * -1), _
App.Path & "\RealEstate.ini")

'Toolbox
lngErr = WritePrivateProfileString _
("Toolbox", "MaxClose", _
CStr(chkToolbox(ntsMaxClose).Value * -1), _
App.Path & "\RealEstate.ini")
lngErr = WritePrivateProfileString _
("Toolbox", "RestoreOpen", _
CStr(chkToolbox(ntsRestoreOpen).Value * -1), _
App.Path & "\RealEstate.ini")

'Color
lngErr = WritePrivateProfileString _
("Color", "MaxClose", _
CStr(chkColor(ntsMaxClose).Value * -1), _
App.Path & "\RealEstate.ini")
lngErr = WritePrivateProfileString _
```

```
("Color", "RestoreOpen", _
CStr(chkColor(ntsRestoreOpen).Value * -1), _
App.Path & "\RealEstate.ini")
Unload Me
```

Handling the Menu Events

Step 16

The only part of our add-in left to construct is the menu handling routines. These routines all fire in response to user interaction with the Visual Basic menu structure. In particular, we want to receive notification when the menu items we added to VB are clicked. The add-in tracks the currently active options in memory through several variables we must add before writing our menu routines. Add these variables to the [general][declarations] section of the Connect class:

```
Private m_Immediate As Boolean
Private m_Locals As Boolean
Private m_Watch As Boolean
Private m_Project As Boolean
Private m_Properties As Boolean
Private m_Object As Boolean
Private m_Layout As Boolean
Private m_Toolbox As Boolean
Private m_Color As Boolean
```

Step 17

To open or close the windows, the user interacts with our menu item under the View menu. This item is tracked by the objMenuhandler object. Select the objMenuHandler object from the Object box in the code window of class Connect. You should then see the Click event for this object. This is the event that will fire when the user selects the menu item. Add the following code to objMenuHandler_Click to open or close the designated windows.

```
'Open or Close Windows
  Dim objWindow As VBIDE.Window

  For Each objWindow In m_VBIDE.Windows
    If objMenu.Caption = "&Max Work Area" Then
      If CloseWindow(objWindow.Type) Then
        objWindow.Close
      End If
    Else
      If OpenWindow(objWindow.Type) Then
        objWindow.Visible = True
      End If
    End If
```

```
Next
'Change Menu Item
If objMenu.Caption = "&Max Work Area" Then
  objMenu.Caption = "&Restore Work Area"
Else
  objMenu.Caption = "&Max Work Area"
End If
```

Step 18

To change the selected options, the user interacts with our menu item under the Add-Ins menu. This menu item is tracked by the objForm-MenuHandler object, which also has a Click event. Add the following code to this event to show the Options dialog:

```
frmOptions.Show
```

Testing the Add-In

Step 19

Make sure all of your work is saved. While still inside the VB design environment, select Run\Start from the menu. You add-in will start, but nothing will happen. Enter Break mode by selecting Run\Break. Now register your add-in by typing directly into the immediate debug window:

This calls the routine you created earlier and writes to the VBADDIN.INI file. Now continue running your add-in by selecting Run\Continue. Minimize this copy of VB 5 and start another copy from the system Start menu. When the second copy starts, your add-in will automatically be loaded for use. You may have to debug any mistakes you made in creating the code.

Once you have successfully connected, use the View menu to try and maximize the work area. Also try the options window under the Add-Ins menu.

Increasing Your Knowledge

Help File Search Strings

- Add-Ins
- IDTExtensibility
- OnConnection
- OnDisconnection

Books OnLine References

▪ "Extending the Visual Basic Environment with Add-Ins"

Mastery Questions

1. Which are methods of the IDTExtensibility interface?
   a. OnConnection
   b. OnStartupComplete
   c. OnDisconnection
   d. OnAddInsUpdate

2. Which object represents the current instance of Visual Basic?
   a. VBIDE
   b. VBE
   c. VB
   d. VBD

3. What object allows VBIDE events to be trapped?
   a. ActiveEvents
   b. Events
   c. VBEvents
   d. VBEEvents

4. What ActiveX component contains toolbar and menu objects?
   a. VBIDE
   b. ADODB
   c. ODBC
   d. Office

5. What file contains the information for all registered add-ins?
   a. VBADDIN.INI
   b. ADDIN.INI
   c. ADD.INI
   d. VBADD.INI

6. What object allows access to the code in a project?
   a. CodeWindow
   b. CodePanes
   c. CodeModules
   d. CodeModule

7. What VB projects can perform as add-ins?
   a. ActiveX Documents
   b. ActiveX EXE
   c. ActiveX DLL
   d. ActiveX Control

8. What code can be used to access the main VB toolbar?
   a. CommandBars("Main")
   b. ComandBars("Standard Menu")
   c. CommandBars("Main")
   d. CommandBars("Standard")

9. Which object is used to generate project-level events?
   a. VBEvents
   b. ProjectEvents
   c. VBProjectsEvents
   d. VBProjectEvents

10. Which project must be modified when creating an add-in setup?
   a. SETUP132.VBP
   b. SETUP.VBP
   c. SETUP1.VBP
   d. SETUP32.VBP

Answers to Mastery Questions

1. a, b, c, d
2. b
3. c
4. d
5. a
6. d
7. a, b, c
8. d
9. c
10. b

INDEX

SOFTWARE AND INFORMATION LICENSE

The software and information on this diskette (collectively referred to as the "Product") are the property of The McGraw-Hill Companies, Inc. ("McGraw-Hill") and are protected by both United States copyright law and international copyright treaty provision. You must treat this Product just like a book, except that you may copy it into a computer to be used and you may make archival copies of the Products for the sole purpose of backing up our software and protecting your investment from loss.

By saying "just like a book," McGraw-Hill means, for example, that the Product may be used by any number of people and may be freely moved from one computer location to another, so long as there is no possibility of the Product (or any part of the Product) being used at one location or on one computer while it is being used at another. Just as a book cannot be read by two different people in two different places at the same time, neither can the Product be used by two different people in two different places at the same time (unless, of course, McGraw-Hill's rights are being violated).

McGraw-Hill reserves the right to alter or modify the contents of the Product at any time.

This agreement is effective until terminated. The Agreement will terminate automatically without notice if you fail to comply with any provisions of this Agreement. In the event of termination by reason of your breach, you will destroy or erase all copies of the Product installed on any computer system or made for backup purposes and shall expunge the Product from your data storage facilities.

LIMITED WARRANTY

McGraw-Hill warrants the physical diskette(s) enclosed herein to be free of defects in materials and workmanship for a period of sixty days from the purchase date. If McGraw-Hill receives written notification within the warranty period of defects in materials or workmanship, and such notification is determined by McGraw-Hill to be correct, McGraw-Hill will replace the defective diskette(s). Send request to:

Customer Service
McGraw-Hill
Gahanna Industrial Park
860 Taylor Station Road
Blacklick, OH 43004-9615

The entire and exclusive liability and remedy for breach of this Limited Warranty shall be limited to replacement of defective diskette(s) and shall not include or extend to any claim for or right to cover any other damages, including but not limited to, loss of profit, data, or use of the software, or special, incidental, or consequential damages or other similar claims, even if McGraw-Hill has been specifically advised as to the possibility of such damages. In no event will McGraw-Hill's liability for any damages to you or any other person ever exceed the lower of suggested list price or actual price paid for the license to use the Product, regardless of any form of the claim.

THE McGRAW-HILL COMPANIES, INC. SPECIFICALLY DISCLAIMS ALL OTHER WARRANTIES, EXPRESS OR IMPLIED, INCLUDING BUT NOT LIMITED TO, ANY IMPLIED WARRANTY OF MERCHANTABILITY OR FITNESS FOR A PARTICULAR PURPOSE. Specifically, McGraw-Hill makes no representation or warranty that the Product is fit for any particular purpose and any implied warranty of merchantability is limited to the sixty day duration of the Limited Warranty covering the physical diskette(s) only (and not the software or information) and is otherwise expressly and specifically disclaimed.

This Limited Warranty gives you specific legal rights; you may have others which may vary from state to state. Some states do not allow the exclusion of incidental or consequential damages, or the limitation on how long an implied warranty lasts, so some of the above may not apply to you.

This Agreement constitutes the entire agreement between the parties relating to use of the Product. The terms of any purchase order shall have no effect on the terms of this Agreement. Failure of McGraw-Hill to insist at any time on strict compliance with this Agreement shall not constitute a waiver of any rights under this Agreement. This Agreement shall be construed and governed in accordance with the laws of New York. If any provision of this Agreement is held to be contrary to law, that provision will be enforced to the maximum extent permissible and the remaining provisions will remain in force and effect.